FUNCTIONAL FOOD - IMPROVE HEALTH THROUGH ADEQUATE FOOD

Edited by **María Chávarri Hueda**

Functional Food - Improve Health through Adequate Food
http://dx.doi.org/10.5772/66263
Edited by María Chávarri Hueda

Contributors

Jailane De Souza Aquino, Francisca Nayara Menezes, Kamila Batista, Priscilla Lins, Jessyca Gomes, Laiane Da Silva, Maira Segura Campos, Yvonne Maphosa, Rosa María Martínez-Espinosa, Noelia García Uribe, Manuel Reig García-Galbis, Greta Krešić, Nikolina Mrduljaš, Tea Bilušić, Oluwafemi Ayodeji Adebo, Janet Adebiyi, Sefater Gbashi, Judith Phoku, Patrick Berka Njobeh, Eugenie Kayitesi, Sonia Ancuta Socaci, Anca Corina Farcas, Zorita Diaconeasa, Oana Lelia Pop, Adriana Paucean, Maria Tofana, Dumitrita Rugina, Adela Pintea, Deshanie Rai, Gyan Rai, Pablo Bautista, Lorena González, Berenice González-Esparza, Camila Del Castillo-Rosas, Hiroko Watanabe, Tomoko Miyake, Elisabeta Botez, Oana-Viorela Nistor, Georgeta Doina Andronoiu, Gabriel - Danut Mocanu, Ioana Otilia Ghinea, Fernando Pérez-Rodríguez, Maria Luisa Poyatos-Guerrero, Silvia Del Valle Alonso, Marina Marsanasco, Nadia S. Chiaramoni

Published by InTech
Janeza Trdine 9, 51000 Rijeka, Croatia

Publishing Process Manager Marijana Francetic
Technical Editor SPi Global
Cover InTech Design team

First published August, 2017
Printed in Croatia
Legal deposit, Croatia: National and University Library in Zagreb

Additional hard copies can be obtained from orders@intechopen.com

Functional Food - Improve Health through Adequate Food, Edited by María Chávarri Hueda
p. cm.
Print ISBN 978-953-51-3439-8
Online ISBN 978-953-51-3440-4

For EU product safety concerns:
IN TECH d.o.o., Prolaz Marije Krucifikse Kozulić 3, 51000 Rijeka, Croatia,
info@intechopen.com or visit our website at intechopen.com.

Meet the editor

Maria Chávarri Hueda has received her MS degree in Biological Sciences from Universidad de Navarra, Spain, in 1997. She obtained her PhD degree from Nutrition and Food Science Area, Department of Pharmacy and Food Sciences, Faculty of Pharmacy, University of the Basque Country. Chávarri has experience in biotechnology and food science, acting on the following subjects: bioactive molecules and functional activity, probiotics, and nutritional status. She worked on the "Influence of the lipid source of the diet on various aspects of hepatic metabolism of triglycerides and cholesterol." Over the last few decades, Chávarri worked as a senior researcher at TECNALIA R&I, Technological Development Center, in food and health area, and she has focused her studies on bioactive molecules of food and plant origin and their functional activities, as well as to deepen her knowledge on probiotics, with the objective of developing functional foods.

Contents

Preface

Society is increasingly aware that food can help prevent the development of certain diseases. This, together with the increase in life expectancy, is changing the trend of food consumption. For this reason, it is important to know which bioactive compounds possess a functional activity; that is to say, they produce a beneficial effect in the organism and can improve the health of the people, and they can be used for the development of functional foods.

This book comprehensively reviews and compiles information on bioactive ingredients and functional foods in 14 chapters, which cover the impact of bioactive ingredients (vitamins, antioxidants, compounds of the pulses, etc.) on nutrition through food, how functional foods can prevent disease, and tools used to evaluate the effects of bioactive ingredients, functional foods, and diet.

Section 1—The Impact of Bioactive Ingredients on Nutrition through Food, which includes Chapters 1–8—showed that bioactive compounds can help prevent the development of certain diseases.

Section 2—Functional Foods to Help Prevent Diseases of a Society Increasingly Aware of Its Feeding, which includes Chapters 9 and 10—showed that functional foods based on meat and drinkable foods may be a suitable route for use in the development of functional foods.

Section 3—Tools Used to Evaluate the Effects of Bioactive Ingredients, Functional Foods, and Diet, which includes Chapters 11–14—showed the use of metabolomics in fermented foods, evaluation of prebiotics in the microbiota, and diet quality indices for nutrition assessment.

This book is written by authors from America, Europe, Asia, and Africa; in addition, between the editor and the authors, it has been sought to delve into each chapter to provide information on the benefits of functional foods for both food companies and consumers.

The scientists involved in the writing of this book were selected and invited because of their recognized expertise and important contribution on the field in which they are acting. Thanks to the involvement of these scientists in this work, the publication of this book was made possible.

This book will hopefully be of help to many scientists, doctor, pharmacists, and chemical and other experts in a variety of disciplines, both academic and industrial. This book in addition to supporting research and development also wants to be a support material to be used in teaching.

Finally, I would like to thank my daughters Paula and Lucia and my husband Alex for their patience and love. I extend my apologies for many hours spent on the editing of this book, which kept me away from them.

Dr. María Chávarri Hueda
Health and Food Area
Health Division
TECNALIA Research & Innovation
Miñano (Alava), Spain

Section 1

Impact of Bioactive Ingredients on Nutrition through Food

Chapter 1

Antioxidant Compounds Recovered from Food Wastes

Sonia Ancuța Socaci, Dumitrița Olivia Rugină,
Zorița Maria Diaconeasa, Oana Lelia Pop,
Anca Corina Fărcaș, Adriana Păucean,
Maria Tofană and Adela Pintea

Additional information is available at the end of the chapter

http://dx.doi.org/10.5772/intechopen.69124

Abstract

The increase awareness of nowadays consumers regarding the food they purchase and consume and the health has led to an increase demand of foods containing biologically active compounds, namely antioxidants, which can help the body to fight against oxidative stress. As a consequence finding, new or nonconventional sources of antioxidants are a priority for food and also pharmaceutical industries. Wastes from fruits and vegetable processing are shown to contained valuable molecules (antioxidants, dietary fibers, proteins, natural colorants, aroma compounds, etc.) which can be extracted, purified and valorized in value-added products. The present chapter is underlying the great potential of food wastes to be exploited as sources of antioxidants based on the scientific evidences regarding the possibilities of extraction and purification, health benefits and envisaged applications of antioxidants recovered from these wastes.

Keywords: bioactive compounds, antioxidants, food waste exploitation, functional ingredients, health benefits

1. Introduction

Statistics announced by the Food and Agriculture Organization (FAO) of United Nation showed that approximately one-third of food produced for human consumption is wasted globally. These statistics indicated that even though the quantity of wastes differs between regions, all regions have major losses at production level. Fruits and vegetables, plus roots and tubers, have the highest wastage rates of any food. The same organization reported that a global quantitative loss and waste of root crops, fruits and vegetables per year is 40–50% [1]. The disposal of such amounts of wastes not only represents a challenge for the food processors, but it is a matter of crucial importance at international level due to both environmental pollution and economical aspects [2, 3]. Studies showed that plant-derived wastes should be reconsidered and regarded as renewable sources of valuable molecules which can be extracted, purified and valorized in different fields, including food industry, cosmetics, pharmaceutical and chemical industry and so on [4, 5]. For example, the search for efficient and nontoxic natural compounds with antioxidant activity has gained increased attention, especially due to the consumers' awareness regarding the direct relation between food (diet) and health [6]. The introduction into the diet of the antioxidant compounds, like polyphenols, is an efficient way to combat the negative effects caused by the excess of reactive oxygen species (ROS) in the body. The oxidative stress, caused by the ROS, is considered to be one of the main triggers of chronic diseases, such as cancer, diabetes, cardiovascular or neurodegenerative disorders [7]. In the case of fruits and vegetables, usually a high amount of antioxidant compounds is found in peels, kernels or seeds, namely in parts that are removed during processing and become wastes [8–13]. Thus, these compounds could be extracted from fruit and vegetable wastes and reused in other food products, as functional ingredients able to confer some characteristic quality criteria and at the same time to exert human health benefits due to their antioxidant properties.

The aim of this chapter is to emphasize existing studies on fruit and vegetable wastes regarding their potential as sources of bioactive compounds (antioxidants) with health-promoting benefits that can be exploited as functional ingredients.

2. Extraction and identification of antioxidants from food wastes

Nowadays, the growing interest of consumers toward the relation between the ingested food and the effects on health has led to an increase demand of foods without what they perceive harmful chemicals (e.g., synthetic preservatives, antioxidants, colorants) and with high nutritional and functional properties. This demand, in the scientific field, was translated by intensifying the research focused on finding new sources of bioactive molecules (antioxidants), optimizing the extraction and purification methods as well as developing innovative functional foods that promote health. In this conjuncture, the exploitation of food wastes (by-products) for the recovery and reuse of valuable bioactive compounds is one of the most sustainable approaches. Thus, efficient extraction techniques can be implemented for the separation and isolation of naturally occurring compounds with antioxidant characteristics from food wastes, such as polyphenols, carotenoids, glucosinolates, dietary fibers and so on.

There is no universal method for the extraction of bioactive compounds, but in order for a method to be suitable it has to fulfil several requirements, including selectivity toward the analyte, high extraction yields, possibility of solvent recovery (e.g., environmental friendly) or using "green solvents," maintaining the functionality of the recovered molecules, low-cost reagents, possibility to be implemented from laboratory scale to industrial scale and so on [14–17]. Among the classical methods used for the isolation of bioactive compounds, the most common ones are solid-liquid extraction (maceration), Soxhlet extraction and liquid-liquid extraction [18]. Depending on the type of matrix (fruit and vegetable waste) and on the type of compounds that are to be recovered, solvents with different polarities may be used (e.g., methanol, ethanol, methanol-water mixtures, water, acetone, ethyl-acetate and so on) [16, 19–21]. In the case of phenolic compounds such as flavonoids or proanthocyanidins (condensed tannins), improved extraction yields were noticed when the organic solvent was used in combination with water, while for the methoxylated compounds recovered from mango peels, a higher yield was achieved when less polar solvents such as acetone were used [16, 22]. Choosing the appropriate extraction solvent is of utmost importance, because it significantly influences the yield and the composition of the extract. Nevertheless, the enhancement of the extraction procedure may be also achieved by optimizing the sample-to-solvent ratio, extraction temperature and time, agitation degree and particle size [18, 23, 24]. Although conventional methods were optimized, there are still some limitations in their use mainly due to the high amount of solvent, time-consuming, difficulty to scaled-up. Thus, to overcome these limitations and in accordance with the "zero waste" desiderate, the current researches are focused on developing greener, sustainable and viable extraction processes. The modern extraction techniques comprise microwave-assisted extraction, ultrasound-assisted extraction, pressurized liquid extraction (e.g., pressurized hot water extraction), enzyme-assisted extraction, supercritical CO_2-based extraction and other emerging techniques [18, 25–27]. For maximum valorization, several integrated extraction systems were developed (e.g., biorefineries), in which the wastes are subjected to sequential extraction steps for the recovery of different classes of bioactive compounds which can be further used such as or as raw materials for value-added chemicals production [17, 28, 29]. Recently, a new integrated extraction-adsorption process has been developed for production of large quantities of extracts rich in antioxidants. This process was proposed for a selective recovery of antioxidants from black chokeberry wastes at pilot scale, by applying a scale-up factor of 50, but the results were similar to those obtained at laboratory scale [30].

The identification and quantification of the recovered antioxidant compounds are generally achieved using high-pressure liquid chromatography (HPLC) and hyphenated techniques (e.g., LC-MS), in particular spectrophotometric methods (e.g., UV-VIS). The bioactivities of the antioxidant compounds are evaluated using methods for the assessment of their antioxidant activity (2,2-diphenyl-1-picrylhydrazyl (DPPH), ferric reducing antioxidant power (FRAP), 2,2′-azino-bis 3-ethylbenzothiazoline-6-sulphonic acid (ABTS), cupric reducing antioxidant capacity (CUPRAC), Oxygen radical absorbance capacity (ORAC)), inhibition of lipid oxidation (peroxide value, Thiobarbituric acid reactive substances (TBARs)), antimicrobial activity, antiproliferative activity and so on. **Table 1** summarizes some of the techniques generally used for the separation and isolation of antioxidant compounds as well as the analytical methods applied for their bioactivity evaluation.

Waste source	Antioxidant compounds	Extraction techniques	Evaluation methods	References
Onion waste	Phenolics Flavonoids	Solid-liquid extraction	Total phenolic content (UV-VIS) Total flavonoids (UV-VIS) Total flavonols (HPLC) Antioxidant activity (FRAP)	[31]
Apple pomace	Phenolics	Solvent extraction	Phenolics (UV-VIS, HPLC) Total flavonoids (UV-VIS) Antioxidant activity (DPPH, FRAP)	[32]
Macadamia skin	Phenolics Flavonoids Proanthocyanidins	Ultrasound-assisted extraction	Total phenolics (UV-VIS) Total flavonoids (UV-VIS) Proanthocynidins (UV-VIS) Antioxidant activity (ABTS, DPPH, CUPRAC, FRAP)	[16]
Potato peels	Phenolics Flavonoids Ferulic acid Chlorogenic acid	Hydroalcoholic solution extraction	Phenolics (UV-VIS, HPLC) Total flavonoids (UV-VIS) Antioxidant activity (DPPH, β-carotene bleaching assay) Lipid oxidation inhibiting potential (peroxide value, p-anisidine value, TOTOX, TBARs, conjugated dienes, volatile compounds)	[20]
	Phenolics	Green ultrasound-assisted extraction	Phenolics (UV-VIS, LC-DAD-MS) Antioxidant activity (DPPH, reducing power)	[24]
	Phenolics	Solvent extraction	Total phenolics (UV-VIS) Total flavonoids (UV-VIS) Antioxidant activity (ABTS, DPPH) Antimicrobial activity (antibacterial and antifungal activity)	[21]
Pomegranate peels	Phenolics Flavonoids	Solvent extraction	Total phenolics (UV-VIS) Total flavonoids (UV-VIS) Antioxidant activity (DPPH)	[33]
	Carotenoids	Ultrasound assisted extraction	Carotenoid content (UV-VIS, HPLC) Antioxidant activity (DPPH)	[25]
Passion fruit rinds	Phenolics	Ethanolic-water pressurized liquid extraction	Total phenolics (UV-VIS) Phenolic composition (UPLC-MS/MS) Antioxidant activity (DPPH, FRAP, ORAC)	[34]
Acerola peels and seeds	Phenolics	Sequential solvent extraction	Total phenolics (UV-VIS) Antioxidant activity (DPPH, ABTS) Lipid oxidation inhibiting potential (thiocyanate method, Schaal oven test)	[35]
Mango seeds	Phenolics (tannins and proanthocyanidins)	Microwave assisted extraction	Lipid oxidation inhibiting potential (β-carotene bleaching assay) Antioxidant activity (DPPH, ABTS) Total phenolic content, tannins content and proanthocyanidine content (UV-VIS)	[22]

Waste source	Antioxidant compounds	Extraction techniques	Evaluation methods	References
Guava seeds and pomace	Phenolics	Solvent extraction	Total phenolics (UV-VIS) Total flavonoids (UV-VIS) Antioxidant activity (ABTS, DPPH) Antimicrobial activity (antibacterial and antifungal activity)	[21]
Grape pomace	Phenolics	Solvent extraction	Phenolics (UV-VIS, HPLC) Antioxidant activity (DPPH, peroxide value, rancimat method)	[36]
	Phenolics	Supercritical fluids extraction (CO_2) Soxhlet extraction	Total phenolics (UV-VIS)	[37]
Chestnut and hazelnut shells	Phenolics	Solvent extraction	Phenolics (UV-VIS, HPLC) Antioxidant activity (FRAP)	[19]
Hazelnut waste	Phenolics	Supercritical fluids extraction (CO_2) Soxhlet extraction	Total phenolics (UV-VIS)	[37]
Spent filter coffee	Phenolics chlorogenates Flavonoids	Glycerol-based extraction	Phenolics (HPLC) Antioxidant activity (DPPH, ferric reducing power)	[17]
Spent ground coffee	Phenolics	Supercritical fluids extraction (CO_2) Soxhlet extraction	Total phenolics (UV-VIS)	[37]
Olive leaves and pomace	Phenolics	Solvent extraction	Total phenolics (UV-VIS) Total flavonoids (UV-VIS) Antioxidant activity (ABTS, DPPH) Antimicrobial activity (antibacterial and antifungal activity)	[21]
Broccoli leaves	Glucosinolates	Microwaved assisted extraction	Glucosinolate composition (LC-DAD-ESI-MS)	[38]
Tomato waste (skin and seeds)	Carotenoids (lycopene)	Enzyme and high pressure assisted extraction	Total carotenoid content (UV-VIS) Lycopene content (HPLC)	[39]
	Carotenoids	Ultrasound and manosonication assisted extraction	Total carotenoid content (UV-VIS) Carotenoid composition (HPLC)	[40]
Artichoke waste (internal and external bracts)	Phenolics	Ultrasound-assisted extraction and nanofiltration	Total phenolics (UV-VIS) Antioxidant activity (DPPH, FRAP) Chlorogenic acid content (HPLC)	[41]
Immature fruits	Phenolics	Reflux extraction (water) Pressurized hot water extraction	Total phenolics (UV-VIS) Antioxidant activity (ORAC) Cell viability (3-(4,5-dimethylthiazol-2-yl)-2,5-diphenyltetrazolium bromide MTT assay)	[27]

Table 1. Some techniques used for the separation and isolation of antioxidant compounds and the analytical methods applied for their bioactivity evaluation.

3. Potential health benefits of recovered antioxidants

3.1. Berries

Blueberries, ribes, chokeberries, raspberries, and blackberries are used to obtain food products such as juices, jams, and jellies. A high amount of wastes are released during industry manufacturing of these fruits. Hence, valuable compounds from wastes, such as anthocyanins, phenolic acids, and flavonoids, could be successfully recovered and used for different industries.

Seed pomace, wastes of blackberry (*Rubusfruticosus* L.) and raspberry (*Rubusidaeus* L.), is generated in large quantities, being a good raw material for oil extraction. Besides linoleic (omega-6) and α-linolenic (omega-3) (2–4:1 ratio) content, these oils are also rich in bioactive compounds, such as tocopherols, phenols, sterols, and carotenoids, which are known to exert antioxidant properties. Therefore, the composition of the oil resulted from blackberry and raspberry seed pomace proved to be stable despite a long-term frozen, due to the presence of natural antioxidants [42]. Consequently, these seed oils can be considered value-added products and could be used as functional or nutraceutical food products.

Leaves could also be a potential source of health-promoting compounds. Leaves and pomace of cranberry (*Vacciniummacrocarpon* L.) contained more polyphenols and exhibited higher antioxidant activity than fruit and juices. Therefore, leaves and pomace could be another excellent source for the production of foods with high health-promoting value [43].

Among polyphenols, anthocyanins and ellagitannins from berries are known for their antitumor potential [44, 45]. A waste of black raspberry seeds applied on colon cancer HT-29 cells inhibited cellular proliferation and induced apoptosis, both through the extrinsic apoptotic pathway (activation of caspase 3, 8) and through intrinsic apoptotic pathway (activation of caspase 9 and poly(ADP-ribose) polymerase (PARP)) [46].

3.2. Apples

The apple waste generally refers to a heterogeneous mixture of peels, pomace, and seeds. Apple waste resulted after juice processing was tested on tumor colon HT29, HT115, and CaCo-2 cell lines. Results showed that waste compounds are able to confer protection against DNA damage, to improve barrier function and to inhibit cell invasion [47]. Comparing the inhibitory effects of nonextractable antioxidants with extractable antioxidants from a freeze-dried apple waste on HeLa, HepG2, and HT-29 human cancer cells, the nonextractable antioxidants were more efficient [48].

Apple peel waste could also be an excellent source of natural antioxidants and bioactive compounds that may improve the human health [49]. Apple peel extract showed a significant dose response reduction in cell proliferation in the HT-29 colon cancer cells but not on MCF-7 breast cancer cells, from ten different extracts of fruits and berries which have been tested [50].

3.3. Citrus

The production of citrus fruits, the most widely cultivated fruits, is increasing every year due to a high market demand. Orange is the main citrus fruit that dominates the global customer requests. Unfortunately, 50–60% of the fruits including seed, peel and segment membrane resulted from juice production ends up as waste [51]. Among these wastes, citrus peel is the major constituent accounting 50% of the wet fruit mass. It contains flavonoids, carotenoids, polyphenols, ascorbic acids, pectin, dietary fibers and essential oils [52]. Orange (*Citrus auranthium*) flesh waste has a higher antioxidant activity than the peel. Although both of the extracts used in a study on human leukocytes showed protection against H_2O_2-induced DNA damage [53].

3.4. Exotic fruits

Pomegranate fruit gained a lot of interest due to multiple beneficial effects on human health. A recent study demonstrated that the antioxidant potential of pomegranate extract is directly related to the phenolic content, whereas its antiproliferative activity is mainly attributed to ellagic acid [54]. The ability of ellagitanins from *Punicagranatum* L. to reduce breast MCF-7 and prostate LNCaP cancer cell proliferation was proved [55].

Juice industry underuses large amounts of passion fruit residues. The seeds of passion fruit are used for oil production, but the residue remained after the seed cold pressing (cake seed) still contains compounds of interest, like fatty acids and/or others polyphenols. Certainly, the antioxidant and the antimicrobial activities of passion fruit residue contribute to its adding value [56]. Similarly, the wastes of mango, peel and kernel contain a noteworthy amount of bioactive components such as xanthones (mangiferin), flavonoids, flavanols, and phenolic acids with therapeutic effects [57]. The *Antidesma thwaitesianum Müll. Arg.* fruit waste was tested on six human normal and cancer (COR-L23, A549, LS174T, PC-3, MCF7 and HeLa) cell lines. Interesting is that extracts of fresh fruits exhibited moderate cytotoxicity against human breast MCF7 cells, while the extract obtained by decocting the residue left after maceration of dried fruits showed the highest cytotoxicity on COR-L23 carcinoma lung cells [58]. The waste resulted from *Myracrodruon urundeuva* seeds, containing steroids, alkaloids and phenols, was twofold more cytotoxic on leukemia HL-60 line than on glioblastoma SF-295 and Sarcoma 180 cells [59]. All these data are strong evidence that exotic fruits wastes are a valuable source of antioxidants with potential health benefits.

3.5. Potatoes and tomatoes

Industrialization of potatoes and tomatoes generates by-products rich in antioxidants. There are scientific evidences that wastes of potatoes and tomatoes could be used as natural antioxidant additives in the protection of vegetable oils, effectively limiting the oxidation of oils [60, 61]. The main antioxidant compounds that have been identified in potato waste were caffeic acid, chlorogenic acid, protocatechuic acid, para-hydroxybenzoic acid and gallic acid [62].

The antioxidant and antiproliferative activity of tomato waste were strongly correlated with its concentration in β-carotene and lycopene [63]. The waste obtained during the production of tomato juice scavenged hydroxyl and superoxide anion radicals and exerted anticancer properties, by inhibiting HeLa, MCF7 and MRC-5 tumor cell growth [64].

4. Applications of recovered antioxidants

Fruits, vegetables, and plant-derived wastes are commonly composed of peels, stems, seeds, kernels, shells, bran, and trimmings residues being a promising source of functional compounds due to their favorable nutritional and rheological properties. The most important bioactive compounds found in these types of wastes are fibers, phenolic compounds, vitamin E, C, carotenoids, and other antioxidants, which are found to have beneficial effects for human health. Trying to comply with the consumers' demand for healthier products, the modern food industry is presently focused on one hand on designing and producing food products with bioactive ingredients—the so-called "functional foods" and "super foods"—for which health claims are made and on the other hand on finding suitable natural compounds that can replace the synthetic food additives (preservatives, antioxidants, colorants, aromas) [65]. Although a lot of investigations studied the antioxidant potentials of plant-derived wastes and by-products, the studies regarding their incorporation in food products are in early stages. Some examples of applications of recovered antioxidant compounds in foods are presented in the next paragraphs.

Carotenoids are a group of natural pigments beneficial for the health of humans due to their antioxidant properties but they are also used as food colorants. Most utilized in the food industry, for their antioxidant and coloring effect, are lycopene and β-carotene. These compounds, together with phytoene, phytofluene, lutein, ξ-carotene, γ-carotene and neurosporene, are found in tomato peel in considerable quantities. Besides the fact that the tomato peels contain up to five times more lycopene than the pulp, some studies also showed that the bioavailability of lycopene from processed tomato (submitted to heating and trituration) is greater than that from raw tomatoes [65–67]. Other fruit wastes (peels and seeds), sources of carotenoids, are avocado peel, banana peel, and mango peel. Carotenoids may be incorporated in different food products due to their antioxidant properties (improving the product shelf life), and colorant properties but also as nutritional constituents acting as precursor of vitamin A. Thus, some examples of products in which recovered carotenoids from wastes were incorporated include macaroni (nutritional, improving sensorial attributes before and after processing) [68], refined vegetable oils (antioxidant, increasing thermal stability) [69], and antioxidant edible films (improving shelf life) [9].

Another big class of natural pigments is represented by the polyphenols. They have a high capacity of scavenging reactive oxygen species (e.g., free radicals), thus being suitable to be used in food products as antioxidants. There are many fruits and vegetable wastes from which polyphenols can be recovered (see **Table 1**). A recent study evaluated the use of a polyphenol-rich extract from olive oil waste to act as a natural antioxidant in lamb meat patties [70]. The results were promising, showing that the polyphenolic extract could improve the product shelf life by preventing the discoloration and oxidative processes. Adding antioxidants from potato peel extracts at concentrations ranging from 2.4 to 4.8 g/kg in minced horse mackerel had also positive impact on the product preservation. In the mackerel treated with polyphenolic extracts, the oxidation of proteins and lipids was prevented, considerably reducing peroxide value, tocopherol degradation, and generation of volatile secondary oxidation substances [71]. Similar results were obtained when polyphenolic extract from carob seeds peel was used as antioxidant in minced horse mackerel [72]. The polyphenolic extracts from potato peels were proved to have similar antioxidant capacity as the synthetic ones (butylated hydroxyanisole (BHA), butylated hydroxytoluene (BHT)) when incorporated in sunflower and soybean. The inhibition of thermal degradation of the oils may be attributed to the main polyphenolic compounds identified in potato peel extract: chlorogenic and gallic acids [73]. Brewers' spent grain—a by-product from brewing process—is a potentially valuable source of natural antioxidant compounds derived from the barley husk [74]. Ferulic acid, *p*-coumaric acids, and caffeic acid are in the highest concentrations, and they have been found with an excellent antioxidant potential, anti-inflammatory, and anticancer activities [75]. Brewers' spent grain flour or extracts can be added in bakery products, like enhancing their nutritional value [76]. Grape pomace, the winery waste, is particularly rich in polyphenols. The polyphenolic extract from muscadine grape pomace was tested *in vitro* to evaluate its capability to reduce the acrylamide formation. Acrylamide, a human carcinogen is a by-product of Maillard reaction, formed during the thermal treatment in different starchy food products (e.g., bread, potato chips). The results showed that the grape polyphenols (especially fractions recovered from skin and seed) significantly reduced the acrylamide level (by 60.3%) in potato chip model, even though there was no significant correlation between polyphenol antioxidant capacity and their potential for acrylamide inhibition [77].

Grape pomace is also an important source of fibers. Dietary fibers are generally known as being a health-promoting component of a diet. The consumption of this kind of fibers is connected with prevention, amelioration, and reductions in risks associated with cardiovascular disease, cancer, and diabetes [78]. Additionally, in the grape pomace, besides the dietary fiber, flavonoids are also present. The investigation of the antioxidant activity of flavonoids extracted from grape pomace has led to the elaboration of a new idea of antioxidant dietary fiber [79]. The presence of antioxidant compounds in the dietary fibers enhances their health benefits and their applications in pharmacological, cosmetic and food industries [80, 81]. Thus, for example, incorporating antioxidant dietary fibers into meat products could improve both their nutritional value and stability to oxidation. Grape pomace-added beef sausages (1% w/w) had a decreased rate of lipid oxidation and better sensorial attribute (taste and color) [82], while yogurt and salad dressings fortified with grape pomace likewise showed increased lipid oxidation stability without negatively influencing the consumers' acceptance of the products [83]. Another source of antioxidant dietary fiber is the apple pomace. Obtained as a by-product after fruit processing, it is composed mainly of skin and pulp tissues which consist of pectin, cellulose, hemicellulose, lignin, gums, and phenolic compounds [32]. Among phenolic compounds found in apple pomace, phlorizin is used as a basic structure for a new class of oral antidiabetic drugs [84]. Other health benefits of apple polyphenols are antioxidant, antihypertensive, anticancer, antidiabetic, and hypolipidemic activities, thus making them appropriate to be used as nutraceutical [29, 85]. Many dietary polyphenolic components derived from plants have more efficient antioxidant activity *in vitro* than vitamins E or C and thus have the ability to lead significantly to the protective results *in vivo*. Several studies consider that fruit and vegetable dietary fiber could have better nutritional properties due to the synergistic effect of associated bioactive compounds such as flavonoids and carotenoids [86, 87].

Some of the antioxidant compounds recovered from vegetable wastes are already valorized in food products that can be found on the market. Thus, for example, some of the patented applications of recovered antioxidants include: the "sugar syrup" extracted with solvent from citrus peels which is used as food natural sweetener (AU1983/0011308D); lycopene from tomato waste used as food antioxidant and supplement (PCT/EP2007/061923); proanthocyanidines from grape and cranberry seeds used as coloring additive in soy sauce (JP1998/0075070); polyphenols from grape pomace or seeds used in food supplements (WO/1999/030724); ellagic acid (40%) and punicalagin (40%) from pomegranate rind and seedcase residues used as food antioxidants (CN2010/1531940); hydroxytyrosol from olive leaves extract as natural antioxidant in food stuff (EP 1582512 A1); and bioactive silverskin extract from coffee silverskin with potential applications in cosmetic, nutrition and health (WO2013/004873) [88].

5. Re-evaluation of food wastes as a source of valuable molecules

The interest of the research community in finding new or nonconventional sources of antioxidants is triggered by the numerous scientific evidences regarding the health effects of the dietary intake of antioxidants. Thus, by fortifying food products with antioxidant compounds, a supplementation of the daily diet with bioactive compounds may be achieved, therefore helping the human body to fight against damaging factors.

The key point for the recovery of natural compounds from fruits and vegetable wastes is to develop flexible strategies for each stage in which wastes are produced. Implementation of a modern technology by using green solvents and safer materials is strongly recommended. Obtaining purified active compounds is rather demanding for food industry and consumers, although this procedure involves an accurate safety assessment and long and sophisticated tests. From the laboratory scale and testing, the procedures used for the recovery of bioactive compounds are now facing the challenges for the scaled-up and further commercialization. The industrial recovery of antioxidants from food wastes, on one hand, is sustained by the numerous studies which have demonstrated their health benefits and, on the other hand, by the food companies which have foreseen the manifold applications of these bioactive compounds. Even though the scaled-up recovery processes may encounter some limitations (e.g., the variability in the composition of vegetable waste, waste collection and preservation method, purity of the isolated antioxidants, functionality of recovered antioxidants), with a proper management, a company could economically benefit by exploiting the recovered compounds to develop new functional food that meet the consumers not only organoleptic criteria but also their demand for healthier food products and at the same time addressing their concern for the environment [2, 6, 46, 48, 88].

Taking in consideration the health and food issues in the actual economical and environmental context, food wastes should no longer be regarded as a waste to be disposed but as a renewable source of valuable molecules that should be fully exploited. Still nowadays, despite their potential, food wastes remain often underexploited. So instead of the classical "waste to waste" perspective, new "waste to health" or "waste to food" perspectives should be considered especially because functional foods or nutraceuticals can be obtained by utilizing low-cost sources of bioactive compounds, ranging from antioxidants to dietary fibers, proteins, dietary lipids, natural colorants, or aroma compounds (e.g., essential oils). Health benefits of bioactive compounds from wastes will open up new research directions not only in functional food innovation but also in the medicine, pharmacy, or chemistry research fields.

Acknowledgements

This work was supported by a grant of the Romanian National Authority for Scientific Research, CNCSIS–UEFISCDI, project number PN-II-RU-TE-2014-4-0842.

Author details

Sonia Ancuța Socaci[1]*, Dumitrița Olivia Rugină[2], Zorița Maria Diaconeasa[1], Oana Lelia Pop[1], Anca Corina Fărcaș[1], Adriana Păucean[1], Maria Tofană[1] and Adela Pintea[2]

*Address all correspondence to: sonia.socaci@usamvcluj.ro

1 Faculty of Food Science and Technology, University of Agricultural Sciences and Veterinary Medicine, Cluj-Napoca, Romania

2 Faculty of Veterinary Medicine, University of Agricultural Sciences and Veterinary Medicine, Cluj-Napoca, Romania

References

[1] SAVE FOOD: Global Initiative on Food Loss and Waste Reduction. Food and Agriculture Organization of the United Nations. Available from: http://www.fao.org/save-food/resources/keyfindings/en/ [Accessed: February 2, 2017]

[2] Oreopoulou V, Tzia C. Utilization of plant by-products for the recovery of proteins, dietary fibers, antioxidants, and colorants. In: Oreopoulou V, Russ W, editors. Utilization of By-Products and Treatment of Waste in the Food Industry. Vol. 3. Springer US, United States; 2007. pp. 209-232. DOI: 10.1007/978-0-387-35766-9_11

[3] Baiano A. Recovery of biomolecules from food wastes—a review. Molecules. 2014;**19**: 14821-14842. DOI: 10.3390/molecules190914821

[4] Ravindran R, Jaiswal AK. Exploitation of food industry waste for high-value products. Trends in Biotechnology. 2016;**34**(1):58-69. DOI: 10.1016/j.tibtech.2015.10.008

[5] Gil-Chavez GJ, Villa JA, Ayala-Zavala JF, Heredia JB, Sepulveda D, Yahia EM, Gonzalez-Aguilar GA. Technologies for extraction and production of bioactive compounds to be used as nutraceuticals and food ingredients: An overview. Comprehensive Reviews in Food Science and Food Safety. 2013;**12**:5-23. DOI: 10.1111/1541-4337.12005

[6] Varzakas T, Zakynthinos G, Verpoort F. Plant food residues as a source of nutraceutical and functional food. Foods. 2016;**5**:88. DOI: 10.3390/foods5040088

[7] Maritim AC, Sanders RA, Watkins JB III. Diabetes, oxidative stress, and antioxidants: A review. Journal of Biochemical and Molecular Toxicology. 2003;**17**(1):24-38. DOI: 10.1002/jbt.10058

[8] Arogba SS. Mango (*Mangifera indica*) kernel: Chromatographic analysis of the tannin, and stability study of the associated polyphenol oxidase activity. Journal of Food Composition and Analysis. 2000;**13**:149-156. DOI: 10.1006/jfca.1999.0838

[9] Gomez-Estaca J, Calvo MM, Sanchez-Faure A, Montero P, Gomez-Guillen MC. Development, properties, and stability of antioxidant shrimp muscle protein films incorporating carotenoid-containing extracts from food by-products. LWT—Food Science and Technology. 2015;**64**:189-196. DOI: 10.1016/j.lwt.2015.05.052

[10] Lante A, Nardi T, Zocca F, Giacomini A, Corich V. Evaluation of red chicory extract as a natural antioxidant by pure lipid oxidation and yeast oxidative stress response as model systems. Journal of Agricultural and Food Chemistry. 2011;**59**:5318-5324. DOI: 10.1021/jf2003317

[11] Moure A, Cruz JM, Franco D, Dominguez JM, Sineiro J, Dominguez H, Nunez MJ, Parajo JC. Natural antioxidants from residual sources. Food Chemistry. 2001;**72**:145-171. DOI: 10.1016/S0308-8146(00)00223-5

[12] Puravankara D, Boghra V, Sharma RS. Effect of antioxidant principles isolated from mango (*Mangifera indica* L.) seed kernels on oxidative stability of buffalo ghee (butter-fat). Journal of the Science of Food and Agriculture. 2000;**80**:522-526. DOI: 10.1002/(SICI)1097-0010(200003)80:4<522::AID-JSFA560>3.0.CO;2-R

[13] Schieber A, Stintzing FC, Carle R. By-products of plant food processing as a source of functional compounds—recent developments. Trends in Food Science & Technology. 2001;**12**:401-413. DOI: 10.1016/S0924-2244(02)00012-2

[14] Tunchaiyaphum S, Eshtiaghi MN, Yoswathana N. Extraction of bioactive compounds from mango peels using green technology. International Journal of Chemical Engineering and Applications. 2013;**4**(4):194-198. DOI: 10.7763/IJCEA.2013.V4.293

[15] Azabou S, Abid Y, Sebii H, Felfoul I, Gargouri A, Attia H. Potential of the solid-state fermentation of tomato by products by Fusarium solani pisi for enzymatic extraction of lycopene. LWT—Food Science and Technology. 2016;**68**:280-287. DOI: 10.1016/j.lwt.2015.11.064

[16] Dailey A, Vuong QV. Effect of extraction solvents on recovery of bioactive compounds and antioxidant properties from macadamia (*Macadamia tetraphylla*) skin waste. Cogent Food & Agriculture. 2015;**1**:1115646. DOI: 10.1080/23311932.2015.1115646

[17] Manousaky A, Jancheva M, Grigorakis S, Makris DP. Extraction of antioxidant phenolics from agri-food waste biomass using a newly designed glycerol-based natural low-transition temperature mixture: A comparison with conventional eco-friendly solvents. Recycling. 2016;**1**:194-204. DOI: 10.3390/recycling1010194

[18] Banerjee J, Singh R, Vijayaraghavan R, MacFarlane D, Patti AF, Arora A. Bioactives from fruit processing wastes: Green approaches to valuable chemicals. Food Chemistry. 2017;**225**:10-22. DOI: 10.1016/j.foodchem.2016.12.093

[19] Nazzaro M, Mottola MV, La Cara F, Del Monaco G, Aquino RP, Volpe MG. Extraction and characterization of biomolecules from agricultural wastes. Chemical Engineering Transactions. 2012;**27**:331-336. DOI: 10.3303/CET1227056

[20] Rodríguez Amado I, Franco D, Sánchez M, Zapata C, Vázquez JA. Optimisation of antioxidant extraction from Solanum tuberosum potato peel waste by surface response methodology. Food Chemistry. 2014;**165**:290-299. DOI: 10.1016/j.foodchem.2014.05.103

[21] Khalifa I, Barakat H, El-Mansy HA, Soliman SA. Optimizing bioactive substances extraction procedures from guava, olive and potato processing wastes and evaluating their antioxidant capacity. Journal of Food Chemistry and Nanotechnology. 2016;**2**(4):170-177. DOI: 10.17756/jfcn.2016-027

[22] Dorta E, Lobo MG, González M. Optimization of factors affecting extraction of antioxidants from mango seed. Food and Bioprocess Technology. 2013;**4**(6):1067-1081. DOI: 10.1007/s11947-011-0750-0

[23] Paleologou I, Vasiliou A, Grigorakis S, Makris DP. Optimisation of a green ultrasound-assisted extraction process for potato peel (*Solanum tuberosum*) polyphenols using bio-solvents and response surface methodology. Biomass Conversion and Biorefinery. 2016;**3**(6):289-299. DOI: 10.1007/s13399-015-0181-7

[24] Prado JM, Vardaneg R, Debien ICN, de Almeida Meireles MA, Gerschenson LN, Sowbhagya HB, Chemat S. Conventional extraction. In: Galanakis CM, editor. Food Waste Recovery. San Diego: Academic Press; 2015. pp. 127-148

[25] Goula AM, Ververi M, Adamopoulou A, Kaderides K. Green ultrasound-assisted extraction of carotenoids from pomegranate wastes using vegetable oils. Ultrasonics Sonochemistry. 2017;**34**:821-830. DOI: 10.1016/j.ultsonch.2016.07.022

[26] Angiolillo L, Del Nobile MA, Conte A. The extraction of bioactive compounds from food residues using microwaves. Current Opinion in Food Science. 2015;**5**:93-98. DOI: 10.1016/j.cofs.2015.10.001

[27] Heng MY, Katayama S, Mitani T, Ong ES, Nakamura S. Solventless extraction methods for immature fruits: Evaluation of their antioxidant and cytoprotective activities. Food Chemistry. 2017;**221**:1388-1393. DOI: 10.1016/j.foodchem.2016.11.015

[28] Matharu AS, de Melo EM, Houghton JA. Opportunity for high value-added chemicals from food supply chain wastes. Bioresource Technology. 2016;**215**:123-130. DOI: 10.1016/j.biortech.2016.03.039

[29] Yates M, Gomez MR, Martin-Luengo MA, Ibañez VZ, Martinez Serrano AM. Multivalorization of apple pomace towards materials and chemicals. Waste to wealth. Journal of Cleaner Production. 2017;**143**:847-853. DOI: 10.1016/j.jclepro.2016.12.036

[30] Vauchel P, Galván D'Alessandro L, Dhulster P, Nikov I, Dimitrov K. Pilot scale demonstration of integrated extraction–adsorption eco-process for selective recovery of antioxidants from berries wastes. Journal of Food Engineering. 2015;**158**:1-7. DOI: 10.1016/j.jfoodeng.2015.02.023

[31] Benítez V, Mollá E, Martín-Cabrejas MA, Aguilera Y, López-Andréu FJ, Cools K, Terry LA, Esteban RM. Characterization of industrial onion wastes (Allium cepa L.): Dietary fibre and bioactive compounds. Plant Foods for Human Nutrition. 2011;**66**:48-57. DOI: 10.1007/s11130-011-0212-x

[32] Rana S, Gupta S, Rana A, Bhushan S. Functional properties, phenolic constituents and antioxidant potential of industrial apple pomace for utilization as active food ingredient. Food Science and Human Wellness. 2015;**4**(4):180-187

[33] Sood A, Gupta M. Extraction process optimization for bioactive compounds in pomegranate peel. Food Bioscience. 2015;**12**:100-106. DOI: 10.1016/j.fbio.2015.09.004

[34] Viganó J, Zaboti Brumer I, de Campos Braga PA, da Silva JK, Maróstica MR Júnior, Reyes Reyes FG, Martínez J. Pressurized liquids extraction as an alternative process to readily obtain bioactive compounds from passion fruit rinds. Food and Bioproducts Processing. 2016;**100**(**Part A**):382-390. DOI: 10.1016/j.fbp.2016.08.011

[35] da Silva Caetano AC, de Araújo CR, de Lima VLAG, Maciel MIS, Melo EA. Evaluation of antioxidant activity of agro-industrial waste of acerola (Malpighia emarginata D.C.) fruit extracts. Food Science and Technology (Campinas). 2011;**31**:769-775

[36] Lafka TI, Sinanoglou V, Lazos ES. On the extraction and antioxidant activity of phenolic compounds from winery wastes. Food Chemistry. 2007;**3**(104):1206-1214. DOI: 10.1016/j.foodchem.2007.01.068

[37] Manna L, Bugnone CA, Banchero M. Valorization of hazelnut, coffee and grape wastes through supercritical fluid extraction of triglycerides and polyphenols. The Journal of Supercritical Fluids. 2015;**104**:204-211. DOI: 10.1016/j.supflu.2015.06.012

[38] Ares AM, Nozal MJ, Bernal JL, Bernal J. Optimized extraction, separation and quantification of twelve intact glucosinolates in broccoli leaves. Food Chemistry. 2014;**152**:66-74. DOI: 10.1016/j.foodchem.2013.11.125

[39] Strati IF, Gogou E, Oreopoulou V. Enzyme and high pressure assisted extraction of carotenoids from tomato waste. Food and Bioproducts Processing. 2015;**94**:668-674. DOI: 10.1016/j.fbp.2014.09.012

[40] Luengo E, Condón-Abanto S, Condón S, Álvarez I, Raso J. Improving the extraction of carotenoids from tomato waste by application of ultrasound under pressure. Separation and Purification Technology. 2014;**36**:130-136. DOI: 10.1016/j.seppur.2014.09.008

[41] Rabelo RS, Machado MTC, Martínez J, Hubinger MD. Ultrasound assisted extraction and nanofiltration of phenolic compounds from artichoke solid wastes. Journal of Food Engineering. 2016;**178**:170-180. DOI: 10.1016/j.jfoodeng.2016.01.018

[42] Radočaj O, Vujasinović V, Dimić E, Basić Z. Blackberry (*Rubus fruticosus* L.) and raspberry (*Rubus idaeus* L.) seed oils extracted from dried press pomace after longterm frozen storage of berries can be used as functional food ingredients. European Journal of Lipid Science and Technology. 2014;**116**:1015-1024. DOI: 10.1002/ejlt.201400014

[43] Oszmiański J, Wojdyło A, Lachowicz S, Gorzelany J, Matłok N. Comparison of bioactive potential of cranberry fruit and fruit-based products versus leaves. Journal of Functional Foods. 2016;**22**:232-242. DOI: 10.1016/j.jff.2016.01.015

[44] Rugina D, Diaconeasa Z, Coman C, Bunea A, Socaciu C, Pintea A. Chokeberry anthocyanin extract as pancreatic beta-cell protectors in two models of induced oxidative stress. Oxidative Medicine and Cellular Longevity. 2015;**2015**:429075. DOI: 10.1155/2015/429075

[45] Rugină D, Sconţa Z, Leopold L, Pintea A, Bunea A, Socaciu C. Antioxidant activities of chokeberry extracts and the cytotoxic action of their anthocyanin fraction on HeLa human cervical tumor cells. Journal of Medicinal Food. 2012;**15**:700-706

[46] Cho H, Jung H, Lee H, Yi HC, Kwak HK, Hwang KT. Chemopreventive activity of ellagitannins and their derivatives from black raspberry seeds on HT-29 colon cancer cells. Food & Function. 2015;**6**:1675-1683. DOI: 10.1039/c5fo00274e

[47] McCann MJ, Gill CI, O'Brien G, Rao JR, McRoberts WC, Hughes P, et al. Anti-cancer properties of phenolics from apple waste on colon carcinogenesis in vitro. Food and Chemical Toxicology. 2007;**45**:1224-1230. DOI: 10.1016/j.fct.2007.01.003

[48] Tow WW, Premier R, Jing H, Ajlouni S. Antioxidant and antiproliferation effects of extractable and nonextractable polyphenols isolated from apple waste using different extraction methods. Journal of Food Science. 2011;**76**:T163-T172. DOI: 10.1111/j.1750-3841.2011.02314.x

[49] Wolfe K, Wu X, Liu RH. Antioxidant activity of apple peels. Journal of Agricultural and Food Chemistry. 2003;**51**:609-614. DOI: 10.1021/jf020782a

[50] Olsson ME, Gustavsson KE, Andersson S, Nilsson A, Duan RD. Inhibition of cancer cell proliferation in vitro by fruit and berry extracts and correlations with antioxidant levels. Journal of Agricultural and Food Chemistry. 2004;**52**:7264-7271. DOI: 10.1021/jf030479p

[51] Marín FR, Soler-Rivas C, Benavente-García O, Castillo J, Pérez-Alvarez JA. By-products from different citrus processes as a source of customized functional fibres. Food Chemistry. 2007;**100**:736-741. DOI: 10.1016/j.foodchem.2005.04.040

[52] Sharma K, Mahato N, Cho MH, Lee YR. Converting citrus wastes into value-added products: Economic and environmentally friendly approaches. Nutrition. 2017;**34**:29-46. DOI: 10.1016/j.nut.2016.09.006

[53] Jae-Hee P, Minhee L, Eunju P. Antioxidant activity of orange flesh and peel extracted with various solvents. Preventive Nutrition and Food Science. 2014;**19**:291-298. DOI: 10.3746/pnf.2014.19.4.291

[54] Masci A, Coccia A, Lendaro E, Mosca L, Paolicelli P, Cesa S. Evaluation of different extraction methods from pomegranate whole fruit or peels and the antioxidant and antiproliferative activity of the polyphenolic fraction. Food Chemistry. 2016;**202**:59-69. DOI: 10.1016/j.foodchem.2016.01.106

[55] Orgil O, Schwartz E, Baruch L, Matityahu I, Mahajna J, Amir R. The antioxidative and anti-proliferative potential of non-edible organs of the pomegranate fruit and tree. LWT—Food Science and Technology. 2014;**58**:571-577. DOI: 10.1016/j.lwt.2014.03.030

[56] Oliveira DA, Angonese M, Gomes C, Ferreira SRS. Valorization of passion fruit (*Passiflora edulis* sp.) by-products: Sustainable recovery and biological activities. The Journal of Supercritical Fluids. 2016;**111**:55-62. DOI: 10.1016/j.supflu.2016.01.010

[57] Asif A, Farooq U, Akram K, Hayat Z, Shafi A, Sarfraz F, et al. Therapeutic potentials of bioactive compounds from mango fruit wastes. Trends in Food Science & Technology. 2016;**53**:102-112. DOI: 10.1016/j.tifs.2016.05.004

[58] Hansakul P, Dechayont B, Phuaklee P, Prajuabjinda O, Juckmeta T, Itharat A. Cytotoxic and antioxidant activities of Antidesma thwaitesianum Müll Arg (Euphorbiaceae) fruit and fruit waste extracts. Tropical Journal of Pharmaceutical Research. 2015;**14**:627-634

[59] Ferreira PMP, Farias DF, Viana MP, Souza TM, Vasconcelos IM, Soares BM, et al. Study of the antiproliferative potential of seed extracts from Northeastern Brazilian plants. Anais da Academia Brasileira de Ciências. 2011;**83**:1045-1058

[60] del Carmen Robles-Ramírez M, Monterrubio-López R, Mora-Escobedo R, del Carmen Beltrán-Orozco M. Evaluation of extracts from potato and tomato wastes as natural antioxidant additives. Archivos Latinoamericanos de Nutrición. 2016;**66**:66-73

[61] Amado IR, Franco D, Sanchez M, Zapata C, Vazquez JA. Optimisation of antioxidant extraction from Solanum tuberosum potato peel waste by surface response methodology. Food Chemistry. 2014;**165**:290-299. DOI: 10.1016/j.foodchem.2014.05.103

[62] Zeyeda NN, Zeitoun MAM, Barbary OM. Utilization of some vegetables and fruits waste as natural antioxidants. Alexandria Journal of Food Science and Technology. 2008;**5**:1-11

[63] Stajčić S, Ćetković G, Čanadanović-Brunet J, Djilas S, Mandić A, Četojević-Simin D. Tomato waste: Carotenoids content, antioxidant and cell growth activities. Food Chemistry. 2015;**172**:225-232. DOI: 10.1016/j.foodchem.2014.09.069

[64] Ćetković G, Savatović S, Čanadanović-Brunet J, Djilas S, Vulić J, Mandić A, et al. Valorisation of phenolic composition, antioxidant and cell growth activities of tomato waste. Food Chemistry. 2012;**133**:938-945. DOI: 10.1016/j.foodchem.2012.02.007

[65] Martins N, Ferreira ICFR. Wastes and by-products: Upcoming sources of carotenoids for biotechnological purposes and health-related applications. Trends in Food Science & Technology. 2017;**62**:33-48. DOI: 10.1016/j.tifs.2017.01.014

[66] Britton G, Gambelli L, Dunphy P, Pudney P, Gidley M. Physical state of carotenoids in chromoplasts of tomato and carrots: Consequences and bioavailability. In: Proceedings of the Second International Congress on Pigments in Foods; Lisbon, Portugal. Sociedade Portuguesa de Quimica 2002. pp. 151-154

[67] Horvitz MA, Simon PW, Tanumihardjo SA. Lycopene and [beta]-carotene are bioavailable from lycopene 'red' carrots in humans. European Journal of Clinical Nutrition. 2004;**5**(58):803-811

[68] Ajila CM, Aalami M, Leelavathi K, Rao UJSP. Mango peel powder: A potential source of antioxidant and dietary fiber in macaroni preparations. Innovative Food Science and Emerging Technologies. 2010;**11**(1):219-224. DOI: 10.1016/j.ifset.2009.10.004

[69] Benakmoum A, Abbeddou S, Ammouche A, Kefalas P, Gerasopoulos D. Valorisation of low quality edible oil with tomato peel waste. Food Chemistry. 2008;**110**:684-690. DOI: 10.1016/j.foodchem.2008.02.063

[70] Muíno I, Díaz MT, Apeleo E, Perez-Santaescolastica C, Rivas-Canedo A, Perez C, Caneque V, Lauzurica S, de la Fuente J. Valorisation of an extract from olive oil waste as a natural antioxidant for reducing meat waste resulting from oxidative processes. Journal of Cleaner Production. 2017;**140**:924-932. DOI: 10.1016/j.jclepro.2016.06.175

[71] Farvin SKH, Grejsen HD, Jacobsen C. Potato peel extract as a natural antioxidant in chilled storage of minced horse mackerel (Trachurus trachurus): Effect on lipid and protein oxidation. Food Chemistry. 2012;**131**:843-851. DOI: 10.1016/j.foodchem.2011.09.056

[72] Albertos I, Jaime I, Diez AM, Gonzalez-Arnaiz L, Rico D. Carob seed peel as natural antioxidant in minced and refrigerated (4°C) Atlantic horse mackerel (Trachurus trachurus). LWT—Food Science and Technology. 2015;**64**:650-656. DOI: 10.1016/j.lwt.2015.06.037

[73] Mohdaly AAA, Sarhan MA, Mahmoud A, Ramadan MF, Smetanska I. Antioxidant efficacy of potato peels and sugar beet pulp extracts in vegetable oils protection. Food Chemistry. 2010;**123**:1019-1026. DOI: 0.1016/j.foodchem.2010.05.054

[74] Fărcaş AC, Socaci SA, Dulf FV, Tofană M, Mudura E, Diaconeasa Z. Volatile profile, fatty acids composition and total phenolics content of brewers' spent grain by-product with potential use in the development of new functional foods. Journal of Cereal Science. 2015;**64**:34-42. DOI: 10.1016/j.jcs.2015.04.003

[75] McCarthy AL, O'Callaghan YC, Piggott CO, FitzGerald RJ, O'Brien NM. Brewers' spent grain; bioactivity of phenolic component, its role in animal nutrition and potential for incorporation in functional foods: A review. Proceedings of the Nutrition Society. 2013;**72**:117-125

[76] Farcas A, Socaci S, Tofana M, Muresan C, Mudura E, Salanta L, Scrob S. Nutritional properties and volatile profile of brewer's spent grain supplemented bread. Romanian Biotechnological Letters. 2014;**19**(5):9705-9714

[77] Xu C, Yagiz Y, Marshall S, Li Z, Simonne A, Lu J, Marshall MR. Application of muscadine grape (*Vitis rotundifolia* Michx.) pomace extract to reduce carcinogenic acrylamide. Food Chemistry. 2015;**182**:200-208. DOI: 10.1016/j.foodchem.2015.02.133

[78] Anderson JW, Baird P, Davis RH Jr, Ferreri S, Knudtson M, Koraym A, Waters V, Williams CL. Health benefits of dietary fiber. Nutrition Reviews. 2009;**4**(67):188-205. DOI: 10.1111/j.1753-4887.2009.00189.x

[79] Saura-Calixto F. Antioxidant dietary fiber product: A new concept and a potential food ingredient. Journal of Agricultural and Food Chemistry. 1998;**46**:4303-4306. DOI: 10.1021/jf9803841

[80] Teixeira A, Baenas N, Dominguez-Perles R, Barros A, Rosa E, Moreno DA, Garcia-Viguera C. Natural bioactive compounds from winery by-products as health promoters: A review. International Journal of Molecular Sciences. 2014;**15**:15638-15678

[81] Zhu F, Du B, Zheng L, Li J. Advance on the bioactivity and potential applications of dietary fibre from grape pomace. Food Chemistry. 2015;**186**:207-212. DOI: 10.1016/j.foodchem.2014.07.057

[82] Riazi F, Zeynali F, Hoseini E, Behmadi H, Savadkoohi S. Oxidation phenomena and color properties of grape pomace on nitrite-reduced meat emulsion systems. Meat Science. 2016;**121**:350-358. DOI: 10.1016/j.meatsci.2016.07.008

[83] Tseng A, Zhao Y. Wine grape pomace as antioxidant dietary fibre for enhancing nutritional value and improving storability of yogurt and salad dressing. Food Chemistry. 2013;**138**:356-365. DOI: 10.1016/j.foodchem.2012.09.148

[84] Helkar PB, Sahoo AK, Patil NJ. Review: Food industry by-products used as a functional food ingredients. International Journal of Waste Resources. 2016;**6**(3):1-6. DOI: 10.4172/2252-5211.1000248

[85] Rabetafikaa HN, Bchirb B, Bleckerb C, Richela A. Fractionation of apple by-products as source of new ingredients: Current situation and perspectives. Trends in Food Science & Technology. 2014;**40**:99-114

[86] Vergara-Valencia N, Granados-Perez E, Agama-Acevedo E, Tovar J, Ruales J, Bello-Perez LA. Fibre concentrate from mango fruit: Characterization; associated antioxidant capacity and application as a bakery product ingredient. LWT—Food Science and Technology. 2007;**40**:722-729

[87] Fernandez-Lopez J, Sendra-Nadal E, Navarro C, Sayas E, Viuda-Martos M, Alvarez JAP. Storage stability of a high dietary fibre powder from orange by-products. International Journal of Food Science and Technology. 2009;**44**:748-756

[88] Galanakis CM, Martinez-Saez N, del Castillo MD, Barba FJ, Mitropoulou VS. Patented commercialized applications. In: Galankis C, editor. Food Waste Recovery: Processing Technologies and Industrial Techniques. 1st ed. Academic Press Elsevier, US; 2015. pp. 337-357. DOI: 10.1016/B978-0-12-800351-0.00015-8

Chapter 2

Polyphenols: Food Sources and Health Benefits

Nikolina Mrduljaš, Greta Krešić and Tea Bilušić

Additional information is available at the end of the chapter

http://dx.doi.org/10.5772/intechopen.68862

Abstract

The current scientific knowledge on the relationship between diet and human health is greatly focused on the effects of phytochemicals, especially polyphenols, on chronic diseases, due to their preventive effect as shown by many epidemiological studies. Herbs, cocoa products, and darkly colored berries, such as black elderberries, chokeberries, and black currants, are the richest dietary sources that contribute to the average intake of polyphenols of about 1 g/day. Polyphenols that are the most common in the human diet are not necessarily the most active in the body because their beneficial effects depend on the plant matrix in which they are incorporated and on processing methods and endogenous factors such as microbiota and digestive enzymes. Polyphenol-rich foods are considered as being potential functional foods due to antioxidant, anti-inflammatory, antimicrobial, immunomodulatory, anticancer, vasodilating, and prebiotic-like properties. This review will outline findings on the preventive effects of polyphenols on chronic diseases, the factors affecting polyphenol bioavailability and bioaccessibility, and new trends in functional food production.

Keywords: polyphenols, dietary intake, chronic diseases, bioavailability, functional food

1. Introduction

Polyphenols are the most common phytochemicals in human diet and comprise a variety of compounds with a great diversity of structures, ranging from simple molecules to polymers with high molecular weight. Polyphenols are plant secondary metabolites present in all plant tissues, and their primary role is to protect plants from insects, ultraviolet radiation, and microbial infections and to attract pollinators [1]. According to the chemical structures of aglycones, polyphenols are classified as flavonoids, phenolic acids, lignans, and stilbenes [2]. Fruits, vegetables, whole grains, chocolate, and drinks like tea and wine are good sources of polyphenols, but due to diverse chemical structures, it is difficult to estimate the total polyphenol content in foods. Beneficial health effects of these phytochemicals are directly linked to regular daily intake and bioavailability. The aim of this review is to present current knowledge regarding evidence on chronic disease prevention, factors affecting polyphenol bioavailability and bioaccessibility, and new trends in the production of polyphenol-enriched functional foods.

2. Classification and food sources of polyphenols

Dietary polyphenols comprise a variety of compounds among which flavonoids and several classes of non-flavonoids are usually distinguished. In nature, polyphenols are bound to sugars in the form of glycosides. However, classification of polyphenols in this review will be presented according to the chemical structures of aglycones. These compounds contain at least one aromatic ring and are classified into different groups according to the number of aromatic rings and the structural elements that bind these rings together. Therefore, polyphenols are classified as flavonoids, phenolic acids, lignans, and stilbenes [2].

Flavonoids are the largest group of phenolic compounds and are widely distributed in plants, especially in fruits. Their structures consist of two aromatic rings that are bound together with a three-carbon bridge that form an oxygenated heterocycle (**Figure 1**). Their biological activities, including antioxidant activity, depend considerably on both structural difference

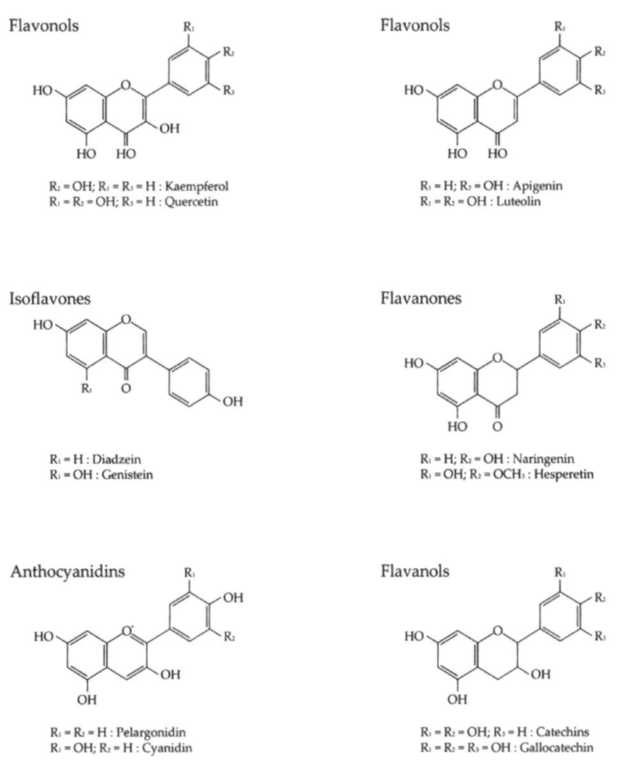

Figure 1. Chemical structures of flavonoids [2].

and glycosylation patterns [3]. According to the degree of oxidation of the central ring and the number and position of –OH groups, flavonoids can be divided in six subclasses: flavonols, flavones, isoflavones, flavanones, anthocyanidins, and flavanols.

Flavonols are one of the most ubiquitous flavonoids in food, and their main representatives are quercetin and kaempferol, typically found as glycosides [2]. Data on the content of flavonols in commonly consumed fruits, vegetables, and drinks can vary significantly due to local growing conditions (microclimate and agrotechnical requirements), seasonal changes, and varietal differences. The most significant dietary sources of this group of flavonoids are yellow and red onion and spinach, but the richest sources are capers, saffron, and dried Mexican oregano (**Table 1**).

The most common *flavones*, such as apigenin and luteolin, are not widely distributed in the plant kingdom although significant amounts are found in celery, parsley, and some herbs (**Table 1**). Tangeretin and nobiletin are polymethoxylated flavones, occurring only in tissues and peels of citrus fruits such as tangerine, grapefruit, and orange. These flavones have methylated hydroxyl groups, which increase their metabolic stability and improve oral bioavailability [4].

The best sources of *isoflavones* are legumes, especially soybeans, and their processed products containing significant amounts of daidzein and genistein (**Table 1**). Although the fermentation of soybeans during the manufacturing of certain foods, such as miso and tempeh, does not cause the loss of isoflavones, they are, however, in the form of aglycones due to bacterial hydrolysis of glycosides [2]. Unlike fermentation, the use of high temperature (the production of soy milk or tofu) can significantly reduce the concentration of isoflavones. Isoflavones possess pseudohormonal properties because of their structural similarity to estrogen, and they are consequently classified as phytoestrogens. Due to their ability to bind to estrogen receptors, soy foods and isoflavone supplements can be potential alternatives to conventional hormone therapy [5].

The most important *flavanones* in food are naringenin and hesperetin. The highest concentrations are found in dried herbs and citrus fruits (**Table 1**), and their glycosides are responsible for the bitter taste of grapefruit and some varieties of oranges.

Anthocyanidins are a subgroup of flavonoids that provide color to plant tissues (flowers, leaves, fruits, and roots), ranging from blue, purple, and red, depending on the pH and their structural composition. Anthocyanidins are considered the most important group of flavonoids in plants, having more than 600 compounds identified in nature [6]. They are widely distributed in colored fruits like berries, plums, and cherries as well as in many dark colored vegetables such as red cabbage, eggplant, red onion, and red radish, while the food content is generally proportional to color intensity. The most common anthocyanidin aglycones are pelargonidin, delphinidin, peonidin, petunidin, malvidin, and cyanidin, which is the most widespread in fruits and vegetables. Being highly unstable in the aglycone form, they are in the form of glycosides (anthocyanins) in plants, enabling them to be resistant to light, pH, and oxidation process [2].

Flavanols are the most complex subclass of flavonoids, ranging from simple monomers (catechin and its isomer epicatechin) to oligomers and polymers (proanthocyanidins) and other derived compounds (e.g., theaflavins and thearubigins) [7]. Catechins and epicatechin are

Flavonoid subgroup	Food source	Content (mg/100 g)
Flavonols	Capers	654.71
	Saffron	509.99
	Mexican oregano (dried)	272.07
	Red onion (raw)	128.51
	Spinach (raw)	119.27
Flavones	Celery seed	2094.00
	Peppermint (dried)	1486.29
	Common verbena (fresh)	790.00
	Mexican oregano (dried)	733.77
	Celery leaves (fresh)	133.38
Isoflavones	Soy (flour)	466.99
	Soy paste (cheonggukang)	264.40
	Soybean (roasted)	246.95
	Soy (tempeh)	147.72
	Soy paste (nato)	103.90
Flavanones	Peppermint (dried)	8739.98
	Mexican oregano (dried)	1049.67
	Grapefruit/pummelo hybrid (pure juice)	67.08
	Orange (juice from concentrate)	61.29
	Rosemary (fresh)	55.05
Anthocyanidins	Black elderberry	1316.65
	Black chokeberry	878.12
	Black currant (raw)	592.23
	Lowbush blueberry (raw)	187.23
	Blackberry (raw)	172.59
Flavanols	Cocoa (powder)	511.62
	Chocolate (dark)	212.36
	Broad bean pod (raw)	154.45
	Black tea (infusion)	73.30
	Green tea (infusion)	71.17

Table 1. The richest food sources of flavonoid groups determined by liquid chromatography [8].

found in many types of fruits such as strawberry, apple, and peach, but cocoa products and black and green tea are the richest sources (**Table 1**). In contrast to other classes of flavonoids, flavanols are stable and are not glycosylated in foods. The production of black tea decreases the concentration of catechins, mainly due to the action of polyphenol oxidase during fermentation, but at the same time, theaflavins and thearubigins are accumulating agents [1]. The oligomers and polymers of flavanols are also referred to as condensed tannins or proanthocyanidins that mainly consist of (epi)catechin units called procyanidins. They are responsible for the astringent character of some fruits and beverages and for the bitterness of chocolate [2].

Phenolic acids can be divided into two main groups—benzoic and cinnamic acids and their derivatives (**Figure 2**). The most important derivatives of benzoic acids are gallic and ellagic acid, which are found in various types of fruit such as raspberries, cranberries, and pomegranates and in nuts (e.g., chestnut contains 1215.22 mg of hydroxybenzoic acids per 100 g). Hydroxybenzoic acids are also components of complex structures like hydrolyzable tannins (gallotannins in mangoes and ellagitannins in red fruit such as strawberries and raspberries) [2].

The most important derivatives of cinnamic acids are coumaric, caffeic, ferulic, and sinapic acids. In food, they are often in the bound form and can only be released upon acid or alkaline hydrolysis or by enzymes. Caffeic acid is the most abundant phenolic acid and represents about 87% of the total hydroxycinnamic acid content of most fruits [2]. Caffeic and quinic acid together form chlorogenic acid, which makes up about 10% of green Robusta coffee beans. Regular consumption of coffee may provide more than 1 g of chlorogenic acid, which means that for many people it is the main source of dietary polyphenol [1].

Lignans are formed with two phenylpropane units and a four-carbon bridge, leading to many different chemical structures in nature (**Figure 3**). The highest amount of these compounds is found in flaxseeds, and other valuable sources are grains and certain vegetables. Lignans are one of the major classes of phytoestrogens, together with isoflavones mentioned earlier. In plants, they are typically found as glycosides and are converted by intestinal bacteria to give metabolites with estrogen activity like equol, enterodiol, and enterolactone [9].

Stilbenes are phytoalexins produced by plants in response to injury and infections. They are present in human diet in low quantities, and only resveratrol is considered important to human health (**Figure 4**). The most important dietary source of resveratrol is grapes and red wine. Resveratrol is directly linked to the *French paradox*, in which it was observed that the French consume significant amounts of saturated fatty acids while rarely suffering from cardiovascular disease and having a lower mortality rate compared with populations from other European countries. It is believed that their regular consumption of red wine plays a key role in preventing heart disease [10].

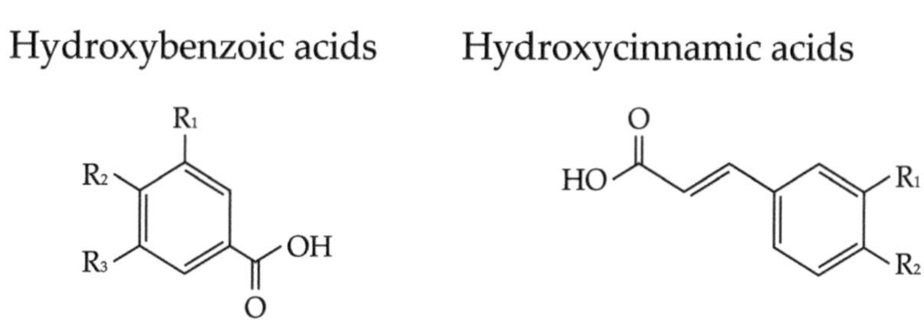

Figure 2. Chemical structure of phenolic acids [2].

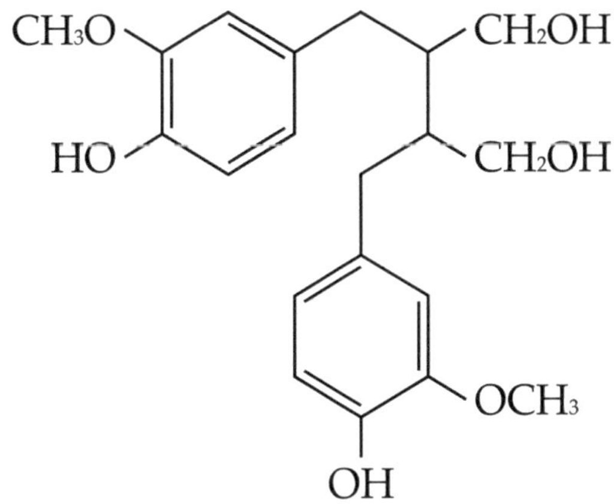

Figure 3. Chemical structure of lignans [2].

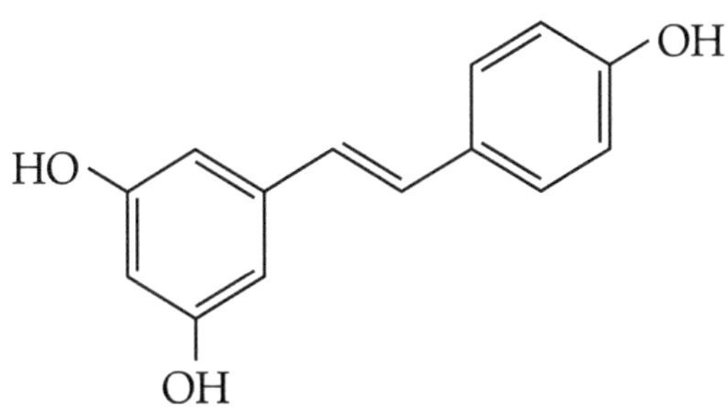

Figure 4. Chemical structure of stilbenes [2].

3. Health benefits

Polyphenols are the most common phytochemicals in human diet and are in the focus of scientific research due to their biological properties, bioavailability, and bioaccessibility, as well as their effects on the prevention of chronic diseases. Epidemiological studies confirm that moderate and prolonged intake of foods rich in polyphenols could prevent the formation of cancer and chronic diseases such as cardiovascular disease, neurodegenerative disease, type 2 diabetes, and obesity, which are the most common in Western populations [1].

A large primary prevention trial tested the long-term effects of the Mediterranean diet, containing polyphenol-rich foods, on the incidence of cardiovascular disease in participants with high risk but free of cardiovascular disease at baseline (the PREDIMED study). Data on their dietary habits were collected with a validated food frequency questionnaire, and the polyphenol content in foods was obtained from the Phenol-Explorer database. Results showed a significant reduction of cardiovascular events and cardiovascular mortality with a higher intake of total polyphenols, especially flavanols, lignans, and hydroxybenzoic acids [11]. The aim of this study was also to investigate the effect of polyphenol intake on all-cause mortality. Among high-risk subjects, those with higher polyphenol intake showed a 37% lower mortality risk, compared with those with lower intake. Subgroups of polyphenols with the strongest inverse association were stilbenes and lignans, while flavonoids and phenolic acids had no significant effect on mortality reduction [12]. However, the European Prospective Investigation into Cancer and Nutrition (EPIC) reported that higher flavonoid intake in the diet was associated with a 29% reduction in all-cause mortality, in particular for the subclasses of flavanones and flavonols, which decreased the incidence of cardiovascular disease by 40 and 41%, respectively [13]. Although a beneficial effect has been proven, more controlled trials are needed to definitively clarify the benefits of different polyphenol subgroups and to define minimum levels of dietary intake. Beneficial effects of polyphenols on cardiovascular disease have been attributed to their antioxidant activities, but recent evidence suggests that vasodilatory, anti-inflammatory, and anti-atherogenic properties may also contribute to cardiovascular risk reduction, indicate their ability to improve lipid profile, and modulate apoptotic processes in the vascular endothelium [14].

Growing evidence also indicates that polyphenols may prevent neurodegenerative diseases such as Alzheimer's disease and Parkinson's disease by decreasing inflammatory stress signaling, leading to the expression of genes that encode antioxidant enzymes and cytoprotective proteins [15]. A study conducted by Schmidt et al. [16] showed that green tea extracts can increase the number of connections between neurons of frontal and parietal brain regions which positively correlated with the improvement in the task performance. A double-blind study included 12 healthy volunteers who received either a milk solution with 27.5 g of green tea extract or a milk solution without the extract. The effect of green tea extract on working memory was visualized with functional magnetic resonance imaging (MRI) while performing memory test. Another intervention study confirmed the beneficial effect of blueberries. During 12 weeks of blueberry juice consumption, cognitive function (paired associate learning and word list recall) was significantly improved in older patients with early symptoms of dementia. In addition, symptoms of depression and blood glucose levels were reduced [17].

Many studies investigated the impact of polyphenols on carbohydrate metabolism and possible prevention of diabetes type 2. Polyphenols have the potential to inhibit key enzymes that are responsible for the digestion of dietary carbohydrates (α-amylase and α-glucosidase) and thus modify the postprandial glycemic response [18]. In vitro studies have shown that polyphenol-rich extracts from berries are effective in the inhibition of α-amylase and α-glucosidase at low levels. Tannin-like components (ellagitannins and proanthocyanidins) from raspberry and rowanberry were the most effective for amylase inhibition. A rowanberry fraction rich in proanthocyanidins was as equally strong an inhibitor as the whole rowanberry extract for α-amylase inhibition but was considerably less effective for α-glucosidase inhibition which suggests that tannins are poor inhibitors of α-glucosidase. Among the tested berry extracts, black currants rich in anthocyanins and flavonols had the strongest inhibitory effect on α-glucosidase [19]. The aim of an interesting study conducted by Yang and Kong [20, 21] was to investigate the effect of green tea polyphenols and green, black, and oolong tea extracts on α-amylase and α-glucosidase activity. All tested samples showed a strong inhibitory effect on α-glucosidase, and their inhibitory potency is mainly attributed to tea polyphenols. In contrast, all three types of tea extract significantly enhanced α-amylase activity, whereas green tea showed the highest activation effect. Green tea polyphenols significantly increased α-amylase activity in low concentrations. A high concentration, however, resulted in a mild inhibitory effect, suggesting that other constituents in the tea counteract the inhibitory effect of polyphenols. A large prospective EPIC-InterAct study examined the association between dietary flavonoid and lignan intake and the risk of developing diabetes type 2 in eight European countries. High intake of flavonoids was associated with a significant risk reduction, while the intake of lignans had no effect. Among flavonoid subclasses, flavonols and flavanols were associated with a significantly reduced risk of diabetes [22]. A comprehensive review by Kim et al. [18] summarizes epidemiological and clinical studies that investigated the relationship between food rich in polyphenols and risk of diabetes type 2. Despite promising data from in vitro and animal studies, the number of intervention surveys conducted on human beings is small. Most studies showed that polyphenols were associated with a lower risk of diabetes type 2, but this association was not entirely consistent. Potential mechanisms of the action of polyphenols in preventing diabetes type 2 include the stimulation of insulin secretion and protection of pancreatic β-cells against glucose toxicity, in addition to the inhibition of salivary and pancreatic α-amylase and α-glucosidase.

Obesity is considered one of the most serious health problems that have assumed the character of a global epidemic. According to the data published by Eurostat in 2014, 51.6% of adults in the European Union are overweight (35.7% pre-obese and 15.9% obese). The in vitro and some in vivo studies suggested that consumption of particular polyphenols (such as catechin in green tea, anthocyanins in blueberries, resveratrol in wine, and curcumin in turmeric) may facilitate weight loss and prevent weight gain due to changes in lipid and energy metabolism [23]. A survey conducted by Basu et al. [24] showed that using a freeze-dried blueberry beverage in obese people with metabolic syndrome for 8 weeks decreased blood pressure and the concentrations of oxidized LDL cholesterol and products of lipid peroxidation. Some researchers suggested that polyphenols may inhibit lipase activity and consequently reduce lipid absorption [25, 26]. Uchiyama et al. [27] have shown that black tea polyphenols in rats with diet-induced obesity can inhibit intestinal lipase activity and suppress the increase of triglyceride levels.

The cause of the aforementioned chronic disease can be associated with oxidative stress resulting from reactive oxygen and nitrogen species. Many in vitro studies have demonstrated that polyphenols can decrease inflammatory markers, reduce oxidative stress, and improve cancer biomarkers, but intervention studies have not always confirmed these positive effects. The reasons which could explain these differences include different doses of administered compounds, polyphenol instability in food and in the gastrointestinal system, a synergistic effect with other antioxidants from the whole food, differences in bioavailability as a result of release from the food matrix, and the presence of food components in the matrix which may enhance or reduce polyphenol bioavailability [28].

4. Dietary intake

The beneficial effects of polyphenols on human health depend considerably on dietary intake. Due to the great diversity of their chemical structures, it is difficult to estimate the total polyphenol content in foods. Hence, a comprehensive database was developed to help estimate the polyphenol content in certain foods and has been available online since 2009 [8]. Data summarized there were derived from more than 1300 scientific publications. According to this database, Pérez-Jiménez et al. [29] established a list of the 100 richest dietary sources of polyphenols per 100 g of food and in a food serving, using common serving sizes. Data on the total content of polyphenols were calculated based on the sum of all individual polyphenol contents determined by chromatography. In addition, the results were compared with data obtained by the Folin-Ciocalteu method, one of the most commonly used method for estimating total phenolic content. The results showed that the richest sources per 100 g of foods are various herbs and cocoa products (as shown in **Table 1**), while at the top of the list, expressed per serving size, are various darkly colored berries such as black elderberry, chokeberry, black currant, and blueberry. Comparison of the data obtained by different methods showed that the values obtained by the Folin-Ciocalteu method systematically exceed the total amount of polyphenols because this method is not specific and interference with other antioxidants present in the food is possible.

With the aim of estimating polyphenol intake, a large European cohort study was recently conducted in ten countries on more than 36,000 subjects. The results showed that the largest intake of phenolic compounds is in Denmark (1706 mg/day), while the lowest is in Greece (664 mg/day). Similar findings were observed after comparison of intake according to regions; the total polyphenol intake in the non-Mediterranean countries was higher compared with the Mediterranean countries. The most significant sources of phenolic compounds are coffee, tea, and fruit, with phenolic acids contributing to the total intake with more than 50% [30]. This was the first study that applied retention factors from the Phenol-Explorer database to assess the effects of cooking and processing on polyphenol contents in foods. Although the usual cooking of common plant foods causes substantial losses of polyphenols, in this study it did not have a high impact on the estimated total polyphenol intake because vegetables and legumes were not major contributors to polyphenol intake [31].

Research on the dietary intake of phenolic compounds has been conducted also in certain European countries, and the results show that the average intake in France is 1193 mg/day [32], in Poland 1756.5 mg/day [33], and in Spain 820 mg/day [34]. The main dietary sources of the total polyphenols in Spain and France are fruits and nonalcoholic beverages (principally coffee and tea). In Spain, fruits accounted for 44% of the total polyphenol intake and nonalcoholic beverages for 23%, whereas in France fruit accounted for only 17% and nonalcoholic beverages for 55% of the total polyphenol intake. Considering individual foods, the main source of total dietary polyphenols is coffee with 18 and 44% of contribution in Spain and France, respectively. In Spain, in contrast to other countries, olives and olive oils are important sources of polyphenols, accounting for 11% of the total polyphenol intake. Nonalcoholic beverages were the main food contributors to polyphenol intake in Poland and accounted for fully 67% of the total polyphenol intake due to high consumption of coffee and tea. The third main contributor to total polyphenol intake is chocolate, whereas fruits accounted for a lower percentage of intake.

5. Bioavailability and bioaccessibility

The beneficial effects of phenolic compounds on health depend not only on food sources but also on their stability, which can vary depending on the method of raw material processing, the matrix in which they are incorporated, and endogenous factors such as microbiota and digestive enzymes. The fraction of the phenolic compounds that can be released from the food matrix by digestive enzymes or intestinal bacterial flora in the colon is bioaccessible and, therefore, potentially bioavailable for absorption [28]. The FDA has defined bioavailability as the rate and extent to which the active substances or therapeutic moieties contained in a drug are absorbed and become available at the site of the action [35].

Understanding the effects of food processing on polyphenol content and bioavailability is important since most of the food consumed on a daily basis is in a processed form. Conventional methods of thermal processing, such as pasteurization that is still most commonly used, provide microbiological stability and extend shelf life but also cause some undesirable changes such as degradation of polyphenols and other bioactive compounds. The possibility of ensuring food safety and at the same time preserving biologically active compounds has resulted in increased interest in the minimal processing of foods using nonthermal methods, such as high-pressure processing and ultrasound. Studies have demonstrated that in comparison with high-pressure processing, pasteurization causes more degradation of polyphenol, anthocyanins, vitamin C, and the color of strawberry puree [36]. Treatment with high-intensity ultrasound, due to the cavitation effect, can break down cell walls and facilitate the extraction of bioactive compounds, thus increasing their bioavailability. Additionally, increased antioxidant capacity and monomeric anthocyanin content in red raspberry puree treated with high-intensity ultrasound were achieved by Golmohamadi et al. [37].

Food matrix composition and other food components significantly influence bioaccessibility, uptake, and further metabolism of polyphenols. Before becoming bioavailable, polyphenols must be released from the food matrix and hydrolyzed by intestinal enzymes or microflora to aglycones. In vitro gastrointestinal digestion models are a useful tool for assessing the impact of the food matrix and other endogenous factors on the stability and biological activity of phenolic compounds and can be well correlated with results from human studies and animal models [38]. Simulation of the physiological parameters, such as variation in the enzymes, acid and bile salt excretion, availability of the substrate, and the transit time of food through the stomach and duodenum, is challenging in all in vitro digestion models. Gastric digestion is simulated by pepsin-HCl at pH 2 and small intestinal digestion with pancreatin-bile mixture at pH 7, while the absorption step can be simulated with polarized human colon carcinoma cell line (Caco-2 cells) [39]. Commercial digestive enzymes, collected or extracted from omnivorous animals, are most commonly used, but their role in the simulation of the human digestion process is still questionable. On the other hand, human digestive juices contain a complex mixture of different enzymes, enzyme inhibitors, and bile salts, which together contribute to the digestion process of food; therefore, the use of human digestive juices may represent a great advantage over commercial digestive enzymes [40]. Phenolic acids and flavonoids with small molecular weight such as gallic acid, catechins, and quercetin glucosides are easily absorbed through the tract, whereas large polyphenols such as proanthocyanidins are poorly absorbed [41]. In most of conducted studies, gastric digestion did not have a significant effect on polyphenol stability. In fact, the majority of polyphenols appear to be released in the stomach. Bouayed et al. [38] observed that approximately 65% of apple total phenolics and flavonoids were released in the stomach and only an additional 10% in the small intestine. Results of the study conducted by Correa-Betanzo et al. [42] showed a high stability of total polyphenols and anthocyanins (7 and 1% of reduction, respectively) during simulated gastric digestion, while intestinal digestion caused a significant decrease of 51 and 83%, respectively, in comparison with the non-digested wild blueberry samples. Similar results were obtained by Bermúdez-Soto et al. [43] who reported a significant reduction of anthocyanins (43%) and flavonols (26%) after intestinal digestion of chokeberry. Mild alkaline intestinal environment was shown to influence all phenolic compounds, especially anthocyanins, and it is generally accepted that their bioavailability is low (<1%). An interesting study was conducted by Czank et al. [44] who proved that bioavailability of anthocyanins has been underestimated. The participants consumed an isotopically labeled anthocyanin tracer (cyanidin-3-glucoside), and the concentration was determined in blood, urine, breath, and feces samples. Results showed a high combined recovery from urine and breathe, which was approximately 12%. To date, a little research has been conducted in investigating polyphenol stability by using human gastrointestinal enzymes. Zorić et al. [45] conducted a study on the stability of rosmarinic acid in an aqueous extract of thyme, lemon balm, and winter savory using human digestive juices of the stomach and small intestine. The results showed lower gastrointestinal stability of rosmarinic acid in comparison with similar studies with commercial digestive enzymes.

In the food matrix, polyphenols are usually mixed with different macromolecules such as proteins, lipids, and carbohydrates. Large polyphenols and those with a high number of hydroxyl groups have a high affinity for proteins, which can result in a complex formation that reduces polyphenol absorption [28]. Food rich in polyphenols, such as coffee or tea, is usually consumed with milk. Studies have shown that interactions between polyphenols and milk proteins, especially casein, can decrease the antioxidant activity of coffee and tea [46]. The effect of milk was confirmed in an intervention study by Serafini et al. [47] whose aim was to determine the total antioxidant capacity and (−)epicatechin content in blood plasma after consumption of plain dark chocolate, dark chocolate with full-fat milk, and milk chocolate. Results have shown that the addition of milk, either during ingestion or in the manufacturing process, caused a significant reduction in total antioxidant activity and absorption of (−) epicatechin in the bloodstream. The explanation was in the formation of a complex between chocolate flavonoids and milk proteins. However, not all studies showed the negative impact of milk addition to food on polyphenol absorption. Keogh et al. [48] monitored the concentration of catechin and epicatechin in the blood after consumption of chocolate polyphenols with and without milk proteins. Results showed that milk protein did not influence the average plasma polyphenol concentration after ingestion. Contradictory results of these and many other studies were explained by the influence of polyphenol concentration. Milk could inhibit absorption in the case of lower polyphenol concentration, while it could have only minimal impact if the concentration is high [35]. In addition to food proteins, polyphenols can also bind to digestive enzymes and act as effective inhibitors as previously described in Chapter 3.

Only a few studies have investigated the interactions between polyphenols and dietary lipids. Since most polyphenols are water soluble, dietary lipids are considered to have a limited influence. Some studies, however, have observed a positive relationship. Ortega et al. [49] found that higher fat content has a positive effect on the stability of cocoa polyphenols in an in vitro digestion model.

Interactions between polyphenols and dietary fibers are important since these interactions have a significant role in the human body. Most non-extractable polyphenols with higher molecular weight (such as tannins and proanthocyanidins) are usually attached with covalent bounds to dietary fibers [28]. The bioavailability of polyphenols depends on the release of polyphenols from such a complex, which, in turn, depends on the polyphenols' structure, the complexity of the polyphenol-carbohydrate structure, and the possibility of enzymes to reach the carbohydrates [35]. According to Ortega et al. [50], soluble dietary fibers, in the in vitro digestion model, enhanced the stability of phenolic compounds during duodenal digestion. Since dietary fibers act as an entrapping matrix and restrict the diffusion of the enzymes to their substrates in the stomach and small intestine, many polyphenols reach the large intestine [51]. Regardless of their bioavailability, polyphenols, as strong antioxidants, may contribute to a healthy antioxidant environment, thus protecting the colonic lumen from oxidative stress, and, furthermore, polyphenols and carbohydrates that have reached the large intestine can have a beneficial effect on colon microflora growth.

6. Polyphenols as functional food components

Today's consumers' expectations of food, besides appropriate taste, appearance, and price, are more focused on positive health effects. Since consumers' awareness of health benefits associated with the consumption of food rich in polyphenols and preferences of herbal over synthetic products are increasing, meeting the consumers' expectations is a key to success.

The global polyphenol market was valued USD 757 million in 2015, and it is estimated to exceed USD 1 billion by 2022 [52]. The most successful applications of plant extracts containing polyphenols are fortification of beverages, while the most popular plant extracts used in beverages and other types of functional food are grape seed, green tea, and apple extract. The market for functional food and the number of studies focused on functional food with a positive effect on health beyond basic nutrition are constantly growing. The bioavailability of functional food components and the levels required in humans are critical factors necessary to optimize health benefits [53]. Polyphenols are the most numerous and widely distributed group of functional molecules. Studies have shown that products enriched with polyphenols could be useful for the dietary management of diabetes and cardiovascular disease prevention. Blueberry polyphenol-enriched defatted soybean flour was incorporated into a very high-fat diet of obese and hyperglycemic mice for 13 weeks. Compared with the control group (very high-fat diet containing defatted soybean flour), the diet supplemented with blueberry polyphenols reduced weight gain, improved glucose tolerance, and lowered fasting blood glucose levels and serum cholesterol [54]. The aim of an intervention study conducted by Sarriá et al. [55] was to evaluate the effect of two cocoa functional products (one rich in dietary fibers and the other rich in polyphenols) on the markers of cardiovascular health. The most significant finding observed after consumption of both products was an increase in HDL cholesterol which was attributed to flavanols, the most common flavonoids in cocoa, while the fiber-rich product was associated with the hypoglycemic and anti-inflammatory effect. As recently reviewed by Tomé-Carneiro and Visioli [56], polyphenol-based nutraceuticals and functional food might be used as adjunct therapy for cardiovascular disease.

Since it is generally accepted that the bioavailability of polyphenols is rather low, recent scientific studies are focused on the enhancement of polyphenol bioaccessibility and the bioavailability rate in the body using encapsulation techniques such as spray-drying, freeze-drying, emulsions, and liposomes. Encapsulated polyphenols are more stable and are protected from light, oxygen, temperature, and moisture. Spray-drying is the most commonly applied encapsulation method in the food industry, transforming liquids into stable and easily applied powders, and can help in the controlled release of phenolic functional ingredients in the human body for more efficient nutraceutical usage [57]. Idham et al. [58] studied the degradation kinetics and color stability of spray-dried encapsulated anthocyanins with four different encapsulation agents (maltodextrin, gum Arabic, a combination of maltodextrin and gum Arabic, and soluble starch). Results have shown that the combination of maltodextrin and gum Arabic resulted in the highest encapsulation efficiencies as well as the longest shelf life and the smallest change in pigment color.

Emulsions are considered one of the most promising techniques for the protection and delivery of polyphenols, due to high-efficiency encapsulation, maintenance of chemical stability, and controlled release [59]. An emulsion is a mixture of two immiscible liquids, usually oil and water, with one of the liquids (the dispersed phase) being dispersed as small droplets in the other liquid (the continuous phase). Ru et al. [60] have shown that epigallocatechin-3-gallate (EGCG), the most abundant polyphenol in green tea, encapsulated in oil-in-water (O/W) emulsions demonstrated an improved anticancer effect, compared with free EGCG, on human hepatocellular carcinoma cell lines. The unpleasant bitter taste of flavanol monomers (catechin and epicatechin) could be successfully masked by using encapsulation, thus increasing flavanol delivery in the gut [61].

7. Conclusion

Polyphenols comprise a large group of phytochemicals with very diverse chemical structures and are considered as being the most common antioxidants in the diet. Since many foods and beverages contain a diversity of polyphenols, it is difficult to determine which specific compounds are directly responsible for beneficial health effects in vivo. The health effects of polyphenols depend on both dietary intake and bioavailability, which can vary greatly. The strongest evidence for the beneficial effects of polyphenols with regard to chronic disease, cardiovascular diseases in particular, exists for flavanol-rich foods. Most dietary polyphenols have relatively short half-lives once ingested, due to rapid metabolism, so it is important that their consumption is maintained throughout the life span. More detailed knowledge on the relationship between the food matrix, processing, and bioavailability of polyphenols should lead to a better understanding of their role in human health and to the development of novel functional foods.

Author details

Nikolina Mrduljaš[1], Greta Krešić[1]* and Tea Bilušić[2]

*Address all correspondence to: greta.kresic@fthm.hr

1 Faculty of Tourism and Hospitality Management, University of Rijeka, Opatija, Croatia

2 Faculty of Chemistry and Technology, University of Split, Split, Croatia

References

[1] Del Rio D, Rodriguez-Mateos A, Spencer JPE , Tognolini M, Borges G, Crozier A. Dietary (Poly)phenolics in human health: Structures, bioavailability, and evidence of protective effects against chronic diseases. Antioxidants & Redox Signaling. 2013;**18**(14):1818-1892. DOI: 10.1089/ars.2012.4581

[2] Manach C. Polyphenols: Food sources and bioavailability. American Journal of Clinical Nutrition. 2004;**79**:727-747

[3] Tsao R. Chemistry and biochemistry of dietary polyphenols. Nutrients. 2010;**2**(12): 1231-1246. DOI: 10.3390/nu2121231

[4] Evans M, Sharma P, Guthrie N. Bioavailability of citrus polymethoxylated flavones and their biological role in metabolic syndrome and hyperlipidemia. In: Noreddin A, editor. Readings in Advanced Pharmacokinetics – Theory, Methods and Applications. Rijeka, Croatia: InTech; 2012. pp. 267-285. DOI: 10.5772/1982

[5] Messina M. Soy foods, isoflavones, and the health of postmenopausal women. The American Journal of Clinical Nutrition. 2014;**100**(suppl):423S–430S. DOI: 10.3945/ajcn.113.071464

[6] Zia Ul Haq M, Riaz M, Saad B. Anthocyanins and Human Health: Biomolecular and therapeutic aspects. 1st ed. Switzerland: Springer International Publishing; 2016. 138 p. DOI: 10.1007/978-3-319-26456-1. ISBN 978-3-319-26456-1

[7] Mena P, Domínguez-Perles P, Gironés-Vilaplana A, Baenas N, García-Viguera C, Villaño D. Flavan-3-ols, anthocyanins, and inflammation. IUBMB Life. 2014;**66**(11):745-758. DOI: 10.1002/iub.1332

[8] INRA, Unité de Nutrition Humaine. Phenol-Explorer Database [Internet]. August 2009. [Updated: June 2015]. Available from: http://phenol-explorer.eu [Accessed: January 10, 2017]

[9] Aehle E, Müller U, Eklund PC, Willför SM, Sippl W, Dräger B. Lignans as food constituents with estrogen and antiestrogen activity. Phytochemistry. 2011;**72**(18):2396-2405. DOI: 10.1016/j.phytochem.2011.08.013

[10] Haminiuk CWI, Maciel GM, Plata-Oviedo MSV, Peralta RM. Phenolic compounds in fruits – An overview. International Journal of Food Science and Technology. 2012;**47**(10):2023-2044. DOI: 10.1111/j.1365-2621.2012.03067.x

[11] Tresserra-Rimbau A, Rimm EB, Medina-Remón A, Martínez-González MA, de la Torre R, Corella D et al. Inverse association between habitual polyphenol intake and incidence of cardiovascular events in the PREDIMED study. Nutrition, Metabolism and Cardiovascular Diseases. 2014;**24**(6):639-647. DOI: 10.1016/j.numecd.2013.12.014

[12] Tresserra-Rimbau A, Rimm EB, Medina-Remón A, Martínez-González MA, López-Sabater MC, Covas MI et al. Polyphenol intake and mortality risk: A re-analysis of the PREDIMED trial. BMC Medicine. 2014;**12**:77. DOI: 10.1186/1741-7015-12-77

[13] Zamora-Ros R, Jimenez C, Cleries R, Agudo A, Sanchez M-J, Sanchez-Cantalejo E et al. Dietary flavonoid and lignan intake and mortality in a Spanish cohort. Epidemiology. 2013;**24**(5):726-733. DOI: 10.1097/EDE.0b013e31829d5902

[14] Quiñones M, Miguel M, Aleixandre A. Beneficial effects of polyphenols on cardiovascular disease. Pharmacological Research. 2013;**68**(1):125-131. DOI: 10.1016/j.phrs.2012.10.018

[15] Vauzour D. Dietary polyphenols as modulators of brain functions: Biological actions and molecular mechanisms underpinning their beneficial effects. Oxidative Medicine and Cellular Longevity. 2012;**2012**:914273. DOI: 10.1155/2012/914273

[16] Schmidt A, Hammann F, Wölnerhanssen B, Meyer-Gerspach AC, Drewe J, Beglinger C et al. Green tea extract enhances parieto-frontal connectivity during working memory processing. Psychopharmacology. 2014;**231**(19):3879-3888. DOI: 10.1007/s00213-014-3526-1

[17] Krikorian R, Shidler MD, Nash TA, Kalt W, Vinqvist-Tymchuk MR, Shukitt-Hale B et al. Blueberry supplementation improves memory in older adults. Journal of Agricultural and Food Chemistry. 2010;**58**(7):3996-4000. DOI: 10.1021/jf9029332

[18] Kim YA, Keogh JB, Clifton PM. Polyphenols and glycemic control. Nutrients. 2016;**8**(1):17. DOI: 10.3390/nu8010017

[19] Boath AS, Grussu D, Stewart D, McDougall GJ. Berry polyphenols inhibit digestive enzymes: A source of potential health benefits. Food Digestion. 2012;**3**(1-3):1-7. DOI: 10.1007/s13228-012-0022-0

[20] Yang X, Kong F. Evaluation of the in vitro alpha-glucosidase inhibitory activity of green tea polyphenols and different tea types. Journal of the Science of Food and Agriculture. 2016;**96**(3):777-782. DOI: 10.1002/jsfa.7147

[21] Yang X, Kong F. Effects of tea polyphenols and different teas on pancreatic α-amylase activity in vitro. LWT – Food Science and Technology. 2016;**66**:232-238. DOI: 10.1016/j.lwt.2015.10.035

[22] Zamora-Ros R, Forouhi NG, Sharp SJ, González CA, Buijsse B, Guevara M et al. The association between dietary flavonoid and lignan intakes and incident type 2 diabetes in european populations: The EPIC-InterAct study. Diabetes Care. 2013;**36**(12):3961-3970. DOI: 10.2337/dc13-0877

[23] Meydani M, Hasan ST. Dietary polyphenols and obesity. Nutrients. 2010;**2**(7):737-751. DOI: 10.3390/nu2070737

[24] Basu A, Du M, Leyva MJ, Sanchez K, Betts NM, Wu M et al. Blueberries decrease cardiovascular risk factors in obese men and women with metabolic syndrome. The Journal of Nutrition. 2010;**140**(9):1582-1587. DOI: 10.3945/jn.110.124701

[25] Worsztynowicz P, Napierała M, Białas W, Grajek W, Olkowicz M. Pancreatic α-amylase and lipase inhibitory activity of polyphenolic compounds present in the extract of black chokeberry (*Aronia melanocarpa L.*). Process Biochemistry. 2014;**49**(9):1457-1463. DOI: 10.1016/j.procbio.2014.06.002

[26] Nakai M, Fukui Y, Asami S, Toyoda-Ono Y, Iwashita T, Shibata H et al. Inhibitory effects of oolong tea polyphenols on pancreatic lipase in vitro. Journal of Agricultural and Food Chemistry. 2005;**53**(11):4593-4598. DOI: 10.1021/jf047814+

[27] Uchiyama S, Taniguchi Y, Saka A, Yoshida A, Yajima A. Prevention of diet-induced obesity by dietary black tea polyphenols extract in vitro and in vivo. Nutrition. 2011;**27**(3):287-292. DOI: 10.1016/j.nut.2010.01.019

[28] Bohn T. Dietary factors affecting polyphenol bioavailability. Nutrition Reviews. 2014;**72**(7): 429-452. DOI: 10.1111/nure.12114

[29] Pérez-Jiménez J, Neveu V, Vos F, Scalbert A. Identification of the 100 richest dietary sources of polyphenols: An application of the phenol-explorer database. European Journal of Clinical Nutrition. 2010;**64**(suppl 3):S112-S120. DOI: 10.1038/ejcn.2010.221

[30] Zamora-Ros R, Knaze V, Rothwell JA, Hémon B, Moskal A, Overvad K et al. Dietary polyphenol intake in Europe: The European prospective investigation into cancer and nutrition (EPIC) study. European Journal of Nutrition. 2016;**55**(4):1359-1375. DOI: 10.1007/s00394-015-0950-x

[31] Rothwell JA, Medina-Remón A, Pérez-Jiménez J, Neveu V, Knaze V, Slimani N et al. Effects of food processing on polyphenol contents: A systematic analysis using phenol-explorer data. Molecular Nutrition and Food Research. 2015;**59**(1):160-170. DOI: 10.1002/mnfr.201400494

[32] Pérez-Jiménez J, Fezeu L, Touvier M, Arnault N, Manach C, Hercberg S et al. Dietary intake of 337 polyphenols in French adults. American Journal of Clinical Nutrition. 2011;**93**(6):1220-1228. DOI: 10.3945/ajcn.110.007096

[33] Grosso G, Stepaniak U, Topor-Madry R, Szafraniec K, Pajak A. Estimated dietary intake and major food sources of polyphenols in the Polish arm of the HAPIEE study. Nutrition. 2014;**30**(11-12):1398-1403. DOI: 10.1016/j.nut.2014.04.012

[34] Tresserra-Rimbau A, Medina-Remón A, Pérez-Jiménez J, Martínez-González MA, Covas MI, Corella D et al. Dietary intake and major food sources of polyphenols in a Spanish population at high cardiovascular risk: The PREDIMED study. Nutrition, Metabolism and Cardiovascular Diseases. 2013;**23**(10):953-959. DOI: 10.1016/j.numecd.2012.10.008

[35] Jakobek L. Interactions of polyphenols with carbohydrates, lipids and proteins. Food Chemistry. 2015;**175**(May):556-567. DOI: 10.1016/j.foodchem.2014.12.013

[36] Marszałek K, Mitek M, Skąpska S. The effect of thermal pasteurization and high pressure processing at cold and mild temperatures on the chemical composition, microbial and enzyme activity in strawberry purée. Innovative Food Science and Emerging Technologies. 2015;**27**:48-56. DOI: 10.1016/j.ifset.2014.10.009

[37] Golmohamadi A, Möller G, Powers J, Nindo C. Effect of ultrasound frequency on antioxidant activity, total phenolic and anthocyanin content of red raspberry puree. Ultrasonics Sonochemistry. 2013;**20**(5):1316-1323. DOI: 10.1016/j.ultsonch.2013.01.020

[38] Bouayed J, Hoffmann L, Bohn T. Total phenolics, flavonoids, anthocyanins and antioxidant activity following simulated gastro-intestinal digestion and dialysis of apple varieties: Bioaccessibility and potential uptake. Food Chemistry. 2011;**128**(1):14-21. DOI: 10.1016/j.foodchem.2011.02.052

[39] Sensoy I. A review on the relationship between food structure, processing, and bioavailability. Critical Reviews in Food Science and Nutrition. 2014;**54**(7):902-909. DOI: 10.1080/10408398.2011.619016

[40] Furlund CB, Ulleberg EK, Devold TG, Flengsrud R, Jacobsen M, Sekse C et al. Identification of lactoferrin peptides generated by digestion with human gastrointestinal enzymes. Journal of Dairy Science. 2013;**96**(1):75-88. DOI: 10.3168/jds.2012-5946

[41] Carbonell-Capella JM, Buniowska M, Barba FJ, Esteve MJ, Frígola A. Analytical methods for determining bioavailability and bioaccessibility of bioactive compounds from fruits and vegetables: A review. Comprehensive Reviews in Food Science and Food Safety. 2014;**13**(2):155-171. DOI: 10.1111/1541-4337.12049

[42] Correa-Betanzo J, Allen-Vercoe E, McDonald J, Schroeter K, Corredig M, Paliyath G. Stability and biological activity of wild blueberry (*Vaccinium angustifolium*) polyphenols during simulated in vitro gastrointestinal digestion. Food Chemistry. 2014;**165**:522-531. DOI: 10.1016/j.foodchem.2014.05.135

[43] Bermúdez-Soto MJ, Tomás-Barberán FA, García-Conesa MT. Stability of polyphenols in chokeberry (*Aronia melanocarpa*) subjected to in vitro gastric and pancreatic digestion. Food Chemistry. 2007;**102**(3):865-874. DOI: 10.1016/j.foodchem.2006.06.025

[44] Czank C, Cassidy A, Zhang Q, Morrison DJ, Preston T, Kroon PA et al. Human metabolism and elimination of the anthocyanin, cyanidin-3-glucoside: A 13 C-tracer study. The American Journal of Clinical Nutrition. 2013;**97**:995-1003. DOI: 10.3945/ajcn.112.049247

[45] Zorić Z, Markić J, Pedisić S, Bučević-Popović V, Generalic-Mekinić I, Grebenar K et al. Stability of rosmarinic acid in aqueous extracts from different lamiaceae species

after in vitro digestion with human gastrointestinal enzymes. Food Technology and Biotechnology. 2016;**54**(1):97-102. DOI: 10.17113/ftb.54.01.16.4033

[46] Zhang H, Yu D, Sun J, Liu X, Jiang L, Guo H et al. Interaction of plant phenols with food macronutrients: Characterisation and nutritional–physiological consequences. Nutrition Research Reviews. 2013;**27**(1):1-15. DOI: 10.1017/S095442241300019X

[47] Serafini M, Bugianesi R, Maiani G, Valtuena S, De Santis S, Crozier A. Plasma antioxidants from chocolate. Nature. 2003;**424**(August):1013. DOI: 10.1038/4241013a

[48] Keogh JB, McInerney J, Clifton PM. The effect of milk protein on the bioavailability of cocoa polyphenols. Journal of Food Science. 2007;**72**(3):S230-S233. DOI: 10.1111/j.1750-3841.2007.00314.x

[49] Ortega N, Reguant J, Romero MP, Macià A, Motilva MJ. Effect of fat content on the digestibility and bioaccessibility of cocoa polyphenol by an in vitro digestion model. Journal of Agricultural and Food Chemistry. 2009;**57**(13):5743-5749. DOI: 10.1021/jf900591q

[50] Ortega N, Maciá A, Romero MP, Reguant J, Motilva MJ. Matrix composition effect on the digestibility of carob flour phenols by an in-vitro digestion model. Food Chemistry. 2011;**124**(1):65-71. DOI: 10.1016/j.foodchem.2010.05.105

[51] Palafox-Carlos H, Ayala-Zavala JF, González-Aguilar GA. The role of dietary fiber in the bioaccessibility and bioavailability of fruit and vegetable antioxidants. Journal of Food Science. 2011;**76**(1):6-15. DOI: 10.1111/j.1750-3841.2010.01957.x

[52] Prasad E. Polyphenol Market – Global Opportunity Analysis and Industry Forecast, 2014-2022 [Internet]. January 2011. Available from: https://www.alliedmarketresearch.com/polyphenol-market [Accessed: 06-03-2017]

[53] Abuajah CI, Ogbonna AC, Osuji CM. Functional components and medicinal properties of food: a review. Journal of Food Science and Technology. 2014;**52**(May):2522-2529. DOI: 10.1007/s13197-014-1396-5

[54] Roopchand DE, Kuhn P, Rojo LE, Lila MA, Raskin I. Blueberry polyphenol-enriched soybean flour reduces hyperglycemia, body weight gain and serum cholesterol in mice. Pharmacological Research. 2013;**68**(1):59-67. DOI: 10.1016/j.phrs.2012.11.008

[55] Sarriá B, Martínez-López S, Sierra-Cinos JL, Garcia-Diz L, Goya L, Mateos R et al. Effects of bioactive constituents in functional cocoa products on cardiovascular health in humans. Food Chemistry. 2015;**174**:214-218. DOI: 10.1016/j.foodchem.2014.11.004

[56] Tomé-Carneiro J, Visioli F. Polyphenol-based nutraceuticals for the prevention and treatment of cardiovascular disease: Review of human evidence. Phytomedicine. 2016;**23**(11): 1145-1174. DOI: 10.1016/j.phymed.2015.10.018

[57] Yousuf B, Gul K, Wani AA, Singh P. Health benefits of anthocyanins and their encapsulation for potential use in food systems: A review. Critical Reviews in Food Science and Nutrition. 2016;**56**:2223-2230. DOI: 10.1080/10408398.2013.805316

[58] Idham Z, Muhamad II, Sarmidi MR. Degradation kinetics and color stability of spray-dried encapsulated anthocyanins from *Hibiscus sabdariffa L.* Journal of Food Process Engineering. 2012;**35**(4):522-542. DOI: 10.1111/j.1745-4530.2010.00605.x

[59] Lu W, Kelly AL, Miao S. Emulsion-based encapsulation and delivery systems for polyphenols. Trends in Food Science and Technology. 2016;**47**(October):1-9. DOI: 10.1016/j.tifs.2015.10.015

[60] Ru Q, Yu H, Huang Q. Encapsulation of epigallocatechin-3-gallate (EGCG) using oil-in-water (O/W) submicrometer emulsions stabilized by ι-carrageenan and β-lactoglobulin. Journal of Agricultural and Food Chemistry. 2010;**58**(19):10373-10381. DOI: 10.1021/jf101798m

[61] Vitaglione P, Lumaga RB, Ferracane R, Sellitto S, Morello JR, Miranda JR, Shimoni E et al. Human bioavailability of flavanols and phenolic acids from cocoa-nut creams enriched with free or microencapsulated cocoa polyphenols. British Journal of Nutrition. 2013;**109**(10):1832–1843. DOI: 10.1017/S0007114512003881

Chapter 3

Folic and Folate Acid

Hiroko Watanabe and Tomoko Miyake

Additional information is available at the end of the chapter

http://dx.doi.org/10.5772/intechopen.69383

Abstract

Folate is a water-soluble B vitamin, also known as vitamin B9 or folacin. It is found naturally in a wide variety of foods, including vegetables, fruits, nuts, beans, dairy products, meats, eggs, seafood, and grains. However, only about 50% of the folate naturally present in food is bioavailable. Folate is critical in the metabolism of nucleic acid precursors and several amino acids, as well as in methylation reactions. Folic acid helps our bodies produce and maintain new cells, and it helps prevent DNA changes that may lead to cancer. Folate deficiency can cause anemia, insomnia, irritability, depression, Alzheimer's disease, cardiovascular disease, and more serious health problems. An inadequate folate status during early pregnancy increases the risk of congenital anomalies, such as neural tube defects (NTDs), which are life-threatening and cause life-long disabilities. Therefore, it has been recommended by the U.S. Public Health Service that even before becoming pregnant, women should consume 400 μg of synthetic folic acid daily, whether in the form of foods or supplements, as well as maintain a healthy diet of folate-rich foods to reduce NTD risk.

Keywords: folate, folic acid, homocysteine, health and outcomes, nutritional education

1. Introduction

Folate is a group of small water-soluble molecules that form one of the so-called B complex vitamins, also known as vitamin B9 or folacin. It is found naturally in a wide variety of foods, including vegetables, fruits, nuts, beans, dairy products, meats, eggs, seafood, and grains. However, only approximately 50% of the folate naturally present in food is bioavailable [1]. Folate is critical for the metabolism of nucleic acid precursors and several amino acids, as well as to methylation reactions. Folic acid helps our bodies produce and maintain new cells, and it helps prevent DNA changes that may lead to cancer. DNA methylation is an epigenetic mechanism that evidently plays a role in Alzheimer's disease [2]. An increase in the risks of depression and cardiovascular disease was observed independent of folic acid and vitamin B12 status [3, 4]. In this review, we will discuss recent issues related to the impact of folate and folic acid on cognitive and reproductive functions.

2. Folate metabolism in humans

Folate metabolism is closely linked to homocysteine (Hcy) metabolism, where Hcy is as an important factor in arteriosclerosis and aging. After the discovery of Hcy in 1932, it was demonstrated to be an important intermediate in the metabolism of amino acids. The folate metabolite 5-methyltetrahydrofolate (5-MTHF) is a substrate of methionine synthase, which remethylates Hcy to form methionine and links the folate cycle with Hcy metabolism (**Figure 1**) [5].

The substrate 5-methultetrahydrofolate requires vitamin B12 as a cofactor of methionine synthase. The effect of vitamin B12 is diminished by the larger role of folate status in determining total Hcy. Pyridoxal phosphate, the active form of vitamin B, is a cofactor for enzymes involved in amino acid metabolism. These enzymes include cystathionine β-synthase, the first enzyme in the transsulfuration pathway that breaks down Hcy to sulfate.

2.1. Folate metabolism and neurodegenerative and neuropsychiatric diseases

Insufficient amounts of folate and vitamin B12 limit the conversion of Hcy into methionine, which is a direct precursor of S-adenosylmethionin (SAM). SAM plays an important role in the methylation of neurotransmitters involved in depression [6]. Lower concentrations of

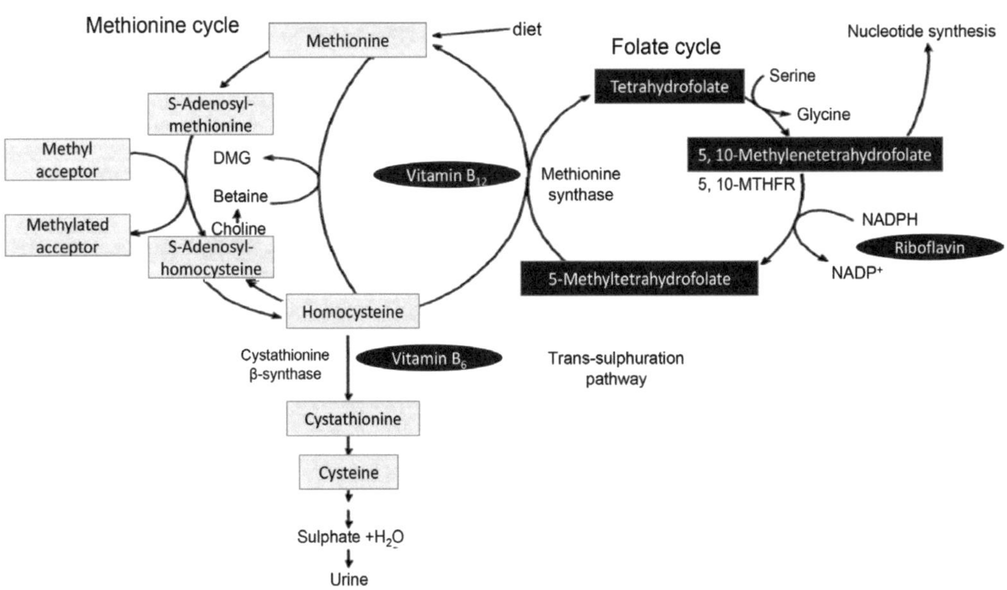

Figure 1. Pathways for the folate cycles and homocysteine metabolism. *Source*: Ref. [5].

http://dx.doi.org/10.5772/intechopen.69383

SAM and monoamine neurotransmitter metabolites were observed in the cerebrospinal fluid of severely depressed patients with high Hcy levels, compared to similar patients with normal Hcy levels [7].

MTHF is able to cross the blood-brain barrier into the cerebrospinal fluid [8]. One important function of folate is its role in the one-carbon cycle. In this pathway, folate is converted by methylenetetrahydrofolate reductase into MTHF, which combines with the amino acid Hcy to produce eventually, with the help of vitamin B12, S-adenosylmethionine (SAMe). SAMe is important, because it functions as a methyl donor in a variety of biochemical reactions and has been suggested to be somehow involved in the synthesis of the three neurotransmitters in the brain: serotonin, epinephrine, and dopamine [9]. **Figure 1** illustrates the actions of 5-MTHF and SAMe in methylation and neurotransmitter synthesis [10]. Thus, a folate deficiency could result in a deficiency of these neurotransmitters.

According to epidemiological and biological evidence, depressive disorders among individuals with epilepsy or neurological and psychiatric problems and the elderly could be caused by low folate [11, 12]. Folic acid affects the rate of the synthesis of the neurotransmitters dopamine, norepinephrine, and serotonin and it acts as a cofactor in the hydroxylation of phenylalanine and tryptophan [13]. Biogenic amine metabolism disturbances may lead to various psychiatric disorders, and a deficiency in folic acid may exacerbate neuropsychiatric disorders such as mental confusion, memory changes, cognitive slowing, and mood disorders.

Measurements of folate levels in plasma, serum, and erythrocytes are the most widely used biochemical indices of folate status, in addition to measurements of dietary folate intake. In a previous study of 883 elderly Latina women aged 60–93, the adjusted odds ratio for increased depressive symptoms in women in the lowest tertile of plasma folate was 2.04, which was significantly different from that in women in the highest tertile of folate [3]. Gilbody et al. reported that subjects with low serum levels, red blood cell (RBC) folate levels, and low folate intake had 1.4 times increased risk of depressive symptoms, compared with those with a high folate status [14]. On the other hand, in the Women's Health and Aging Study, serum homocysteine and folic acid levels were not associated with depression status among physically disabled women with a mean age of 77.3 years [15].

The elderly are of particular concern because of age-related declines in vitamin absorption and the extraction of vitamin B12 from protein [16] and age-related increases in autoimmunity against intrinsic factor or the gastric parietal cells that produce it [17]. Elevated plasma Hcy concentrations are common in older age [18]. With advanced age, the prevalence of a low vitamin B12 status increases from 5% at age 65 to 20% at age 80 years [19]. The reviews of population-based studies found that a low folate status is associated with mild cognitive impairment, Alzheimer's disease, and depression in healthy and neuropsychiatric diseased older people [20].

2.2. Folic metabolism and cancer

Several epidemiological studies have suggested an inverse association between folate status and the risk of cancer, including colorectal, lung, pancreatic, esophageal, stomach, cervical, ovarian, and breast cancers [21]. Folic acid helps our bodies produce and maintain new cells, and it helps prevent DNA changes. Folate plays an essential role in one-carbon transfer involving the remethylation of Hcy to methionine, thereby ensuring the provision of SAMe, the primary methyl group donor for most biological methylation reactions. Folic acid linked with conjugating agents only enters cells through the folate receptor (FR) [22], a cell surface glycosylphosphatidylinositol-anchored glycoprotein in humans [23]. Folate might influence the development of cancer through its role in one-carbon metabolism and its subsequent effects on DNA replication and cell division [24]. However, research has not established the precise nature of folate's effect on carcinogenesis.

2.3. Folate metabolism and reproductive function

Maternal nutrition, especially folate, is critical for optimizing pregnancy outcomes. The increase in folate required during pregnancy is due to the growth of the fetus and uteroplacental organs. The demand for folate is increased to support both the normal physiological changes of mothers and the optimal growth and development of the fetus and offspring [25]. Impaired placental perfusion due to hyperhomocysteinemia is implicated in having a negative effect on pregnancy outcomes. Inadequate folate intake before conception and early pregnancy increases the risk of congenital malformations of the brain and spinal cord, such as anencephaly, spina bifida, and neural tube defects (NTDs). NTDs are the most common and severe congenital malformations of the central nervous system, occurring secondary to lack of closure of the neural tube and leading to long-term morbidity. Neurulation, the process of neural tube formation, is completed 28 days after conception, as many women do not realize that they are pregnant at this stage [26]. Das et al. found in the systematic review that folate fortification had a significant impact on reducing neural tube defects (risk ratio; RR: 0.57 (95% CI: 0.45, 0.73)), spina bifida (RR: 0.64 (95% CI: 0.57, 0.71)), and anencephaly (RR: 0.80 (95% CI: 0.73, 0.87)). Folate fortification significantly reduced the incidence of congenital abnormalities [27].

3. Recommended dietary intake of folate in humans

Folate deficiency can cause anemia, insomnia, irritability, and far more serious health problems. In 2000, the D-A-CH societies (Germany [D], Austria [A], and Switzerland [CH]) initiated a recommendation of 400 μg of folate daily among adults [28], a value agreed to by the USA, Canada [29], Australia, and New Zealand [30]. The World Health Organization (WHO) and the Food and Agriculture Organization (FAO) of the United Nations [30] also agreed, setting an estimated average requirement (EAR) of 320 μg of dietary folate equivalents (DFE)/day and a recommended dietary allowance (RDA) of 400 μg DFE/day for adults. **Table 1** indicates the various recommendations. By combining RBC folate, plasma total Hcy, and plasma or serum folate, the Institute of Medicine concluded an EAR for adults with a focus on adequate quantities of folate, via food or food plus folic acid and consumed under controlled conditions, to maintain normal blood concentrations of these indicators [31].

Each country issues an RDA as the mean of estimated requirements for pregnant women that must be increased to meet the demands of increasing maternal tissues, fetal growth, fat store

	Adults		Pregnant women	
	EAR (μg/day)	RDA (μg/day)	EAR (μg/day)	RDA (μg/day)
WHO/FAO[1]	320	400	370–470	600
USA, Canada[2]	320	400		600
Australia and New Zealand[3]	320	400	520	600
Japan[4]	200	240	400	440

Notes: EAR, estimated average requirement (average daily level of intake estimated to meet the requirements of 50% of healthy individuals); RDA, recommended dietary allowance (average daily level of intake sufficient to meet the nutrient requirements of nearly all (97–98%) healthy individuals).

[1] *Source:* The WHO/Food and Agriculture Organization of the United Nations at the Institute of Medicine (IOM) of the National Academies [31].

[2] *Source:* IOM [29].

[3] *Source:* Australian National Health and Medical Research Council and New Zealand Ministry of Health [30].

[4] *Source:* Overview of Dietary Reference Intakes for Japanese by the Minister of Health, Labour and Welfare [59].

Table 1. Reference values for folate/folate equivalents for adults from different international societies and organizations.

growth, and the increase in basal metabolic rate. In 1992, the U.S. Public Health Service recommended that all women of reproductive age in the USA capable of becoming pregnant should consume 400 μg of synthetic folic acid daily from fortified foods or supplements. In addition, they should consume a balanced, healthy diet of folate-rich food to prevent two common and serious birth defects: spina bifida and anencephaly [32, 33]. All women between 15 and 45 years of age should consume folic acid daily because half of U.S. pregnancies are unplanned and because these birth defects occur early in pregnancy (3–4 weeks after conception), before most women know they are pregnant. The Food and Drug Administration mandated the addition of folic acid to all enriched cereal grain products by January 1998 [34]. Experimental and epidemiological evidence has shown that periconceptional dietary supplementation with folic acid can result in an estimated 50–70% decrease in the prevalence of NTDs [35].

4. Global strategies of folic acid fortification for reproductive-age women

In 2009, the U.S. Preventive Services Task Force published updated guidelines reinforcing these recommendations [36]. Recently, the National Institute for Health and Clinical Excellence [37] reinforced this focus on the periconceptional period. The best-known recommendation for women who are planning a pregnancy is to take 400 μg of folic acid a day in supplements to prevent NTDs. As of July 2015, almost 80 countries had fortified their wheat flour with folic acid, and health agencies in many countries have officially recommended the periconceptional consumption of folic acid in the range of 400–500 μg by young women capable of conceiving or planning to conceive [37].

Fortification leads to a decrease in the prevalence of serum deficiency from 30% to less than 1% and a decrease in the prevalence of an RBC folate deficiency from 6% to no measureable deficiency [38]. The number of cases of spina bifida and anencephaly among deliveries occurring during 1995–2011 in 19 population-based birth defect surveillance programs in the US was reported by the Centers for Disease Control and Prevention (**Figure 2**) [39]. Overall, a 28% reduction in prevalence was observed for anencephaly and spinal bifida. The mandatory fortification of standardized enriched cereal grain products in the US has resulted in a substantial increase in blood folate concentrations. In a study based on data from a National Health and Nutrition Examination Survey, the mean serum folate concentration for women aged 15–44 years who did not use supplements increased from 10.7 to 28.6 nmol/L shortly after initiating fortification in the USA, an almost threefold increase [40].

The incidence rate of NTDs was reported to be 0.97 per 1000 births in some European countries [41]. The reported NTD prevalence ranges and medians for each region were: Africa (5.2–75.4; 11.7 per 10,000 births), Eastern Mediterranean (2.1–124.1; 21.9 per 10,000 births), Europe (1.3–35.9; 9.0 per 10,000 births), Americas (3.3–27.9; 11.5 per 10,000 births), South-East Asia (1.9–66.2; 15.8 per 10,000 births), and Western Pacific (0.3–199.4; 6.9 per 10,000 births) [42]. According to the Morbidity and Mortality Weekly Report, if 50–70% of NTDs can be prevented by consuming 400 μg of folic acid per day, assuming a prevalence of 300,000 NTDs per year, worldwide folic acid fortification could prevent 150,000–210,000 NTDs annually [35].

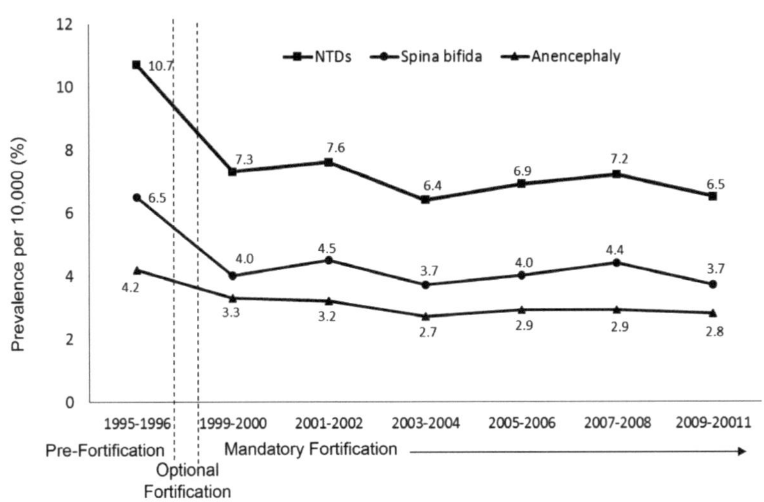

Figure 2. Prevalence of spina bifida and anencephaly in the USA, 1995–2011. *NTDs: Spina bifida + anencephaly. *Source*: Neural tube defect ascertainment project of the National Birth Defects Prevention Network at Centers for Disease Control and Prevention [39].

5. Current trends of worldwide folic acid food fortification

Figure 3 shows the world's industrially milled flour and rice fortification legislation with at least iron and folic acid, from March 2017. According to the food fortification initiative, globally, 86 countries have initiated legislation to mandate the fortification of wheat flour

http://dx.doi.org/10.5772/intechopen.69383

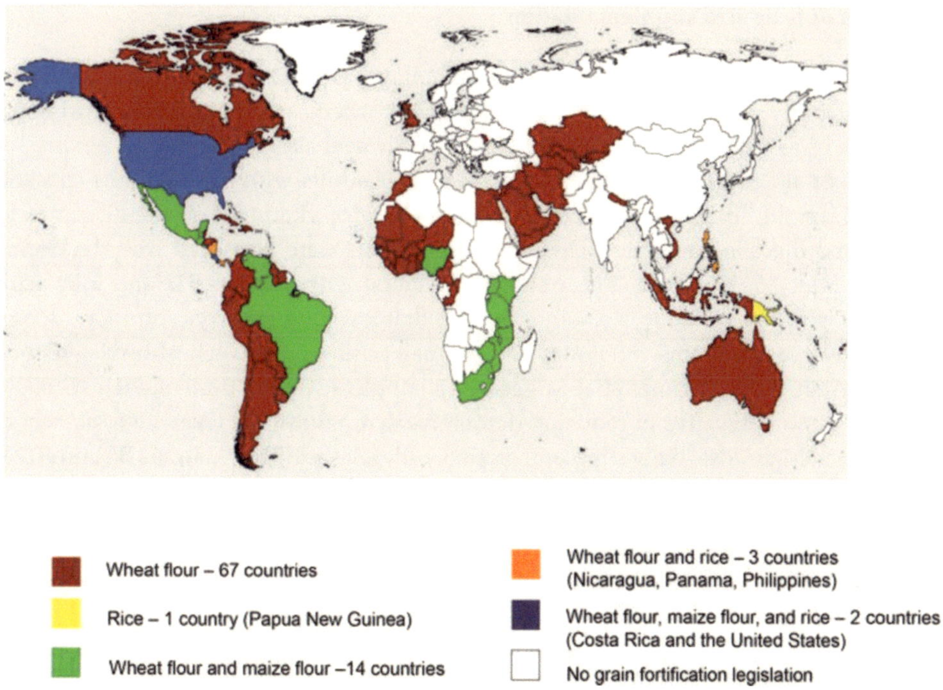

Figure 3. World's map of industrially milled flour and rice fortification with at least iron or folic acid. *Source*: Food fortification initiative in March 2017 [43].

alone or in combination with other grains, while over 100 countries have not introduced mandatory folic acid fortification, including the EU, China, and Japan [43].

The U.S. program adds 140 μg of folic acid per 100 g of enriched cereal grain product and it has been estimated to provide 100–200 μg of folic acid per day to women of childbearing age [44]. In Canada, it is mandatory to fortify white-wheat flour and enriched cornmeal with 150 μg of folic acid/100 g and enriched pasta with 200–270 μg of folic acid/100 g [45]. Berry et al. estimated that in the USA and Canada, the additional intake of about 100–150 μg/day of folic acid through food fortification has been effective in reducing the prevalence of NTDs at birth and in increasing blood folate concentrations in both countries [46].

In Ireland, all bread, including white, wholemeal, and brown, manufactured or marketed in Ireland, with the exception of minor bread products, should be fortified on a mandatory basis with folic acid at a level that provides 120 μg per 100 g of bread consumed. The voluntary folic acid fortification of foods, for example, cereal bars, yogurt, or juice, is permitted [47].

In Australia, all plain, fancy and sweet breads, rolls, and buns, including bagels, focaccia, English muffins made with yeast and flour mixes or flour for domestic bread making must contain folic acid. Organic bread is not required to contain folic acid. Some manufacturers also voluntarily choose to fortify other foods with folic acid, for example, breakfast cereal. Manufacturers must list folic acid in the ingredients list on the labels of foods fortified with folic acid. Currently, some cereals and cereal products, bread, and fruit juice have folic acid voluntarily added by food manufacturers [48].

5.1. Benefits of folic acid supplementation

Folate requirements can be affected by bioavailability, nutrient interactions, and smoking. The bioavailability of folates in food is about 50–60%, whereas that of the folic acid used to fortify foods or as a supplement is about 85% [1]. Folic acid as a supplement is almost 100% bioavailable on an empty stomach. Among 2919 older adults with elevated Hcy concentrations of ≥12 μmol/L, participants received either 500 μg of vitamin B12 and 400 μg of folic acid daily or a placebo for 2 years. Depressive symptoms were measured with the Geriatric Depression Scale-15. However, 2-year supplementation with vitamin B12 and folic acid in older adults with hyperhomocysteinemia showed that lowering Hcy concentrations does not reduce depressive symptoms, but it may have a small positive effect on health-related quality of life [49]. A study by Lachner et al. suggested a supplementation dose of at least 1000 μg/day might be more effective in reducing depressive symptoms [50]. Okereke et al. reported that long-term, high-dose, daily supplementation with folic acid and vitamins B6 and B12 did not reduce overall depression risk in 4331 older women (mean age 63.6 years), without prior depression [51].

Nguyen et al. [52] conducted a randomized controlled trial designed to assess the impact of supplementation in Guatemala. In total, 459 women aged 15–49 years were assigned to four groups at random to receive weekly (5000 or 2800 μg) or daily (400 or 200 μg) folic acid plus iron, zinc, and vitamin B12 for 12 weeks. Depression was measured using the Center for Epidemiologic Studies Depression Scale. Women in the lowest tertile of RBC folate were 1.7 times more likely to be depressed than those were in the highest tertile (OR = 1.71; 95%CI: 0.91, 3.18) at baseline. However, this relationship disappeared after adjustments for potential confounding factors. Mean depression scores and the prevalence of depression decreased postintervention, with no differences in the degree of improvement by group. It is difficult to evaluate the effect of supplementation on depressive symptoms, because this study had no placebo control group. This is because a number of reports have suggested that folate supplementation may enhance the effectiveness of certain antidepressant regimens [53, 54].

6. Education and technical consultation on folate deficiency

The nutritional intake of reproductive-age women appears inadequate during the preconceptional period. Among almost all women, folate intake is less than the RDA. Promoting women's health during preconception is a key public health strategy. Thiele et al. [55] observed in Germany that better educated women had higher indices of qualitatively beneficial diets than did lesser educated women. Adolescents at universities and colleges are potentially important targets for the promotion of healthy lifestyles, including physical, psychological, and eating habits. However, little is known about nutritional and health-related behaviors.

Questions arise as to how pregnant women show concern for their consumed diets and whether pregnant women get appropriate nutrient information during their routine antenatal checkups. Bookari et al. [56] reported that 65% of pregnant women were not familiar with the healthy eating recommendations. Nearly 80% of pregnant women would have liked education about nutrition and dietary advice [57], but Anya et al. [58] reported that women spend

3 min or less with their antenatal care providers and less than 40% had been informed or educated about diet and nutrition. These results suggest pregnant women lack opportunities to receive adequate and appropriate nutrition education during antenatal care. Most women expect advice on general dietary improvements, with the remainder seeking advice on how to promote the quality and quantity of nutritional intake. A critical goal for women should be to make behavior changes to ensure a good nutritional status before, during, and beyond pregnancy, which may lead to improved birth outcomes. More effective education campaigns should be set up by health care providers to improve women's awareness. Health care providers should educate reproductive-age women about careful food selection and meal planning and preparation at clinics, schools, or offices through mass media.

7. Conclusion

Folate deficiency impairs DNA replication and cell division, which adversely affects rapidly proliferating tissues, such as bone marrow, and results in the production of unusually large macrocytic cells with poorly differentiated nuclei. An increase in the risks of anemia, depression, and cancer was observed independent of folic acid. As the world is aging rapidly, attention on aging-related mental disorders has increased. Malnutrition is common among people aged 65 years and older. In addition, despite the abundance of information concerning folic acid, many women of reproductive age either are still unaware of its importance or do not value this information. The RDA guidelines have not worked effectively to appeal to the public. More effective education campaigns should be set up by health care providers to improve women's awareness.

Author details

Hiroko Watanabe* and Tomoko Miyake

*Address all correspondence to: watanabe@sahs.med.osaka-u.ac.jp

Department of Children and Women's Health, Osaka University Graduate School of Medicine, Suita, Japan

References

[1] Pfeiffer CM, Rogers LM, Bailey LB, Gregory JF 3rd. Absorption of folate from fortified cereal-grain products and supplemental folate consumed with or without food determined by using a dual-label stable-isotope protocol. The American Journal of Clinical Nutrition. 1997;**66**(6):1388-1397

[2] Wang J, Yu J-T, Tan M-S, et al. Epigenetic mechanisms in Alzheimer's disease: Implications for pathogenesis and therapy. Ageing Research Reviews. 2013;**12**:1024-1041. DOI: 10.1016/j.arr.2013.05.003

[3] Gougeon L, Payette H, Morais JA, et al. Intakes of folate, vitamin B6 and B12 and risk of depression in community-dwelling older adults: The Quebec Longitudinal Study on Nutrition and Aging. European Journal of Clinical Nutrition. 2016;**70**(3):380-385. DOI: 10.1038/ejcn.2015.202

[4] Stanhewicz AE, Kenney WL. Role of folic acid in nitric oxide bioavailability and vascular endothelial function. Nutrition Reviews. 2017;**75**(1):61-70. DOI: 10.1093/nutrit/nuw053

[5] Robinson K. Homocysteine, B vitamins, and risk of cardiovascular disease. Heart. 2000; **83**(2):127-130

[6] Papakostas GI, Cassiello CF, Iovieno N. Folates and S-adenosylmethionine for major depressive disorder. The Canadian Journal of Psychiatry. 2012;**57**(7):406-413. DOI: 10.1177/070674371205700703

[7] Bottiglieri T, Laundy M, Crellin R, et al. Homocysteine, folate, methylation, and monoamine metabolism in depression. Journal of Neurology Neurosurgery and Psychiatry. 2000;**69**(2):228-232

[8] Wu D, Pardridge WM. Blood-brain barrier transport of reduced folic acid. Pharmaceutical Research. 1999;**16**(3):415-419

[9] Miller AL. The methylation, neurotransmitter, and antioxidant connections between folate and depression. Alternative Medicine Review. 2008;**13**(3):216-226

[10] Stahl SM. Essential Psychopharmacology. New York: Cambridge University Press; 2008

[11] Brito A, Verdugo R, Hertrampf E et al. Vitamin B-12 treatment of asymptomatic, deficient, elderly Chileans improves conductivity in myelinated peripheral nerves, but high serum folate impairs vitamin B-12 status response assessed by the combined indicator of vitamin B-12 status. Am J Clin Nutr. 2016;**103**(1):250-257. DOI: 10.3945/ajcn.115.116509

[12] Sławek J, Roszmann A, Robowski P. The impact of MRI white matter hyperintensities on dementia in Parkinson's disease in relation to the homocysteine level and other vascular risk factors. Neurodegenerative Diseases. 2013;**12**(1):1-12. DOI: 10.1159/000338610

[13] Hutto BR. Folate and cobalamin in psychiatric illness. Comprehensive Psychiatry. 1997; **38**(6):305-314

[14] Gilbody S, Lightfoot T, Sheldon T. Is low folate a risk factor for depression? A meta-analysis and exploration of heterogeneity. Journal of Epidemiology and Community Health. 2007;**61**(7):631-637

[15] Penninx BW, Guralnik JM, Ferrucci L, et al. Vitamin B (12) deficiency and depression in physically disabled older women: Epidemiological evidence from the Women's Health and Aging Study. American Journal of Psychiatry. 2000;**157**(5):715-721

[16] Selhub J, Bagley LC, Miller J, et al. B vitamins, homocysteine, and neurocognitive function in the elderly. The American Journal of Clinical Nutrition. 2000;**71**(2):614S-620S

[17] Carmel R. Cobalamin, the stomach, and aging. The American Journal of Clinical Nutrition. 1997;**66**(4):750-759

[18] van Wijngaarden JP, Doets EL, Szczecinska A et al. Vitamin B12, folate, homocysteine, and bone health in adults and elderly people: a systematic review with meta-analyses. J Nutr Metab. 2013;**2013**:1-19. DOI: 10.1155/2013/486186

[19] Clarke R, Refsum H, Birks J, et al. Screening for vitamin B-12 and folate deficiency in older persons. The American Journal of Clinical Nutrition. 2003;**77**(5):1241-1247

[20] Araújo JR, Martel F, Borges N, Araújo JM, Keating E. Folates and aging: Role in mild cognitive impairment, dementia and depression. Ageing Research Reviews. 2015;**22**:9-19. DOI: 10.1016/j.arr.2015.04.005

[21] Kim YI. Will mandatory folic acid fortification prevent or promote cancer? The American Journal of Clinical Nutrition. 2004;**80**(5):1123-1128

[22] Lu JY, Lowe DA, Kennedy MD, Low PS. Folate-targeted enzyme prodrug cancer therapy utilizing penicillin-V amidase and a doxorubicin prodrug. Journal of Drug Targeting. 1999;**7**(1):43-53. DOI: 10.3109/10611869909085491

[23] Salazar MD, Ratnam M. The folate receptor: What does it promise in tissue-targeted therapeutics? Cancer and Metastasis Reviews. 2007;**26**(1):141-152. DOI: 10.1007/s10555-007-9048-0

[24] Kim YI. Folate and carcinogenesis: Evidence, mechanisms, and implications. Journal of Nutritional Biochemistry. 1999;**10**(2):66-88

[25] Lamer Y. Folate recommendations for pregnancy, lactation, and infancy. Annals of Nutrition and Metabolism. 2011;**59**:32-37

[26] Sadler TW. Embrology of neural tube development: Embryology of neural tube development. American Journal of Medical Genetics Part C Seminars in Medical Genetics. 2005;**135C**(1):2-8. DOI: 10.1002/ajmg.c.30049

[27] Das JK, Salam RA, Kumar R, Bhutta ZA, et al. Micronutrient fortification of food and its impact on woman and child health: A systematic review. Systematic Reviews. 2013;**2**:67. DOI: 10.1186/2046-4053-2-67

[28] Deutsche Gesellschaft für Ernährung (DGE) DACH-Referenzwerte für die Nährstoffzufuhr. 1. Aufl. Frankfurt am Main: Umschau/Braus; 2000

[29] Institute of Medicine (US) Standing Committee on the Scientific Evaluation of Dietary Reference Intakes and its Panel on Folate, Other B Vitamins, and Choline. Dietary Reference Intakes for Thiamin, Boflavin, Niacin, Itamin B b6 s, Folate, Vitamin B12, Pantothenic acid, Biotin, and Choline. Washington, DC: National Academy Press; 1998

[30] National Health and Medical Research Council (NHMRC). Nutrient Reference Values for Australia and New Zealand: Including Recommended Dietary Intakes. Canberra: Commonwealth of Australia; 2006

[31] FAO/WHO. Human vitamin and mineral requirements. Agriculture and consumer protection. Bangkok, Thailand: FAO/WHO, 2002

[32] O'Keefe CA, Bailey LB, Thomas EA, Hofler SA, Davis BA, Cerda JJ, Gregory JF 3rd. Controlled dietary folate affects folate status in nonpregnant women. Journal of Nutrition. 1995;**125**(10):2717-2725

[33] Amarin ZO, Obeidat AZ. Effect of folic acid fortification on the incidence of neural tube defects. Paediatric and Perinatal Epidemiology. 2010;**24**(4):349-351. DOI: 10.1111/j.1365-3016.2010.01123.x

[34] Food and Drug Administration. Food standards: Amendment of standards of identify for enriched grain products to require addition of folic acid. Federal Register. 1996; **61**(44):8781-8797

[35] Centers for Disease Control and Prevention (CDC). CDC grand rounds: Additional opportunities to prevent neural tube defects with folic acid fortification. MMWR Morbidity and Mortality Weekly Report. 2010;**59**(31):980-984

[36] U.S. Preventive Services Task Force. Folic acid for the prevention of neural tube defects: U.S. Preventive Services Task Force recommendation statement. Annals of Internal Medicine. 2009;**150**(9):626-631

[37] National Institute for Health and Clinical Excellence. Improving the Nutrition of Pregnant and Breastfeeding Mothers and Children in low-income Households. London: NICE Public Health Guidance 11; 2008. Available from: https://www.nice.org.uk/guidance/ph11 [Accessed February 5, 2011]

[38] Pfeiffer CM, Hughes JP, Lacher DA, et al. Estimation of trends in serum and RBC folate in the United States population from pre-to post-fortification using assay-adjusted data from NHANES 1988-2010. Journal of Nutrition. 2012;**142**(5):886-893. DOI: 10.3945/jn.111.156919

[39] Centers for Disease Control and Prevention (CDC). Updated estimates of neural tube prevented by mandatory folic acid fortification-United States, 1995-2011. MMWR Morbidity and Mortality Weekly Report. 2015;**64**(1):1-5

[40] Centers for Disease Control and Prevention (CDC). Folate status in women of childbearing age-United States, 1999. MMWR Morbidity and Mortality Weekly Report. 2000;**49**(42):962-965

[41] De Marco P, Merello E, Calevo MG, Mascelli S, Pastorino D, Crocetti L, et al. Maternal periconceptional factors affect the risk of spina bifida-affected pregnancies: An Italian case-control study. Childs Nervous System. 2011;**27**(7):1073-1081

[42] Zaganjor I, Sekkarie A, Tsang BL, et al. Prevalence of neural tube defects worldwide: A systematic literature review. PLoS One. 2016;**11**(4):e0151586. DOI: 10.1371/journal.pone.0151586

[43] Food Fortification Initiative, Enhancing Grains Healthier Lives [Internet]. 2017. Available from: http://www.ffinetwork.org/global_progress/[Accessed: March 1 2017]

[44] Quinlivan EP, Gregory JF, III. Reassessing folic acid consumption patterns in the United States (1999-2004): Potential effect on neural tube defects and overexposure to folate. The American Journal of Clinical Nutrition. 2007;**86**:1773-1779

[45] The Minister of Public Works and Government Services. Regulatory impact analysis statement. Canada Gazette Part II. SOR/98-550. 1998;**132**(24):3029-3033

[46] Berry RJ, Bailey L, Mulinare J, Bower C, Folic Acid Working Group. Fortification of flour with folic acid. Food and Nutrition Bulletin. 2010;**31**(1 Suppl):S22-35. DOI: 10.1177/15648265100311S103

[47] Food Safety Authority of Ireland . Report of the Implementation Group on Folic Acid Food Fortification to the Department of Health and Children [Internet]. 2008. Available from: http://lenus.ie/hse/bitstream/10147/218011/1/Folic+Acid+Implementation+Report+Final+Bookmarked.pdf. [Accessed: February 20]

[48] Food standards Australia and New Zealand [Internet]. 2016. Available from: http://www.foodstandards.gov.au/consumer/nutrition/folicmandatory/pages/default.aspx [Accessed: March 1 2017]

[49] de Koning EJ, van der Zwaluw NL, et al. Effect of two-year vitamin B12 and folic acid supplementation on depressive symptoms and quality of life in older adults with elevated homocysteine concentrations: Additional results from B-PROOF study, and RCT. Nutrients. 2016;**8**(11):748-753. DOI: 10.3390/nu8110748

[50] Lachner C, Steinle NI, Regenold WT. The neuropsychiatry of vitamin B12 deficiency in elderly patients. The Journal of Neuropsychiatry and Clinical Neurosciences. 2012;**24**(1):5-15. DOI: 10.1176/appi.neuropsych.11020052

[51] Okereke OI, Cook NR, Albert CM, et al. Effect of long-term supplementation with folic acid and B vitamins on risk of depression in older women. British Journal of Psychiatry. 2015;**206**(4):324-331. DOI: 10.1192/bjp.bp.114.148361

[52] Nguyen PH, Grajeda R, Melgar P, Marcinkevage J, DiGirolamo AM, Flores R, Martorell R. Micronutrient supplementation may reduce symptoms of depression in Guatemalan women. Archivos Latinoamericanos de Nutrición. 2009;**59**(3):278-286

[53] Coppen A, Bailey J. Enhancement of the antidepressant action of fluoxetine by folic acid: A randomised, placebo controlled trial. Journal of Affective Disorders. 2000;**60**(2):121-130

[54] Alpert M, Silva RR, Pouget ER. Prediction of treatment response in geriatric depression from baseline folate level: Interaction with an SSRI or a tricyclic antidepressant. Journal of Clinical Psychopharmacology. 2003;**23**(3):309-313

[55] Thiele S, Mensink G, Beitz R. Determinants of diet quality. Public Health Nutrition. 2004;**7**(1):29-37

[56] Bookari K, Yeatman H, Williamson M. Exploring Australian women's level of nutrition knowledge during pregnancy: A cross-sectional study. International Journal of Women's Health. 2016;**8**:405-419. DOI: 10.2147/IJWH.S110072

[57] de Jersey SJ, Nicholson JM, Callaway LK, Daniels LA. An observational study of nutrition and physical activity behaviours, knowledge, and advice in pregnancy. BMC Pregnancy & Childbirth. 2013;**13**:115. DOI: 10.1186/1471-2393-13-115

[58] Anya SE, Hydara A, Jaiteh LE. Antenatal care in The Gambia: Missed opportunity for information, education and communication. BMC Pregnancy & Childbirth. 2008;**8**:1-7. DOI: 10.1186/1471-2393-8-9

[59] Overview of Dietary Reference Intakes for Japanese (2015) [Internet]. 2015. Available from: http://www.mhlw.go.jp/file/06-Seisakujouhou-10900000-Kenkoukyoku/Overview.pdf [Accessed: February 15, 2017]

Chapter 4

New Advances about the Effect of Vitamins on Human Health: Vitamins Supplements and Nutritional Aspects

Noelia García Uribe, Manuel Reig García-Galbis and
Rosa María Martínez Espinosa

Additional information is available at the end of the chapter

http://dx.doi.org/10.5772/intechopen.69122

Abstract

The early twentieth century was a crucial period for the identification and biological-chemical-physical characterisation of vitamins. From then until now, many studies have attempted to clarify into detail the biological role of the vitamins in humans and their direct connection with certain diseases, either in a negative way (appearance of deficiency diseases due to vitamin deficiency) or a positive way (use of vitamins to treat diseases and/or to improve human health). The aim of this work is to analyse, from an integrative point of view, the information about vitamins and their effects on human health, and to identify direct correlations between these compounds and health. The effects of vitamins supplements on diet are also explored. The analysis of the results shows that it is impossible to establish robust and universal conclusions about the benefit of vitamin supplementation on human health beyond the prevention and/or treatment of deficiency states.

Keywords: nutrition, vitamins, human health, antioxidants, dietary supplements, multivitamins

1. Introduction

Human nutrition, as a field of knowledge, had a great impact at the beginning of the twentieth century. From 1912, experiments such as those developed by English biochemist Frederick Hopkins (1861–1947) demonstrated the existence of certain organic substances in food that are essential for health. Hopkins called them 'accessory food factors' [1–3]. Shortly after that discoveries, the Polish biochemist Casimir Funk (1884–1967) proposed the term 'vitamins' to identify the substances previously termed 'accessory food factors' [2, 3]. The etymology of the term vitamin derives from the Latin 'vita' (life) and 'amina'; Funk concluded that these substances were necessary for life and most of them contained an amino group [1, 4]. Thus, in the early sixties, the identification of essential nutrients necessary to support human life and health (macronutrients, micronutrients and trace elements) was almost concluded [4].

In the last half of last century, all vitamins were identified, their chemical structures were determined and natural sources from which vitamins can be obtained were described in detail. The biological role of each vitamin, their connections with several metabolic pathways and human pathologies and their importance in human nutritional processes were also quickly established [2, 4]. Besides, advances in chemical analysis/technologies during the last three decades have provided the tools to produce vitamins *in vitro* (even at large scale). Consequently, vitamins can be currently obtained by chemical synthesis, by isolation of natural sources (fat-soluble vitamins) or by microbial biotechnology (mainly water-soluble vitamins).

Thus, several human pathologies based on vitamins deficiency can be fully eradicated or their prevalence decreases substantially thanks to (i) promotion of good nutrition practices and (ii) use of dietary supplements containing mainly vitamins and trace elements. Even so, malnutrition is still a massive problem, particularly in some geographic regions characterised by poverty, poor nutrition understanding and practices and deficient sanitation and food security.

During the last five decades, several scientific-technical reports have confirmed and/or suggested new biological roles and properties for vitamins in human beings. Despite a large amount of existing information, there are very few integrative studies carried out on the effect of the vitamins on human health. In this sense, the work here presented summarises the main recent evidences that provide an integrated and updated analysis about the effect of vitamins in human health. The main aim is to understand how the use of vitamins (from food or from dietary supplements containing vitamins) can improve human health or the evolution of some specific disease.

2. General aspects of vitamins

2.1. Definition and classification

Vitamins are organic micronutrients mainly synthesised by plants and microorganisms, which do not provide energy. Animals are not able to synthesise them, consequently, these essential micronutrients must be supplied by the diet in small amounts or even trace amounts (micrograms or milligrammes per day) for the maintenance of the metabolic functions of most animal cells [5, 6]. However, some vitamins can be synthesised in varying concentrations by humans. Thus, vitamin D and niacin are endogenously synthesised (in the skin by exposure to the sun or from the amino acid tryptophan, respectively) [7, 8]. On the other hand, vitamins such as K2, B1, B2 and biotin are synthesised by intestinal bacteria [9]. Generally, this

http://dx.doi.org/10.5772/intechopen.69122

endogenous synthesis is not enough to cover daily needs, so dietary intake is required [8, 10]. Most of the vitamins were identified related to the diagnosis of the diseases associated with their deficiency [2, 11]. Thus, these diseases are termed 'deficiency diseases'.

Two groups of vitamins are distinguished based on their solubility (fat-soluble and water-soluble vitamins) [6] (**Table 1**). Each of these two groups exhibit significantly different physical-chemical-biological characteristics. The alphabetic nomenclature indicates the chronology of its discovery; however, the subsequent observation that vitamin B consisted of multiple compounds, gave rise to numerical nomenclature. The gaps in numbering are due to the removal of several substances that were initially described as vitamins [8, 10].

Besides, vitamins are also classified by their biological role, which constitutes a more scientific approach to the current reality (Section 2.3 display details about the biological roles).

2.2. Physical-chemical properties

Each vitamin is a family of chemically related compounds that share qualitatively biological activities and may vary in aspects related to their bioactivity and bio assimilation. Therefore, the common name of the vitamin (i.e. vitamin A) is, in fact, a generic descriptor for all active analogues or relevant vitamin derivatives [12]. **Table 2** summarises the main physical-chemical properties.

2.3. Biological roles

Vitamins play an important role in several metabolic pathways, acting closely associated with many of the enzymes that catalyse the reactions involved in these metabolic processes [10, 13, 14].

Fat-soluble vitamins	Water-soluble vitamins
Vitamin A or Retinol	Vitamin B1 or Thiamine
Vitamin D or Calciferol	Vitamin B2 or Riboflavin
Vitamin E or α-Tocopherol	Vitamin B3 or Niacin
Vitamin K or Phylloquinone	Vitamin B5 or Pantothenic acid
	Vitamin B6 or Pyridoxine
	Vitamin B7 or Biotin
	Vitamin B9 or Folic acid
	Vitamin B12 or Cobalamin
	Vitamin C or Ascorbic acid
Soluble in fats	Soluble in water
They do not contain nitrogen	They contain nitrogen (except vitamin C)
Require bile salts and fats for absorption	Easily absorbed
Normally not excreted in the urine	They present urinary excretion threshold (Unlikely toxicity)
No daily or usual intake is required	Almost daily intake is required
Hypervitaminosis can cause toxicity	Not stored in the body (Exception: vitamin B12 in liver)
Liver and adipose tissue storage	

Note. Underlined: Name mainly used in the scientific literature.

Table 1. Classification and differences of vitamins based on their solubility [6].

Vitamin	Derivatives	Chemical formula	MW	Maximum absorption (nm)	Melting point (°C)	Colour/State
A (retinol)	Retinol	$C_{20}H_{30}O$	286.4	319–328	62–64	Yellow/crystal
	Retinal	$C_{20}H_{28}O$	284.4	373	61–64	Orange/crystal
	Retinoic acid	$C_{20}H_{28}O_2$	300.4	350–354	180–182	Yellow/crystal
D (cholecalciferol)	Cholecalciferol (vitaminD3)	$C_{27}H_{44}O$	384.6	265	84–85	White/crystal
	Ergocalciferol (vitaminD2)	$C_{28}H_{44}O$	396.7	264	115–118	
E (α-tocopherol)	α-tocopherol	$C_{29}H_{50}O_2$	430.7	292	2.5	Yellow/oil
	γ-tocopherol	$C_{28}H_{48}O_2$	416.7	298	−2.4	
K (phylloquinone)	Phylloquinone(K1Menaquinone-s	$C_{31}H_{46}O_2$	450.7	242	-	Yellow/oil
	($K2_n$)	-	444.7–649.2	243–270	35–54	Yellow/crystal
	Menadione(K3)	$C_{11}H_8O_2$	172.2	-	105–107	Yellow/crystal
B1(thiamine)	Thiamine	$C_{12}H_{17}N_4OS^+$	337.3	-	246–250	White/Crystals
B2 (riboflavin)	Riboflavin	$C_{17}H_{20}N_4O_6$	376.4	260	278	Orange-Yellow/ Crystal
B3 (niacin)	Nicotinic acid	$C_6H_5NO_2$	123.1	260	237	White/Crystal
	Nicotinamide	$C_6H_6NO_2$	122.1	261	128–131	
B5 (pantothenic acid)	Pantothenic acid	$C_9H_{17}NO_5$	219.2	204	-	Clear/oil
B6 (pyridoxine)	Pyridoxol	$C_8H_{11}NO_3$	205.6	253	206–208	White/Crystal
	Pyridoxal	$C_{80}H_9NO_3$	203.6	390	165	
	Pyridoxamine	$C_8H_{12}N_2O_2$	241.1	253	226	
B7 (biotin)	Biotin	$C_{10}H_{16}N_2O_3S$	244.3	204	232	Colourless/Crystal
B9 (folic acid)	Folic acid	$C_{19}H_{19}N_7O_6$	441.1	282	-	-
B12 (cobalamin)	Cyanocobalamin	$C_{63}H_{88}CoN_{14}O_{14}P$	1355.4	278	-	Dark red/Crystal
C (ascorbic acid)	Ascorbic acid	$C_6H_8O_6$	176.1	245	190–192	White/Crystals

Table 2. Physic-chemical properties of vitamins and the most relevant derivatives (Adapted from Combs, [12]; https://www.ncbi.nlm.nih.gov/pccompound;http://www.lipidbank.jp/).

Using the 'biological role' as criteria, vitamins are classified into five groups:

- Vitamins acting as coenzymes: B1 (thiamine), B2 (riboflavin), B3 (niacin), B5 (pantothenic acid), B6 (pyridoxine) and B7 (biotin).
- Antioxidant vitamins: E (α-tocopherol) and C (ascorbic acid).
- Vitamins showing hormonal functions: A (retinol) and D (calciferol)
- Vitamins that act in the cellular proliferation: B9 (Folic acid), B12 (cobalamin).
- The vitamins involved in coagulation: K or phylloquinone.

Thus, vitamins belonging to the group B work together at the cellular level and they are essential for neurological functioning and central metabolism [15]. A deficient intake of one or more than one of them may hinder the use of the other vitamins of group B. On the other hand, antioxidant vitamins protect against cell damage caused by the oxidative attack of free radicals reactive nitrogen species (ROS), Reactive nitrogen species (RNS), avoiding the destruction of the body's tissues. This group of vitamins prevent the development of a large number of degenerative diseases, associated with ageing and oxidative stress, such as Alzheimer's disease, Parkinson's disease, multiple sclerosis, cancer and myocardial infarction (heart attack), among others [16, 17]. In addition, some vitamins assume additional endocrine functions [18]. Consequently, the deficiency of a vitamin causes metabolic processes imbalances. This fact results in clinical signs or diseases of different health impact based on the level of deficiency. **Table 3** summarises the main biological roles played by vitamins and anomalies in human health due to vitamin excess (toxic effects in the case of liposoluble vitamins) or vitamin deficiency.

Vitamin	Biological roles	Clinical signs of deficiency	Toxic effects
Vitamin A (retinol)	Cellular repair and maintenance. Immune response. Development of NS. Normal vision. Foetal development. Reproduction. Bone growth. Antioxidant activity.	Xerophthalmia, night blindness, keratinization of the corneal epithelium, dry mucous membranes	Anorexia, weight loss, extreme irritability, diplopia, alopecia, headache, bone abnormalities, liver damages, birth defects
Vitamin D (cholecalciferol)	Bone and dental mineralisation. Absorption and metabolism of calcium and phosphorus.	Rickets (in children), osteomalacia (in adults) and osteoporosis	Hypercalciuria and hypercalcemia with soft tissue calcifications, renal and cardiovascular damage
Vitamin E (α-tocopherol)	Powerful antioxidant. Synthesis of heme group. Antitoxic function.	Peripheral neuropathy, spinocerebellar ataxia and pigmentary retinopathy.	Haemorrhagic toxicity, headache, fatigue, nausea, double vision, muscular pains, creatinurea, gastrointestinal distress
Vitamin K (phylloquinone)	Blood clotting. Protein synthesis. Bone metabolism	Haemorrhages.	Menadione (synthetic form) causes liver damage, jaundice and haemolytic anaemia in newborns

Vitamin	Biological roles	Clinical signs of deficiency	Toxic effects
VitaminB1 (thiamine)	Macronutrient metabolism. Neuronal function.	Beriberi[1]. Wernicke-Korsakoff syndrome. Polyneuritis. Heart failure. Anorexia and gastric atony	Not observed
VitaminB2 (riboflavin)	Energy metabolism. Ocular function. Antibody and red blood cells formation. Mucosal maintenance.	Oral-ocular-genital syndrome[2].	Not observed
VitaminB3 (niacin)	Macronutrient metabolism. Sex hormone production. Glycogen synthesis.	Pellagra[3] (dermatitis, dementia and diarrhoea).	Hepatotoxicity, flushing[4], nausea, blurred vision and IGT
VitaminB5 (pantothenic acid)	Energy metabolism. Antibody synthesis. Corticosteroid synthesis Cholesterol synthesis	Hypertension, gastrointestinal disturbances, muscular cramps, hypersensitivity, neurological disorders	Not observed
VitaminB6 (pyridoxine)	Fat and protein metabolism DNA and RNA synthesis Haemoglobin synthesis. Antibody production. Electrolyte balance. Neuronal function. Conversion of tryptophan to niacin	Neuropathy (paraesthesia). Epileptiform convulsions in infants. Hypochromic anaemia, seborrheic dermatitis and glossitis	Sensory neuropathy and skin disorders.
VitaminB7 (biotin)	Energy metabolism. Cell growth Fatty acids amino acids and glycogen synthesis	Dermatitis, conjunctivitis, alopecia and abnormalities of the CNS (depression, hallucinations and paraesthesia)	Not observed
Vitamin B9 (folic acid)	DNA and RNA synthesis Growth and cell division Leukocytes and erythrocytes formation and maturation. Folic acid metabolism	Macrocytic anaemia	Neurological complications in people with vitamin B12 deficiency
VitaminB12 (cobalamin)	Lipid and protein metabolism Red blood cells maturation. Iron absorption. DNA and RNA synthesis. Neuronal function	Hematologic (macrocytic anaemia), paraesthesia	Not observed
Vitamin C (ascorbic acid)	Multiple functions as coenzyme Iron absorption. Wound healing Antioxidant. Corticosteroid synthesis	Scurvy[5]. Sjögren syndrome, gum inflammation, dyspnoea, oedema y fatigue. Bone abnormalities, haemorrhagic symptoms and anaemia	Diarrhoea and other gastrointestinal disturbances

NS: Nervous system; CNS: central nervous system; IGT: impaired glucose tolerance.

[1]First nutritional deficiency described, typical of populations subsisting on diets in which polished ('white') rice is the major food. The pathology leads to weight loss, heart disorders and neurological dysfunction.

[2]Affectation of the mucous membranes, tongue (glossitis), lips (cheilitis) and hypervascularization of the cornea.

[3]In populations subsisting on diets in which maize is the major food.

[4]Head and neck redness.

[5]Signs and symptoms include: follicular hyperkeratosis, petechial, ecchymosis, coiled broken hairs, swollen and bleeding gums, perifollicular bleeding, joint spasm, arthralgia and altered wound healing (IOM, [18]; Combs, [10]).

Table 3. Main biological functions, clinical signs of deficiency and toxic effects (caused by excessive intake, hypervitaminosis) of vitamins [8, 10, 14, 18–20].

http://dx.doi.org/10.5772/intechopen.69122

3. Recommended dietary intakes

Most foods (exceptions: sucrose, refined grains and alcoholic beverages), provide vitamins in number and variable quantity [6]. However, there is not a single food containing all of them. Therefore, the diets must be mixed and balanced thus supplying the vitamins at the levels required by the body. When a food (or a diet) provides some or all the macronutrients but does not contain the necessary vitamins, it hinders the correct metabolism. Consequently, several official institutions around the world provide guides to recommend the optimum values of daily vitamins intake to promote health and to eradicate deficiency diseases.

The reference values of vitamin intake, allow preventing deficiency states and hypervitaminosis. **Table 4** shows the recommended dietary allowance (RDA) related to vitamins, which are focused on metabolic needs in the general population, and the maximum tolerable daily intake (UL) without risk of adverse health effects for the general population. These may vary between countries.

Vitamin	RDA	UL	Food sources
Vitamin A[1] (retinol)	2900 IU/d* (800 μg/d)	10,000 IU/d (3000 μg/d)	Liver, fish, dairy products, meat, egg yolk, butter, darkly coloured fruits and leafy vegetables
Vitamin D[2] (cholecalciferol)	600 IU/d* (15 μg/d)	2000 IU/d (50 μg)	Fish liver oils, fatty fish, egg yolk, fortified dairy products and fortified cereals
Vitamin E (α-tocopherol)	15 mg/d	1000 mg/d	Vegetable oils, unprocessed cereal grains, nuts, fruits, vegetables, meats
Vitamin K (phylloquinone)	90–120 μg/d	-	Green vegetables, Brussel sprouts, cabbage, plant oils and margarine
Vitamin B1 (thiamine)	1.2 mg/d	-	Enriched, fortified or whole-grain products, bread and bread products, mixed foods whose main ingredient is grain, cereals, potatoes, liver, pork and eggs
Vitamin B2 (riboflavin)	1.2 mg/d	-	Organ meats, milk, bread products and fortified cereals
Vitamin B3 (niacin)	15 mg/d	35 mg/d	Meat, fish, poultry, enriched and whole grain breads and bread products, fortified cereals and mushrooms
Vitamin B5 (pantothenic acid)	5 mg/d	-	Chicken, beef, potatoes, oats, cereals, tomato products, liver, kidney, yeast, egg yolk, broccoli and whole grains
Vitamin B6 (pyridoxine)	1.3 mg/d	100 mg/d	Fortified cereals, organ meats, fortified soy-based meat substitutes and bananas
Vitamin B7 (biotin)	30 μg/d	-	Liver, egg yolk, pork and vegetables

Vitamin	RDA	UL	Food sources
Vitamin B9 (folic acid)	400 μg/d	1000 μg/d (1 mg/d)	Enriched cereal grains, dark leafy vegetables, enriched and whole grain breads, fortified cereals, liver and nuts
Vitamin B12 (cobalamin)	2.4 μg/d	-	Fortified cereals, meat, fish and poultry
Vitamin C (ascorbic acid)	80 mg/d	2000 mg/d	Citrus fruits, tomatoes, potatoes, Brussel sprouts, cauliflower, broccoli, strawberries, cabbage and spinach

*RDAs for vitamins A and D are listed in both International Units (IUs) and micrograms (mg/day) or micrograms (μg/day). The hyphen (-) indicates that the UL is not determined due to lack of data on the adverse effects associated with the excessive intake of these vitamins.

[1] The vitamin A activity in foods is thus currently expressed as retinol equivalents (RE): 1 RE is defined as 1 μg of all-trans retinol, 6 μg of all-trans β-carotene, or 12 μg of another provitamin A carotenoids. Or it is expressed in IU (international units): 1 IU of vitamin A activity has been defined as equal either to 0.30 μg of all-trans retinol or to 0.60 μg of all-trans β-carotene.

[2] In the case of vitamin D, 1 μg calciferol = 40 IU of vitamin D, a value based on a minimum of sun exposure.

Table 4. Recommended dietary allowances (RDAs), tolerable upper intake level (UL) for healthy adults and main food sources containing the vitamins described [18], https://fnic.nal.usda.gov/sites/fnic.nal.usda.gov/files/uploads/DRI_Vitamins.pdf].

Some vitamins can be supplied as provitamins, substances without vitamin activity that when metabolised, give rise to the formation of the corresponding vitamin [8, 12]. In some cases, it is possible to synthesise the vitamin from dietary compounds that apparently have no relation to it. For instance, nicotinic acid (vitamin B3) can be caused by the metabolic transformation of the amino acid tryptophan [8] or retinol (vitamin A), which can be obtained from beta-carotene (a pigment produced by some vegetables and microorganisms) [21].

4. Bibliographic and bibliometric analysis of the selected information.

To identify the main recent scientific-technical works about vitamins and their effect in human beings, a bibliographic/bibliometric review has been made following PRISMA guide [22]. The classical scheme proposed by Vilanova [23] has been used to analyse and to assess the quality of the information obtained. The main aim of this analysis is to understand how the use of vitamins (from food or from dietary supplements containing vitamins) can improve human health or the evolution of some specific diseases.

To do the information search (manuscripts published during the last 27 years in English and Spanish), general and more specific databases were selected (https://scholar.google.es/; PubMed, http://www.ncbi.nlm.nih.gov/pubmed; Scopus, https://www.scopus.com/; Web of Science (WOS), https://apps.webofknowledge.com/). The keywords used to do the search were: all the names of the vitamins, 'vitamins & human health', 'vitamins & biological roles' and 'deficiency diseases'. These terms were previously identified through the

http://dx.doi.org/10.5772/intechopen.69122

database 'MeSH' (medical subject heading) as suitable descriptors for the realisation of this work. Combinations of these keywords with the terms 'diet' and 'nutrition' were also used to identify as many sources as possible. All the following options were selected in the databases previously mentioned: 'Title/Abstract', 'article', 'clinical trial' and 'review'. Search finished in December 2016, the 15th. The research questions used to do the search and to select the information were: What is new about the knowledge of the effect of vitamins on human health? Is human health improving when multivitamin complexes are used?

Figure 1 displays the results of the search just using the combination 'vitamins & human health'. Thanks to this keywords combination, 99,990 publications were identified (32,363 Pubmed; 35,127 WOS; 32,500 Scopus). About 60–77% of these publications are research articles (most of them clinical trials), 13–24% reviews and 5–11% are proceedings. Most of the items consulted (85%) belong to the field of medicine, followed by the fields of biochemistry, genetics and molecular biology (15%). To carry out this work, all the items were analysed by the three authors paying special attention to reviews and clinical trials. As it can be concluded from this figure, the last decade was particularly productive in terms of a number of publications analysing the effect of vitamins in human health or the use of vitamins as part of a treatment against certain pathologies.

To address the detailed analysis of the direct effects of vitamins in human health, described by each item identified (**Figure 1**), four categories or manuscripts were established: 1: experimental studies, clinical trials; 2: analytical observational studies (cohort studies; case-control studies); 3: Descriptive observational studies (series of cases; studies of incidence and prevalence); 4: Reviews, systematic reviews and/or meta-analysis. The main conclusions from this analysis are summarised in the following section.

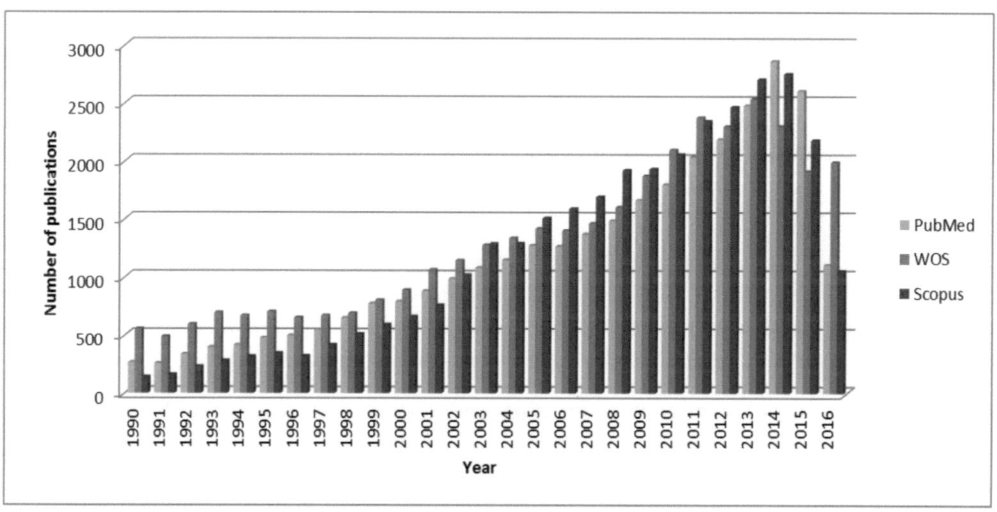

Figure 1. Identification of publications of interest. Number of items reported in the last 27 years. Keywords: Vitamins & Human health.

5. New advances of the effect of the use of vitamins through the diet in human health as well as the treatment of several human diseases

From the database containing the publications of interest previously mentioned, 75% of them were analysed into detail to highlight what is new about the use of vitamins through the diet in human health as well as their use as part of the treatment of several human diseases. Most the publications analysed in this work suggest a possible effect of a vitamin (its derivatives, analogues or precursors), or combinations of vitamins in human health. However, the results presented in the majority these publications are not conclusive. Thus, most of them assume that it is not possible to attribute with certainty the effect observed due to inconsistencies in the design or implementation of the studies. In this sense, there are many aspects to discuss, which are following summarised:

a) The standard method of medical science to establish and to compare the effectiveness of a substance in human beings is the clinical trial [24]. However, despite having strict inclusion criteria, these studies present some features that can affect the results. Some of the main features that may influence the results are: genetic background and style of life of the patient; non-specific effects and bioavailability of the vitamin/molecule tested; selection of the mechanism of action of the molecule tested; validity of the biomarkers used to determine the effect of a compound; the sample size (population) and the duration of the study (especially critical when the pathological condition under study takes decades to develop). All these aspects should be taken into account when interpreting the clinical results; otherwise, the associations observed are inadequately estimated of causality, and consequently, a direct relationship between the administration of a vitamin and effect on human health cannot be properly established.

b) Observational studies are easier to perform in terms of methodology, but they lack the capacity to establish causality of phenomena.

c) The meta-analysis presents a high level of scientific evidence, especially the meta-analysis of randomised controlled trials [24]. Meta-analysis is characterised by the high size of the study population, and therefore, they show better clinical significance. However, as a disadvantage, they usually are not feasible due to the difficulties of finding trials with the homogeneous design.

Therefore, despite a large number of publications on the vitamins and the potential uses of multi/vitamin supplements, there is no scientific evidence of beneficial effects in human health, beyond the prevention and/or treatment of deficiency states.

In this sense, the supplementation of food, as well as strategies to improve nutritional practises, have contributed to the eradication of deficiency diseases [25–28]. The main biological functions, clinical signs of deficiency and toxic effects of vitamins described until the end of the last century were previously discussed in Section 2.3 (**Table 3**). Recently, new correlations between vitamins and human health have been proposed. Details about the best described correlations between the use of vitamins on human health are following summarised:

Vitamin A: Diet supplementation has a positive effect on the blindness and the morbid-mortality in preschool-age children living in developing countries (http://data.unicef.org/nutrition/vitamin-a.html). Since 1960, clinical trials have shown that the disorders caused by vitamin A deficiency in developing countries can be prevented with regular dose and this supplementation significantly reduces infant mortality [29–31]. In relation to the other observed associations between vitamin A and certain diseases (**Table 5**), the evidence obtained do not allow definitive conclusions on the potential benefits of supplementation.

Vitamin D: The role of vitamin D in bone health is probably one of the better-supported relationships (**Tables 3** and **5**). The 'new' properties related to vitamin D are closely linked to the biological function already described. Thus, several meta-analyses of randomised controlled clinical studies conclude that vitamin D supplementation reduces the risk of falls (derived from the bone fragility) in a 19%, the risk of hip fracture in an 18% and the risk of non-vertebral

Name of the vitamin	Diseases or health states	Name of the vitamin	Diseases or health states
Vitamin A (retinol)	Eye diseases Mortality Cancer Anaemia	**Vitamin K (phylloquinone)**	Bone health CVD Cancer Mortality
Vitamin E (α-tocopherol)	CVD, cancer, mortality Alzheimer disease, immunity	**Vitamin D (cholecalciferol)**	Bone health, cancer, CVD Hypertension Autoimmune diseases Pregnancy Quality life Pulmonary infections Mortality
Vitamin B1 (thiamine)	Microalbuminuria in DM Cardiac function	**Vitamin B2 (riboflavin)**	Homocysteine levels in plasma Cancer Migraine
Vitamin B3 (niacin)	Atherosclerosis, Dyslipidaemias, Mortality, Diabetes, Cancer	**Vitamin B5 (pantothenic acid)**	Healing Acne Rheumatoid arthritis
Vitamin B6 (pyridoxine)	TD, Cancer, PMS, CTS Side effects of OCPs, CVA	**Vitamin B7 (biotin)**	DM Multiple sclerosis
Vitamin B9 (folic acid)	Birth defects, Vascular disease Renal disease, Cognitive Function, Cancer, DM, Childhood asthma, Childhood leukaemia	**Vitamin B12 (cobalamin)** **Vitamin C (ascorbic acid)**	Cognitive function Congenital diseases Cancer, CVD, Pulmonary function, Cold, Stress, AMD

AMD: Age-related macular degeneration; CTS: Carpal tunnel syndrome; CVA: stroke (cerebrovascular accident); CVD: cardiovascular disease; DM: diabetes mellitus; OCPs: oral contraceptives; PMS: premenstrual syndrome; TD: tardive dyskinesia.

Table 5. New associations found between vitamins (deficiency or toxicity) and diseases or health states.

fractures in a 20% in older adults. The effect on the prevention of falls or fractures is reached using high doses of at least 700–1000 IU/day or at least 400 IU/day, respectively [32–35]. In addition, supplementation has been shown to have a beneficial effect on the balance and muscle strength [36]. The evidence-based clinical trials suggest that supplementation with vitamin D (1000 IU/day) helps to prevent falls and fractures in the elderly population. However, the studies are not exempt from limitations; in general, these studies were done using supplements of vitamin D combined with calcium, so the effect attributable specifically to the vitamin D is difficult to determine. In addition, in many cases the basal levels of vitamin D and/or calcium uptake is unknown (diet, exposure to the sun, supplements, etc.).

Vitamin B9: intervention trials with folic acid in pregnant women stated that the supplementation reduces the occurrence of neural tube defects (NTD) [37–39]. In USA for instance, the use of folic acid supplements was legally established by the end of 1990, which reduced significantly (20–27%) the prevalence of neural tube defects at birth [19]. Since then, the consumption of 400 μg/day of folic is recommended to women who want to conceive to prevent birth defects in the foetus [40, 41]. In relation to the other observed associations between folic acid and certain diseases (**Table 5**), the evidences obtained do not make possible to attribute potential benefits to supplementation. Besides, for all the statements about the supplementation with vitamins, there are studies that found negative evidence, including the two cases mentioned above (vitamin D and folic acid).

In relation to the other observed associations between individual vitamins and certain diseases (**Table 5**), the evidences do not clearly show direct effects of supplementation, either in a positive way (prevention of chronic diseases and/or improvement of human health) or negative (adverse effects linked to the excessive intake), due to the inadequate methodology of the existing studies [42]. There is a need for new designs of scientific studies to reach valid conclusions. These new designs should consider several aspects such as (i) the initial nutritional status of patients, (ii) the use of homogeneous groups, (iii) the use of control groups and (iv) control of the composition of the ingested food (as it often overestimates the amount of vitamin because it does not consider the bioavailability).

On the other hand, population differences based on genetics could have significant implications in terms of vitamins bio assimilation [43]. The biochemical individuality and the lack of margins for the safety of vitamins sustain the basic premise of the toxicology 'the dose makes the poison'. To evaluate the therapeutic efficacy of a vitamin is essential to analyse the dose to be administered, the form of the vitamin used (solution, microencapsulated or crystallised), the source of the vitamin (synthetic or purified from natural sources), the bioavailability and the interaction of a specific vitamin with other nutrients.

Summarising, the analysis set out in this work shows that 'new' potential benefits have been attributable to several vitamins. However, most of them are not robustly supported by evidences. In addition, the analysis suggests that the information related to individual vitamins for the prevention and/or treatment of diseases is more consistent than that of a multivitamin complex. In this sense, a systematic review carried out in the USA concludes that the evidence is insufficient to support the use of multivitamin supplements to prevent chronic degenerative diseases [42].

Finally, it is not surprising that numerous studies published in more than a decade have related some supplements (including vitamins E, C, D, A, and B) with adverse effects on human health. A meta-analysis of 67 trials showed that supplements of vitamin E, vitamin A and beta-carotene might be associated with a higher incidence of mortality [44]. Another study found a higher incidence (18%) of lung cancer and mortality from all causes (8%) in men who received beta-carotene [45]. In 2008, a large randomised controlled trial was stopped after reporting that supplementation of vitamin E and selenium resulted in an increase in the incidence of prostate cancer [46].

6. Use of multivitamin complexes and potential risk of hypervitaminosis

The rate of use of vitamins, minerals and other bioactive compounds available in food or dietary supplements is increasing significantly in advanced societies, especially in USA population, where the multivitamin complexes are the most commonly used supplements [47–49]. Several works state that currently, more than 47% of men and 59% of the women in the USA use supplements for health benefits, and the number of users is growing significantly [50]. In Europe, the greatest consumption was observed in the countries of the north, especially in Denmark (51.0% among men, 65.8% among women) [51].

Due to this high market demand, the number of companies producing this kind of dietary supplements is increasing around the work (http://biomarket.cat/es/69-vitaminas; http://salud.bayer.es/vitaminas-y-complementos-alimenticios/otras-vitaminas/; http://lifestylemarkets.com/vitamins-and-supplements/multivitamins/).

There are reports indicating that there could be adverse effects on human health attributable to high consumption of multivitamin complexes. Almost 60,000 cases of toxicity by use of vitamins are reported annually USA poison control centres [http://www. aapcc.org/annual-reports/; [52]]. The most common adverse effects associated with excessive intake of vitamins (hypervitaminosis) are shown in **Table 3**, Section 2.3. Fat-soluble vitamins, for instance, due to its ability to accumulate in the body, have a greater potential for toxicity than water-soluble vitamins. However, the overdose of water-soluble vitamins can also cause toxicity affecting several body systems including the nervous system [20, 53]. Between the fat-soluble vitamins, the more toxic are vitamin A and vitamin D. The toxicity of vitamin A can be acute or chronic (IOM, 2006) and high doses cause many toxic manifestations (**Table 3**, Section 2.3). However, there has been no toxic effects of carotenoids (provitamin A), even when eaten in large amounts for weeks or years [41, 54], except for an orange/yellow colouring of the skin [55]. Vitamin D is potentially toxic, especially to small children [56]. In comparison to vitamins A and D, vitamin E is the least toxic when ingested orally [57]. In the case of vitamin K, toxic effects have not been observed even intaking large amounts over a long period [41]; however, a synthetic form of Vitamin K (menadione) has been associated with liver damage, and therefore no longer used therapeutically [18, 41].

The evidence on the safety profile of multivitamin complexes in humans has been established through case reports. However, the data reported from these case reports do not allow the accurate identification of maximum tolerable intake level (U). Besides, the toxicological data show that the margins of safety for multivitamin complexes intake are not yet defined, noting toxic doses significantly different in the scientific literature. This suggests that high doses of vitamins, especially of fat-soluble vitamins, should not be given to any group of the population until the safety of such doses is well established and based on scientific evidence.

7. Conclusion

Despite a large number of research works carried out to study the effects of vitamins in human health during the last decades, evidences to attribute potential benefits of vitamins supplementation on either human health or prevention and/or treatment of chronic degenerative diseases are still scarce. The analysis of the research works published during the last 27 years shows that it is impossible to establish robust and universal conclusions about the benefit of vitamin supplementation on human health beyond the prevention and/or treatment of deficiency states (stated during the second half of the twentieth century).

On the other hand, it is important to highlight the high heterogeneity in the clinical and methodological experiments as well as in the tools used to perform these studies, which contributes to making difficult a comparative analysis at large scale. Clinical trials of high methodological quality and a significant number of patients are yet to come. Due to these reasons, the widespread use of multivitamin complexes as diet supplements is still not fully justified.

The most prudent recommendation and scientifically supported for disease prevention is to eat a balanced diet with an emphasis on fruits and vegetables rich in antioxidants [58], since it is through the diet it is impossible to eat excessive quantities of vitamins. This approach minimises the risk of micronutrient deficiency or excess. However, not all individuals maintain a balanced diet for long periods of time. For this reason, certain circumstances (pregnant women, infants without breastfeeding, vegetarian individuals, elderly, etc.) may require the use of vitamin supplements under control to prevent deficiencies.

Although the potential of the vitamins in the promotion of human health is enormous, it is necessary to assess the risk/benefit ratio in each case. There is much more research to be done to understand the benefits of supplementation in the prevention of diseases and the improvement of human health. Accurate studies about consumption of vitamins by country (including aspects as sex, age, etc.) as well as about food fortification and vitamins protection and stabilisation are yet to come [28]. A greater knowledge in this area of the science of nutrition will have an impact on clinical practice dietetics and nutrition guidelines for public health.

Acknowledgements

This work has not been funding.

Author details

Noelia García Uribe[1], Manuel Reig García-Galbis[2,3] and Rosa María Martínez Espinosa[1]*

*Address all correspondence to: rosa.martinez@ua.es

1 Department of Agrochemistry and Biochemistry, Faculty of Science, Biochemistry and Molecular Biology, University of Alicante, Spain

2 Department of Nursing, Faculty of Health Sciences, University of Alicante, Spain

3 Department of Nutrition, Faculty of Health Sciences, University of Atacama, Chile

References

[1] Rosenfeld L. Vitamine—vitamin. The early years of discovery. Clinical Chemistry. 1997; **43**(4):680-685

[2] Semba RD, Bloem MW. The anemia of vitamin A deficiency: Epidemiology and pathogenesis. European Journal of Clinical Nutrition. 2002;**56**(4):271-281. DOI: 10.1038/sj.ejcn.1601320

[3] Maltz A. Casimer Funk, nonconformist nomenclature, and networks surrounding the discovery of vitamins. Journal of Nutrition. 2013;**143**(7):1013-1020. DOI: 10.3945/jn.112.171827

[4] Varela G. Historia de las Vitaminas. En la alimentación y la nutrición a través de la historia. Salas J, Garcia P, Sanchez JM. ed. Barcelona: Glosa;2005. pp. 374-382

[5] Mora JR, Iwata M y von Andrian UH. Vitamin effects on the immune system: Vitamins A and D take centre stage. Nature Reviews Immunology. 2008;**8**(9):685-698. DOI: 10.1038/nri2378

[6] Mataix J. Vitaminas I. Nutrición para Educadores. 2nd ed. Madrid: Diez de Santos; 2013:117-126

[7] Nair R, Maseeh A. Vitamin D: The "sunshine" vitamin. Journal of Pharmacology and Pharmacotherapeutics. 2012;**3**(2):118-126

[8] Bender DA. Nutritional Biochemistry of the Vitamins. 2nd ed. United Kingdom: Cambridge University Press; 2003. 470

[9] Hill MJ. Intestinal flora and endogenous vitamin synthesis. European Journal of Cancer Prevention. 1997;**6**(1):43-45

[10] Combs GF. The Vitamins: Fundamental Aspects in Nutrition and Health. 3rd ed. Burlington: Academic Press; 2008. 583

[11] Piro A, Tagarelli G, Lagonia P, Tagarelli A, Quattrone A. Casimir Funk: His discovery of the vitamins and their deficiency disorders. Annals of Nutrition and Metabolism. 2010;**57**(2):85-88. DOI: 10.1159/000319165

[12] Combs GF. The Vitamins: Fundamental Aspects in Nutrition and Health. 4th ed. London: Academic Press; 2012. pp. 33-70

[13] Fidanza A, Audisio M. Vitamins and lipid metabolism. Acta Vitaminologica Et Enzymologica. 1982;**4**(1-2):105-114

[14] Berg JM, Tymoczko JL, Stryer L. Biochemistry. 5th ed. New York: WH Freeman; 2002. Section 8.6. Available in: https://www.ncbi.nlm.nih.gov/books/NBK21154/

[15] Kennedy DO. B Vitamins and the brain: Mechanisms, dose and efficacy—A review. Nutrients. 2016;**8**(2):68. DOI: 10.3390/nu8020068

[16] Brambilla D, Mancuso C, Scuderi MR, Bosco P, Cantarella G, Lempereur L. The role of antioxidant supplement in immune system, neoplastic, and neurodegenerative disorders: A point of view for an assessment of the risk/benefit profile. Nutrition Journal. 2008;**7**:29. DOI: 10.1186/1475-2891-7-29

[17] Harrison FE, May JM. Vitamin C function in the brain: Vital role of the ascorbate transporter (SVCT2). Free Radical Biology & Medicine. 2009;**46**(6):719-730. DOI: 10.1016/j.freeradbiomed.2008.12.018

[18] Institute of Medicine (IOM). Dietary references intakes: The essential guide to nutrient requirements. United States: The National Academies Press; 2006. 1344. Available in (access in February 2017): https://www.nal.usda.gov/sites/default/files/fnic_uploads/DRIEssentialGuideNutReq.pdf

[19] Lanska DJ. Chapter 30: Historical aspects of the major neurological vitamin deficiency disorders: The water-soluble B vitamins. Handbook of Clinical Neurology. 2010b;**95**: 445-476. DOI: 10.1016/S0072-9752(08)02130-1

[20] Hammond N, Wang Y, Dimachkie MM, Barohn RJ. Nutritional Neuropathies. Neurologic Clinics. 2013;**31**(2):477-489. DOI: 10.1016/j.ncl.2013.02.002

[21] Tang G. Bioconversion of dietary provitamin A carotenoids to vitamin A in humans. The American Journal of Clinical Nutrition. 2010;**91**(5):1468-1473. DOI: 10.3945/ajcn.2010.28674G

[22] Hutton B, Salanti G, Caldwell DM, Chaimani A, Schmid CH, Cameron C, et al. The PRISMA extension statement for reporting of systematic reviews incorporating network meta-analyses of health care interventions: Checklist and explanations. Annals of Internal Medicine. 2015;**162**(11):777-784

[23] Vilanova JC. Revisión bibliográfica del tema de estudio de un proyecto de investigación. Radiología. 2012;**54**(2):108-114

[24] Baladie E, Basulto J. Sistema de clasificación de los estudios en función de la evidencia científica. Dietética y Nutrición Aplicada Basadas en la Evidencia (DNABE): una herramienta para el dietista-nutricionista del futuro. Act Diet. 2008;**12**(1):11-19. Available in (access in February 2017): http://fedn.es/docs/grep/docs/DNABE_2008.pdf

[25] Hatun Ş, Ozkan B, Bereket A. Vitamin D deficiency and prevention: Turkish experience. Acta Paediatrica. 2011;**100**(9):1195-1199. DOI: 10.1111/j.1651-2227.2011.02383.x

[26] Sherwin JC, Reacher MH, Dean WH, Ngondi J. Epidemiology of vitamin A deficiency and xerophthalmia in at-risk populations. Transactions of The Royal Society of Tropical Medicine and Hygiene. 2012;**106**(4):205-214. DOI: 10.1016/j.trstmh.2012.01.004

[27] Swaminathan S, Thomas T, Kurpad AV. B-vitamin interventions for women and children in low-income populations. Current Opinion in Clinical Nutrition & Metabolic Care. 2015;**18**(3):295-306. DOI: 10.1097/MCO.0000000000000166

[28] Dwyer JT, Wiemer KL, Dary O, Keen CL, King JC, Miller KB, et al. Fortification and health: Challenges and opportunities. Advances in Nutrition. 2015;**6**(1):124-131. DOI: 10.3945/an.114.007443

[29] Lanska DJ. Chapter 29: Historical aspects of the major neurological vitamin deficiency disorders: Overview and fat-soluble vitamin A. Handbook of Clinical Neurology. 2010a;**95**:435-444. DOI: 10.1016/S0072-9752(08)02129-5

[30] Mayo-Wilson E, Yakoob MY, Bhutta ZA. Vitamin A supplementation for preventing morbidity and mortality in children from 6 months to 5 years of age. The Cochrane Database Syst Rev. 2010 Dec 8;(12):CD008524. DOI: 10.1002/14651858.CD008524.pub2. Review

[31] Wilson EM, Imdad A, Herzer K, Yawar MY y Bhutta ZA. Vitamin A supplements for preventing mortality, illness, and blindness in children aged under 5: Systematic review and meta-analysis. BMJ. 2011;**343**:d5094. DOI: https://doi.org/10.1136/bmj.d5094

[32] Bischoff-Ferrari HA, Dawson-Hughes B, Staehelin HB, Orav JE, Stuck AE, Theiler R, et al. Fall prevention with supplemental and active forms of Vitamin D: A meta-analysis of randomised controlled trials. BMJ. 2009;**339**:3692. DOI: 10.1136/bmj.b3692

[33] Bischoff-Ferrari HA, Willett WC, Wong JB, Stuck AE, Staehelin HB, Orav EJ, et al. Prevention of nonvertebral fractures with oral vitamin D and dose dependency: A meta-analysis of randomized controlled trials. Archives of Internal Medicine. 2009;**169**(6):551-561. DOI: 10.1001/archinternmed.2008.600

[34] Michael YL, Whitlock EP, Lin JS, Fu R, O'Connor EA, Gold R. Primary care—Relevant interventions to prevent falling in older adults: A systematic evidence review for the U. S. Preventive services task force. Annals of Internal Medicine. 2010;**153**:815-825. DOI: 10.7326/0003-4819-153-12-201012210-00008

[35] Murad MH, Elamin KB, Abu Elnour NO, Elamin MB, Alkatib AA, Fatourechi MM, et al. Clinical review: The effect of vitamin D on falls: A systematic review and meta-analysis. The Journal of Clinical Endocrinology & Metabolism. 2011;**96**(10): 2997-3006. DOI: 10.1210/jc.2011-1193. Epub 27 Jul 2011

[36] Muir SW, Montero-Odasso M. Effect of vitamin D supplementation on muscle strength, gait and balance in older adults: A systematic review and meta-analysis. Journal of the American Geriatrics Society. 2011;**59**:2291-2300. DOI: 10.1111/j.1532-5415.2011.03733.x

[37] Stamm RA, Houghton LA. Nutrient intake values for folate during pregnancy and lactation vary widely around the World. Nutrients. 2013;**5**(10):3920-3947. DOI: 10.3390/nu5103920

[38] Imbard A, Benoist JF, Blom HJ. Neural tube defects, folic acid and methylation. International Journal of Environmental Research and Public Health. 2013;**10**(9):4352-4389. DOI: 10.3390/ijerph10094352

[39] Liew SC. Folic acid and diseases—Supplement it or not? Revista da Associação Médica Brasileira. 2016;**62**(1):90-100. DOI: 10.1590/1806-9282.62.01.90

[40] Czeizel AE, Dudás I. Prevention of the first occurrence of neural-tube defects by periconceptional vitamin supplementation. The New England Journal of Medicine. 1992;**327**(26):1832-1835. DOI: 10.1056/NEJM199212243272602

[41] Institute of Medicine (IOM) Standing Committee on the Scientific Evaluation of Dietary Reference Intakes and its Panel on Folate, Other B Vitamins, and Choline. Folate. Dietary Reference Intakes for Thiamin, Riboflavin, Niacin, Vitamin B6, Folate, Vitamin B12, Pantothenic Acid, Biotin, and Choline. Washington: National Academies Press; 1998. 196-305. Available in: https://www.ncbi.nlm.nih.gov/pubmed/23193625

[42] Huang HY, Caballero B, Chang S, Alberg AJ, Semba RD, Schneyer CR et al. The efficacy and safety of multivitamin and mineral supplement use to prevent cancer and chronic disease in adults: A systematic review for a National Institutes of Health state-of-the-science conference. Annals of Internal Medicine. 2006;**145**(5):372-385. Available in: http://annals.org/aim/article/728199/efficacy-safety-multivitamin-mineral-supplement-use-prevent-cancer-chronic-disease

[43] Kresevic DM, Denton JE, Burant CJ, Pallaki M. Racial difference in response to Vitamin D supplementation. Journal of the National Medical Association. 2015;**107**(2):18-24. DOI: 10.1016/S0027-9684(15)30020-1

[44] Bjelakovic G, Nikolova D, Gluud LL, Simonetti RG, Gluud C. Antioxidant supplements for prevention of mortality in healthy participants and patients with various diseases. The Cochrane Database Syst Rev. 2008 Apr 16;(2):CD007176. doi: 10.1002/14651858.CD007176. Review. Update in: Cochrane Database Syst Rev. 2012;3:CD007176

[45] Cardenas E, Ghosh R. Vitamin E: A dark horse at the crossroad of cancer management. Biochemical Pharmacology. 2013;**86**(7):845-852. DOI: 10.1016/j.bcp.2013.07.018

[46] Lippman SM, Klein EA, Goodman PJ, Lucia MS, Thompson IM, Ford LG, et al. Effect of selenium and vitamin E on risk of prostate cancer and other cancers: The selenium and vitamin E cancer prevention trial (SELECT). JAMA. 2009;**301**(1):39-51. DOI: 10.1001/jama.2008.864

[47] NIH State-of-the-Science Panel. National Institutes of Health state-of-the-science conference statement: Multivitamin/mineral supplements and chronic disease prevention. The American Journal of Clinical Nutrition. 2007;**85**:257-264

[48] Gahche J, Bailey R, Burt V, Hughes J, Yetley E, Dwyer J, et al. Dietary supplement use among U.S. Adults has increased since NHANES III (1988-1994). NCHS Data Brief. 2011;(**61**):1-8. Available in: http://www.cdc.gov/nchs/data/databriefs/db61.htm

[49] Bailey RL, Gahche JJ, Lentino CV, Dwyer JT, Engel JS, Thomas PR, et al. Dietary supplement use in the United States, 2003-2006. Journal of Nutrition. 2011;**141**:261-266. DOI: 10.3945/jn.110.133025

[50] Soni MG, Thurmond TS, Miller ER, Spriggs T, Bendich A, Omaye ST. Safety of vitamins and minerals: Controversies and perspective. Toxicological Sciences. 2010;**118**(2):348-355. DOI: 10.1093/toxsci/kfq293

[51] Skeie G, Braaten T, Hjartåker A, Lentjes M, Amiano P, Jakszyn P, et al. Use of dietary supplements in the European prospective investigation into cancer and nutrition calibration study. European Journal of Clinical Nutrition. 2009;**63**(4):226-238. DOI: 10.1038/ejcn.2009.83

[52] Mowry JB, Spyker DA, Brooks DE, McMillan N, Schauben JL. 2014 Annual report of the American association of poison control centers' National Poison Data System (NPDS): 32nd Annual Report. Clinical Toxicology. 2015;**53**(10):962-1147. DOI: 10.3109/15563650.2015.1102927

[53] Chawla J, Kvarnberg D. Hydrosoluble vitamins. Handbook of Clinical Neurology. 2014;**120**:891-914. DOI: 10.1016/B978-0-7020-4087-0.00059-0

[54] Hathcock JN, Hattan DG, Jenkins MY, McDonald JT, Sundaresan PR, Wilkening VL. Evaluation of vitamin A toxicity. The American Journal of Clinical Nutrition. 1990;**52**(2):183-202

[55] Bendich A. The safety of beta-carotene. Nutrition and Cancer. 1988;**11**(4):207-214. DOI: 10.1080/01635588809513989

[56] Haussler MR, McCain TA. Basic and clinical concepts related to vitamin D metabolism and action. The New England Journal of Medicine. 1977;**297**:1041-50. DOI: 10.1056/NEJM197711102971906

[57] Bendich A, Machlin LJ. Safety of oral intake of vitamin E. The American Journal of Clinical Nutrition. 1988;**48**(3):612-619

[58] Lock K, Pomerleau J, Causer L, Altmann DR, McKee M. The global burden of disease attributable to low consumption of fruit and vegetables: Implications for the global strategy on diet. Bulletin of the World Health Organization. 2005;**83**(2):100-108. Available in: https://www.ncbi.nlm.nih.gov/pmc/articles/PMC2623811/

Chapter 5

Fermented Pulse-Based Food Products in Developing Nations as Functional Foods and Ingredients

Oluwafemi Ayodeji Adebo, Patrick Berka Njobeh,
Janet Adeyinka Adebiyi, Sefater Gbashi,
Judith Zanele Phoku and Eugenie Kayitesi

Additional information is available at the end of the chapter

http://dx.doi.org/10.5772/intechopen.69170

Abstract

Pulses play a significant and diverse role in the agricultural systems and diets of underprivileged populations worldwide. They are ideal produce for reducing poverty, improving human health and nutrition, and enhancing resilience of the ecosystem. Fermentation is a processing technique that has been used for decades to transform food produce with improved health, functional, and nutraceutical benefits. In tandem with the United Nations' (UN's) sustainable development goal Number 3, fermented food products from pulses with health benefits align with this initiative to end hunger, achieve food security, and improve nutrition. In solidarity with the celebration of International Year of Pulses 2016 (IYP2016) and considering the relative neglect of pulses as compared with other food groups, this chapter would be vital in positioning pulses and fermented products from them as readily available functional foods. With increased interest in fermentation, fermented pulse-based foods have been identified as excellent sources of bioactive and functional foods. Thus, fermented pulse-based products present a viable alternative, relatively available, affordable, and cheap source of foods with properties beyond that of basic nutrition.

Keywords: pulses, fermented foods, functional foods, bioactive compounds

1. Introduction

Following the resolution of the UN on 20 December, 2013, the 68th general assembly declared the year 2016 as the International Year of Pulses (IYP2016) [1], which was celebrated and sponsored by the Department of Science and Technology (DST), South Africa, during the

2016 Autumn International Food Safety and Security Conference hosted by the University of Johannesburg, South Africa. This was designated as such to promote public awareness on the usage of pulses and their potential as critical sources of plant-based proteins. The IYP2016 is rather timely and appropriate considering the relative neglect of pulses when compared to other crops despite its significant role pulses toward ensuring food security and nutrition.

While other processing techniques have been used for the transformation of pulses for food, fermentation is significant because it is known to improve sensory qualities and shelf life, reduce pathogenic microorganisms, and exert functional and health beneficial effects to food [2–5]. Due to its benefits and subsequent findings, developments of novel pulse-based foods through fermentation have been promoted [6]. Fermented pulse-based food contains a number of functional compounds including phytochemicals (phenolic compounds), lectins, polysaccharides, and phytates that confer and play significant role in health [7, 8]. In this regard, this chapter is thus focused on fermented pulse-based foods and the substantiated health-promoting components in them. This is vital considering the fact that these fermented foods form basic sources of diet and primary sources of bioactive compounds in many developing and underdeveloped nations. Furthermore, considerable emerging evidence showing the potential benefits of these fermented pulse-based foods is described.

2. Description of pulses and an overview of their composition

The name "pulses" is generally reserved for crops harvested solely for the dry seed (**Table 1**) and used interchangeably with grain legumes. While all pulses are considered legumes, not all legumes are pulses [9]. The Codex Alimentarius Commission as well as the Food and Agricultural Organization (FAO) of the United Nations defines pulses as dry seeds of leguminous plants, which are usually distinguished from leguminous oilseeds with their low fat content [1–3]. Although different pulse varieties are grown in 173 countries around the world, 11 of them are primarily recognized by FAO [9]. These are presented in **Table 1** along with their documented world production as at 2015.

Pulses are important crops that have a balanced nutritional composition and are among the most important sources of cheap and readily available starch, carbohydrate, protein, dietary fiber, minerals, and vitamins in food [9, 11–15]. Pulses also contain a number of bioactive compounds including phytates, oligosaccharides, enzyme inhibitors, and phenolic compounds that have been reported to positively impact health [7, 8, 16]. For human consumption, pulses are not eaten in its raw state, but typically after subsequent food processing, including boiling, cooking, puffing, grinding, germination (sprouting), and fermentation to increase their sensorial quality, appeal, esthetic value, and use.

Plant proteins are now being regarded as excellent, versatile, and available sources of functional and biologically active food components [9]. The evolution and drive toward the consumption of plant proteins have been influenced by the continued need and drive of health professionals agitating for partial replacement of animal proteins with plants that possess better and cheaper nutritional components. Aside from other components, pulses have been

Common name	Native name	Botanical name	World production[δ]
Dry beans	Kidney bean, navy bean, pinto bean Lima bean Scarlet runner bean Tepary bean Adzuki (azuki) bean Mung bean, golden gram, green gram Black gram, urad Ricebean Moth bean	*Phaseolus vulgaris* *Phaseolus lunatus* *Phaseolus coccineus* *Phaseolus acutifolius* *Vigna angularis* *Vigna radiate* *Vigna mungo* *Vigna umbellate* *Vigna aconitifolia*	27591.35
Dry broad beans	Horse bean Broad bean Field bean	*Vicia faba equina* *Vicia faba* *Vicia faba*	5568.67
Dry peas	Garden pea Protein pea	*Pisum sativum* var. *sativum* *Pisum sativum* var. *arvense*	12536.12
Chickpea	Bengal gram, garbanzo	*Cicer arietinum*	13741
Dry cowpea	Black-eyed pea, black eye bean	*Vigna unguiculata*	5602.72
Pigeon pea	Arhar/toor, cajan pea, Congo bean, gandules	*Cajanus cajan*	4890.10
Lentil		*Lens culinaris*	4952.12
Bambara groundnut	Earth pea	*Vigna subterranea*	160.38
Vetch	Common vetch	*Vicia sativa*	905
Lupins		*Lupinus* sp.	1014.02
Minor pulses	Lablab, hyacinth bean Jack bean Sword bean Winged bean Velvet bean, cowitch Yam bean	*Lablab purpureus* *Canavalia ensiformis* *Canavalia gladiate* *Psophocarpus tetragonolobus* *Mucuna pruriens* var. *utilis* *Pachyrhizus erosus*	NA*

[δ]Production in KT obtained from Ref. [10].
*NA, not available.

Table 1. Commonly consumed pulses and world production (KT).

identified as excellent sources of plant proteins, which are accumulated during their development [9, 17]. Pulses have been incorporated in various forms of traditional and staple diets to supplement basic protein and energy requirements and provide functional properties beneficial to human health [9, 16, 18, 19].

The most important pulses intended for human consumption include adzuki bean, black gram, chickpea, dry broad bean, dry cowpea, field pea, mung bean, green gram kidney bean, lentil, lupin, pigeon pea, lima bean, moth bean, and rice bean [20] with their comparative % protein provided in **Table 2**. Pulses provide between 20 and 30 g of protein per 100 g, twice as

Scientific names	Common names	Protein (per 100g)
Cajanus cajan	Pigeon pea	21.70
Cicer arietinum	Chickpea	21.70
Lens culinaris	Lentils	24.63
Phaseolus lunatus	Lima beans	21.46
Phaseolus vulgaris	Black beans	23.58
	Kidney beans	24.37
	Pinto beans	21.42
Vigna angularis	Adzuki beans	19.87
Vicia faba	Faba beans	26.12
Vigna mungo	Black gram	25.21
Vigna radiata	Mung beans	23.86
Vigna mungo	Vigna mungo	25.21

Table 2. Comparison of protein content of major pulse crops [21].

much as is found in grains and similar to that in meat [20]. Additionally, pulses do not contain residues of hormones and antibiotics like it is the case with animal protein sources such as beef and milk [1]. Nevertheless, while antibiotic and hormones might be absent, they could possibly be contaminated with pesticide and herbicides, used during cultivation. Pulses also possess a considerable amount of vitamins A and B along with iron, phosphorus, and calcium and thus serve as a food of high calorie and nutritive value [1, 7, 9].

3. Fermentation and fermented pulse foods in developing nations

Pulses have been processed in developing nations for centuries using traditional processing techniques of grinding, fermentation, steeping, germination, dehulling, etc. and prior to consumption for further use. Other novel food processing including micronization, microwave processing, high pressure processing (HPP), pulse electric field (PEF), irradiation, and extrusion techniques have found potential use and application for pulse processing. Nevertheless, fermentation remains largely important for pulse processing and gaining increased attention because of its improved functionalities, increase nutritional composition, and production of bioactive compounds [16, 22].

Fermentation can be generally defined as a processing technique used to convert substrates into new products through the action of microorganisms [5]. Fermentation is also used in a broader sense for the intentional use of microorganisms to obtain useful products for humans on an industrial scale. Such industrial products may include biomass, enzymes, primary and secondary metabolites, recombinant, and biotransformation products. The biochemical changes that occur throughout the food fermentation process lead to the modification of the substrate (starch or sugar) and production of other compounds (such as acids and alcohols) [5]. Fermentation improves the texture, appearance, color, flavor, shelf life, and also protein digestibility of pulses [5, 16]. It further decreases the presence of "antinutritional factors" including phytate, lectins, oligosaccharides, and protease inhibitors [5, 16, 23]. Especially in rural and traditional communities, spontaneous fermentation is mostly used for pulse processing. However, better and improved fermentation techniques in terms of specific strain development have been encouraged and introduced to improve product and nutritional quality, microbial safety, and product yield. In addition to fermentation, other processing operations could involve baking, cooking, and compositing, among others.

3.1. Microbiology and biochemistry of pulse fermentation

The microbiota of fermented pulse-based foods is largely dependent on temperature, pH, water activity, type of substrate, and salt levels. The three major types of microorganisms used during fermentation of pulses are bacteria of the genus *Bacillus*, lactic acid bacteria (LABs); some fungal species (**Table 3**); and possibly yeasts. In majority of these pulse-based fermented foods, the fermentation process is spontaneous (natural), and thus a mixture of microorganisms may act parallel or sequentially. This may thus cause changing and non-consistent products and possible production of pathogenic microorganisms and toxins [4, 5]. Nevertheless, LABs are dominant (**Table 3**), normally fastidious and grow willingly in most food substrates reducing the pH rapidly to a point where other competing organisms are no longer able to grow [24]. Several industrial fermentations have also applied LABs for the production of functional foods and the production of enzymes/metabolites. For ages, indigenous or traditional fermented foods have formed an essential part of the diet and can be prepared in the cottage industry using simple techniques and household equipment [25–27]. Fermented pulse-based foods are more abundant and available in developing nations, especially in India where it is passed on as trade secrets in the communities of certain families, a practice protected by custom [25]. The several available fermented pulse-based foods are summarized in **Table 3**.

Fermentation of pulses as with other food crops is associated with reduction of pH; changes in carbohydrates (starch, fibers, saccharides, sugars), proteins (amino acids), and lipids; "antinutritional" factors; and enzymatic degradation of different compounds [4, 5]. It also leads to the improvement of texture, taste, and aroma of the final product. As further described in the later section of this chapter, effect of fermentation on the composition of pulses varies; substantial evidence suggests improvement in nutritional and beneficial composition. Aside from various modifications, fermentation of pulses is also associated with the formation of compounds as a result of microbial actions on endogenous compounds. Such compounds include alcohols, ketones, organic acids, and aldehydes that further contribute to the distinct aroma associated with fermented pulse-based foods.

3.2. Evolvement and market for fermented pulse-based foods

As earlier presented in **Table 3**, fermented pulse-based food products are ubiquitous in developing nations, with some being used as snacks, meal, or spices. From traditional methods of fermentation and preparation of these foods (which are still largely common), there has been

Product	Produce	Country of origin	Microbial group responsible for fermentation	Mode of consumption/form	References
Amriti	Black gram	India	NR	Snack	[28]
Bedvin roti	Black gram, opium seeds, or walnut	India	NR	Breakfast or snack food	[29]
Bhallae	Black gram	India	*Bacillus subtilis, Candida curvata, C. famata, C. membranifaciens, C. variovaarai, Cryptococcus humicola, Debaryomyces hansenii, Enterococcus faecalis, Geotrichum candidum, Hansenula anomala, H. polymorpha, Kluyveromyces marxianus, Lactobacillus fermentum, Leuconostoc mesenteroides, Pediococcus membranaefaciens, Rhizopus marina, Saccharomyces cerevisiae, Trichosporon beigelii, T. pullulans, Wingea robertsii*	Side dish	[30]
Condiment	Pigeon pea	Nigeria	NR	Condiment	[31]
Dalbari (Urad dalbari)	Lentil	India	NR	Snack	[32]
Dawadawa	Local pulses	West and Central Africa	*B. licheniformis, B. subtilis*	Condiment, meat substitute	[16]
Dhokla	Bengal gram	India	*B. cereus, Ent. faecalis, Leuc. mesenteroides, L. fermenti, Tor. candida, Tor. pullulans*	Snack	[33–35]
Dosa	Black gram	India	*Bacillus* sp., *L. fermentum, Leuc. mesenteroides, Streptococcus faecalis,* yeast	Breakfast or snack food	[16]
Idli	Black gram	India, Sri Lanka	*L. delbrueckii, L. fermentum, Lactococcus lactis, Leuc. mesenteroides, Strep. lactic, Ped. cerevisiae,* yeast	Breakfast food	[16]
Khaman	Bengal gram dhal or Chickpeas	India	*Bacillus* sp., *L. fermentum, Leuc. mesenteroides, Lact. lactis, Ped. acidilactici*	Snack	[33, 36, 37]
Maseura	Black gram	Nepal, India	*B. laterosporus, B. mycoides, B. pumilus, B. subtilis, C. castellii, Ent. durans, Ped. acidilactici, Ped. pentosaceus, L. fermentum, L. salivarius, S. cerevisiae, Pichia burtonii*	Dry, ball like, brittle, condiment	[38]
Mashbari	Black gram, spices	India	*Bacillus* sp. A_{94}, *Lactobacillus sp., S. cerevisiae*	Staple food	[39]

Product	Produce	Country of origin	Microbial group responsible for fermentation	Mode of consumption/form	References
Masyaura	Black gram or green gram	Nepal, India	*Aspergillus niger, C. versatilis, Cladosporium* sp., *Lactobacillus* sp., *Ped. acidilactici, Ped. pentosaceus, S. cerevisiae, Penicillium* sp.	Side dish	[40, 41]
Papad	Bengal gram, black gram, lentil, red or green gram	India	*C. krusei, S. cerevisiae.*	Condiment or savory food	[27, 37]
Probiotic food	Mung bean	China	*L. plantarum* B1-6	Beverage	[6]
Sepubari	Black gram, dangal, spices	India	*Bacillus* sp. A_{31}., *Lactobacillus* sp., *S. cerevisiae*	Special dish in marriage feast	[39]
Teliye mah	Black gram	India	NR	Semi solid	[29]
Tempeh	Chickpeas, local pulses	Indonesia, New Guinea, Surinam	*Asp. oryzae, Rhiz. oligosporus*	Breakfast food or snack	[16]
Tempe Benguk	Velvet bean seeds	Indonesia	*Rhiz. arrhizus, Rhiz. oligosporus*	Alkaline, solid, fried cake/ breakfast food	[42]
Tempe Kecipir	Winged bean seed	Indonesia	*Rhiz. achlamydosporus, Rhiz. arrhizus, Rhiz. oligosporus, Rhiz. oryzae*	Alkaline, solid, fried cake/ breakfast food	[43]
Tempe Koro Pedang	Jack bean seed	Indonesia	*Rhiz. achlamydosporus, Rhiz. arrhizus, Rhiz. oryzae*	Alkaline, solid, fried cake/ breakfast food	[43]
Vadai	Black gram	India	*Leuconostoc* sp., *Pediococcus* sp., *Streptococcus* sp.	Paste, side dish	[34]
Wadi	Black gram and oil	India	*L. fermentum, L. mesenteroides*	Spicy condiment or an adjunct for cooking vegetables or rice	[27, 44]
Wari	Bengal gram or Black gram	India, Pakistan	*B. subtilis, Candida.* sp., *Cryptococcus humicolus, Debaryomyces* sp., *Ent. faecalis, G. candidum, H. anomala, Kl. marxianus, L. bulgaricus, S. cerevisiae, Strep. thermophiles, Trich beigelii, Win. robetsii*	Snack, fried balls, brittle, side dish	[16, 45, 46]

NR, not reported.

Table 3. Pulse-based fermented foods in developing countries.

some improvement toward the commercialization of few of these fermented pulse-based food products. Through industrialization and the advent of new technologies, significant developed and commercially available fermented pulse-based foods are *tempeh*, which has evolved to being available as salads and burgers; *dawadawa* (dried and ground form); and *dhokla* flour. Challenges, however, hampering the development and subsequent commercialization have been affordability of starter cultures and inadequate access to appropriate technology. The use of starter cultures in fermentation processes would largely assist in standardizing the fermentation process to ensure consistency, hygiene, and improved sensory quality. The challenge of accessing commercially available starter cultures for use in traditional, rural, and urban homes and small-scale industry is quite significant in developing nations. Related to this is also limited access to necessary technology, equipment, and expertise for production of fermented pulse-based foods, which is needed for development and provision of shelf-stable products.

Considering the ever-growing increasing market for functional foods in the world, with an increase of 25% from 2013, the global functional food market is expected to reach US$54 billion in 2017 [47]. This demand is expected to be largely driven by the need for products with substantiated health benefits, which can address chronic diseases including obesity, diabetes, cardiovascular diseases, and cancer. With such increase in demand coupled with the advent of new and novel processing technologies and the wealth of ongoing research, there is huge potential for the development of new functional products from fermented pulses which could be subsequently commercialized. Although few of these are already available in the market, there is still need for concerted efforts to scale up their production and make them more readily available.

4. Major functional components in fermented pulse-based foods and effects of fermentation on them

As indicated in the earlier sections of this chapter, aside basic nutrition, fermented pulses are sources of important functional components that have been proven critical for human health. These benefits can be attributed to various bioactive and health-promoting components embedded in them [20, 48]. It should also be noted that while fermentation has been used for ages to transform and modify pulses to products with improved benefits, studies have only recently sought a better understanding of the modification and its effects during pulse processing. As one would envisage, fermentation can have an effect on the bioactive components present and subsequent health-promoting benefits derived from fermented pulse-based foods. Examples of such major bioactive components and documented changes are subsequently discussed in the proceeding sections of this chapter.

4.1. Phenolic compounds

Over the years, there has been an increasing interest and desire in phenolic compounds due to their beneficial activity in relation to health. According to Dueñas et al. [49], pulses are excellent sources of phenolic compounds, which are largely accumulated in their hulls. The most essential phase of phenolic metabolism is the accumulation of phenols in plant tissues, as this is responsible for biological activity [50]. Several factors affect the concentration of phenols in pulses, including the degree of maturity at time of harvest, climatic and edaphic conditions, processing (e.g., fermentation), and storage conditions [50, 51].

Phenolic compounds consist of the –OH bonded directly to an aromatic hydrocarbon group and the major ones in pulses include flavonoids, tannins, saponins, and phenolic acids [16, 20]. These compounds impact pigmentation, flavor and taste in foods, and antioxidant activities and interact with proteins as a result of their radical-scavenging capacity [52]. Studies have shown that antioxidants contained in fermented pulses may mitigate the prevalence of some forms of cancer [53–58]. Ademiluyi et al. [57] reported the hypoglycemic and antiacetylcholinesterase activities of fermented bambara in rats and attributed this to the presence of phenolic compounds and other phytochemicals. Phenols in pulses and their fermented products have also been reported to exhibit strong antimutagenic, anti-inflammatory, and anticarcinogenic properties and have the capacity to modulate some important cellular enzyme functions [55, 59, 60]. Reduced levels of oxidative damage to lymphocytic DNA have also been linked to consumption of fermented pulse-based foods rich in antioxidants [50]. Phenolic compounds in fermented pulses have been documented to exhibit antioxidant properties. As reported by Moktan et al. [61], *idli* and *dhokla* exhibited metal chelating, lipid peroxidation, and high free radical-scavenging activities. Likewise, common bean and *tempeh* products exhibited radical-scavenging and antioxidant activities. In an in vivo study using hypercholesterolemic mice, the antioxidants in fermented mung bean were found to reduce the level of serum lipid and liver enzyme profiles [62]. Epidemiological studies have repeatedly shown a positive indication regarding the increased consumption of polyphenolic-rich diets and associated reduction of chronic human diseases [63, 64]. Clinical studies on pulses have also attested that phenolic compounds confer some health benefits in humans, including the reduction of cardiovascular diseases, weight management, cancer prevention, and diabetes control [65–68].

Available literature on the fermentation of pulses has documented both an increase and decrease in the phenolic compounds. An increase in hydroxybenzoic acid and (+)-catechin content was reported in spontaneously fermented lentils [69], while similar increase in free soluble phenols observed during the fermentation of some underutilized pulses [55]. Conversely, a reduction of conjugated forms of ferulic acid, *p*-cumaric, hydroxycinnamic derivatives, and bound phenols was observed during the fermentation of pulses [49, 55]. Surprisingly, same authors reported the synthesis of tyrosol and an increase in free quercetin due to the hydrolysis of quercetin glucosides [49]. Nevertheless, such documented changes have been attributed to the action of glycosidases and esterases from LABs releasing free aglycones, phenolic acids, hydroxyl-cinnamic acids, and less esterified proanthocyanidins and the transformation of bound to free phenolics during fermentation [49, 55, 70–73].

4.2. Protease inhibitors, lectins, and phytates

Proteases, lectins, and phytates are group of compounds normally regarded to as minor components of pulses. As documented by Vasconcelos and Oliveria [74] and Boye et al. [75], they were regarded as antinutrients in the past, because they negatively affect nutrient digestibility and alter glucose transportation. Referring to these minor components as "antinutrient" could however be a misnomer, considering their involvement in health-promoting processes [7, 16].

Protease inhibitors found in pulses act on either or both of the serine proteases chymotrypsin and trypsin and are important from the nutritional point of view [7, 76]. They are found in relatively high quantities in pulses compared to other plant foods and can be broadly classified as either Bowman-Birk or Kunitz type, based on their molecular masses and cystine contents [75]. Inhibitors of the Kunitz type have two disulfide bridges with a molecular mass of approximately 20 kDa and act specifically against trypsin, while the Bowman-Birk type contains seven disulfide bridges, with a molecular mass between 8 and 10 kDa, and inhibits chymotrypsin and trypsin simultaneously at independent binding sites [7, 76]. Although protease inhibitors can block chymotrypsin and trypsin activities, thus reducing protein digestibility, the Boman-Brik family of protease inhibitors has been reported to show anti-inflammatory and anticarcinogenic effects in human colon cancer cells [77–81].

Pulses are the main sources of lectins in everyday human diet, although fermentation is reported to reduce the lectin content of pulses [7, 82]. Lectins are glycoproteins, which have the ability to agglutinate red blood cell in vitro and are thus referred to as phytohaemagglutinins [83]. Like other presumed pulse antinutrients, lectins are now being considered as important in immunological and cell biology, with potentials for clinical applications [7]. They can inhibit tumor growth and exert antimicrobial, immunomodulatory, and HIV-1 reverse transcriptase inhibitory activities [84]. In other studies, lectins are being adopted for the discovery of cancer markers that are proteinaceous in nature via a natural glycoprotein microarray approach [7].

Phytic acid also known as inositol polyphosphate, inositol hexakisphosphate (IP6), or phytate (when in salt form) is found within the hulls of pulses. It is known to be the main storage form of phosphorus in plants [85]. Phytate and some of its secondary products are regarded as antinutrients because of their active role in chelating important minerals such as magnesium, calcium, zinc, and iron, thus contributing to mineral deficiencies [85, 86]. However, the health benefits of phytate have been "rediscovered," thus propelling a gradual change and perspective in its classification as an antinutrient. For example, phytic acid could play a role in regulating DNA repair via nonhomologous end joining [87] and other cellular functions such as nuclear messenger RNA export [7]. In vitro and in vivo studies of fermented pulses have also shown that inositol hexaphosphate (InsP6, phytic acid) exhibits potent anticancer properties (both therapeutic and preventive), tumor abrogation, host defense mechanism, and reduction of cell proliferation [88]. Phytic acids also diminish the bioavailability of toxic heavy metals and demonstrate antioxidant activity [89]. Phytic acid is used as a food additive (preservative) E391 [90], though its exact intracellular physiological roles are still unclear [91].

Fermentation was reported to have reduced trypsin inhibitor activity of mucuna and faba bean [23, 92] and was hypothesized as a consequence of bacterial proteases during the fermentation [92]. In studies conducted by Akpapunam and Achinewhu [93] and Khattab and Arntfield [94], fermentation was observed to reduce the phytic acid and trypsin inhibitor activity in fermented pulses. Such reduction of phytate and phytic acid has been ascribed to the endogenous phytase seeds and that of other microorganisms, which causes hydrolysis of the phytic acid into orthophosphate and inositol and microbial degradation of the phytates [93, 95, 96]. Likewise, a reduction in the lectin content of lentils fermented for 72 h was reported by Cuadrado et al. [82]. This was ascribed by the authors to proteolytic degradation of lectin protein and changes in lectin-protein structure [82].

4.3. Fiber and saccharides

A thorough review of dietary fiber in pulses has been presented in the literature [97] and, accordingly, identified as good sources of both soluble and insoluble dietary fibers. When unprocessed, pulses could contain approximately 15–32% total dietary fiber, of which about one-third to three-quarters is made up of insoluble fiber, while the rest is soluble fiber [66]. Soluble fibers found in pulses comprise of oligosaccharides such as pectin, stachyose, verbascose, and raffinose, whereas, the insoluble ones include lignin, hemicellulose, and cellulose [16, 98, 99]. Major health benefits linked to dietary fiber include laxation and reduced risk of being overweight, cardiovascular diseases, and diabetes [100]. Particularly in fermented pulses, the fiber contents can lower the risk of many diseases such as diabetes, coronary heart diseases, obesity, and some forms of cancer [101]. Fibers (in particular insoluble fibers) provide physicochemical functionality to foods such as fecal bulking via its ability to hold and bind liquids such as water and fat. While soluble fiber ferments in the stomach, thus enhancing colon health via lowered pH, production of short-chain fatty acids (SCFAs), and potential microbiota changes in the colon [66, 99]. Soluble fiber has also been linked with reduction in cholesterol levels, total and low-density lipoprotein, and insulin resistance [102].

Essentially, pulse starches contain higher amylose content with high capacity for retrogradation, thus reducing starch digestion rate [103]. Slowly digestible starches and resistant starches from pulses have been linked to management of diabetes and promotion of satiation [103, 104]. Fermented pulse-based foods such as *tempeh* and *idli* are products that have been recognized as good sources of resistant starches, making them suitable for dietary strategies to manage blood glucose levels [16, 105–107]. Oligosaccharides in pulses and its fermented substrates may also be considered as prebiotics, which could be beneficial to human health [66, 108]. Pulses with their abundance of non-starch polysaccharides, oligosaccharides, and resistant starch are low glycemic index (GI) foods with GI values within 28–52 [109–113]. According to Yeap et al. [114], fermented mung bean products have been recommended for the management of diabetes due to their low GI and have assisted in reducing the prevalence of diabetes in Asia. The cardioprotective effect conferred by fermented pulse-based foods could be due to the synergistic action of the pulse oligosaccharides, resistant starch, protein, minerals, vitamins, and phytochemicals [80, 115, 116]. All these beneficial properties of dietary fiber and saccharides have led to increased interests in its use in food formulations in the food industry [99].

Studies in literature have largely suggested that fermentation increases the digestibility of fiber, starches, and saccharides [117–119]. Reduction or total elimination of raffinose oligosaccharides, verbascose, and stachyose during lactic acid and fungal fermentation of pulses has been reported in in vitro and in vivo studies [16, 120]. Yeast fermentation of peas and kidney beans, however, resulted in increase of oligosaccharides [16]. Adewunmi and Odunfa [121] investigated the effect of fermentation on the oligosaccharide content of two common *Vigna unguiculata* beans (*drum* and *oloyin*) in West Africa and observed that the stachyose content of drum bean slurry decreased by over 50% when fermented for 72 h using *Ped. acidilactici, Lactobacillus plantarum*, and *L. fermentum*. Likewise, a decrease of about 67% of stachyose content of *oloyin* was observed when fermented under similar conditions. However, the sucrose content of both beans was observed to increase significantly for all tested organisms under the same fermentation conditions [121]. They attributed these observations to the α-galactosidase enzyme producing ability of the studied organisms which breaks down the α-1,6-glycosidic bonds. In an earlier study by Odunfa [122], a similar observation was made when stachyose content of locust beans fermented for 24 h decreased. The decrease was attributed to the hydrolyzation of the oligosaccharides to simple reducing sugars by α- and β-galactosidase [122]. In a similar study, Tewari and Muller [45] reported a reduction from 4.4 to 0.6% of total raffinose and stachyose concentration after fermentation of black beans and soybean with *L. bulgaricus* and *Streptococcus thermophiles* [45]. Both increase and decrease in the fiber composition of fermented pulses have been reported in the literature. A decrease in soluble and neutral dietary fiber, cellulose, and hemicellulose in some fermented pulses was reported by Veena et al. [105] and Granito and Alvarez [107], while an increase in total dietary fiber and lignin has equally been reported by Veena et al. [105], Granito and Alvarez [107], and Vidal-Valverde [117].

4.4. Proteins and peptides

Pulses constitute an excellent source of dietary protein, which is accumulated during the growth phase of the plant; hence, pulse seeds that are mature are usually high in protein content and other nutritional components [123]. On dry weight basis, lentil, chickpea, and dry pea contain approximately 28.6, 22, and 23.3% protein, respectively, which may vary slightly depending on growing conditions, maturity, and variety [123, 124]. A greater part of pulse proteins is in the form of storage proteins which fall which are categorized into glutelins, albumins, and globulins depending on their solubility properties. Glutelins are soluble in dilute acid and base and account for between 10 and 20% pulse proteins, albumins (water soluble) also account for 10–20% protein in pulses, and globulins which are soluble in salt water constitute up to 70% of the total proteins found in pulses [123–125].

Peptides on the other had are protein molecules that are smaller than 10 kDa and may occur naturally or are derivatives of cryptic sequences of inherent natural proteins [126, 127]. Essentially, they mainly are derived via hydrolysis by microbial, digestive, and plant proteolytic enzymes [128]. Hydrolysis of pulse proteins occurs during fermentation, which alters protein functionality through the modification of physical size as well as its surface chemical properties [129]. Bioactive peptides formed during this process can show multifunctional characteristics and confer positive effects on human health through various influences on the gastrointestinal, cardiovascular, nervous, and immunological [130]. Peptides and hydrolysates from mug bean, pea, and chickpea have been investigated for various therapeutic activities such as antioxidant capacity, copper-chelating activity, and enhancement of mineral absorption/bioavailability, antiproliferative and antimicrobial properties, and angiotensin-converting enzyme (ACE) activity [131].

Protein and their adhering/conjugated peptides are significant minor components in fermented pulse-based foods. The hydrolysis of these compounds during fermentation can affect protein functionality through a modification of the protein chemical properties and physical size, increase in the number of ionisable amino and carboxylic groups leading to increased protein solubility, water holding capacity, and the formation smaller peptide fragments [66, 129, 132, 133].

In a study conducted by Xiao et al. [133] on solid-state fermentation of chickpea flour with *Cordyceps militaris*, the authors observed increased amounts of true protein, crude protein, and essential amino acids, and further analysis showed that proteins contained in fermented chickpeas were predominantly composed of lower molecular mass than that of the unfermented chickpeas. Results from the same study revealed that protein digestibility, water absorption index, fat absorption capacity, and emulsification capacity were also enhanced by fermentation. Lee et al. [134] observed the formation of bioactive peptides as a result of proteolysis during fermentation. Likewise, the production of angiotensin I-converting enzyme (ACE) was reported during fermentation of mung bean [6]. Jung et al. [135] availed that enhanced emulsification capacity observed in fermented pulses is due to the yield of low-molecular-mass peptides which have the ability to easily migrate to the water-oil interface, hence resulting in a more stable emulsion. These changes in functionalities, however, depend on the degree of hydrolysis and on the nature of the proteins [130].

Other important nutritive and nonnutritive bioactive components of fermented pulses include phytosterols, vitamins, minerals, squalene, saponins, defensins, phytoestrogens, and fatty acids. Detailed description of these other components can be found in documented studies in the literature [7, 16, 136–138]. Nonetheless, other substantiated health benefits of fermented pulses include anticancer activities, reduction of aging and stress, probiotic effects, reduces the risk of chronic diseases, and the general improvement of human well-being [16, 20, 65, 116, 139–142].

5. Development of novel functional foods from fermented pulses

As indicated early on in this chapter, fermentation of pulses to obtain different products generates vital molecules including bioactive peptides, phytochemicals, fibers, saccharides, and other compounds with substantiated health benefits. This thus opens doors for the development of novel foods from these food crops. Although conventional functional fermented foods are saturated with products from cereals and dairy, nondairy foods are gradually gaining global prominence. Coupled with the strict religious/dietary requirements of certain populations in the developing nations and the continued demand and drive for consumption of vegetable proteins, fermented pulse-based foods offer an excellent substitute in this regard. In addition, they should also be explored as technological ingredients for the development of new and novel, healthy foods. As earlier indicated, few of these fermented pulse-based foods commercially exist, but there is still a huge potential and opportunity for the development of novel functional fermented pulse-based food products with improved functionality. With the advent of different innovative technologies, fermented pulse-based foods have enormous prospects and potential for the delivery of functional foods to the populace and intending consumers. With the provision of such, it is envisaged that consumers may be willing to pay for such products with improved functionalities and quality. While these fermented pulse-based functional foods offer considerable market potential, studies and detailed in vivo experiments must be properly done prior to commercialization of such novel products.

6. Conclusion and future prospects

Owing to their relative availability, pulses are recognized as significant sources of food. Nevertheless, they are regarded as "food for the poor" in most developing nations. Fermentation as a food processing technique can improve the quality and other health-promoting benefits of pulses. As evident from earlier studies reviewed herein, consumption of such fermented pulse-based foods would thus be beneficial and largely contribute to nutrition and food security. Although these fermented pulse-based foods are readily available, the daily per capita consumption in traditional settings has been declining in recent years, and this is ironically associated with an increase of chronic diseases plaguing both developing and developed countries. While some of the inherent bioactive compounds in fermented pulse-based foods could possibly inhibit nutrient availability, fermentation can effectively reduce their ability to do this, thus ensuring that the bioactive compounds present confer some functional activities.

There is an existing potential market for functional foods, but the availability of shelf-stable products can hinder their prospects. As such, mechanisms to ensure access to technology and expertise among local and small-scale food processors should be enhanced. Although cost might hinder the provision of commercially available starter cultures, delivery of such starter cultures for improved and effective fermentation could be achieved using dried forms of previous fermented products (with viable fermenting organisms), for subsequent use. Most importantly increasing awareness of pulses and subsequent fermented products from such crops as sources of functional and health-promoting foods would be the role of government, nongovernmental organizations, and other relevant stakeholders within the health and other related sectors. This will to a large extent ensure that developing nations achieve the much-needed and envisaged food and nutrition security.

Acknowledgements

This work was supported via the Global Excellence and Stature (GES) Fellowship of the University of Johannesburg (UJ) granted to the main author (Adebo, O.A). This work was also partly supported by the National Research Foundation (NRF) Center of Excellence (CoE) in Food Security cohosted by the University of Pretoria (UP) and the University of Western Cape (UWC), South Africa.

Author details

Oluwafemi Ayodeji Adebo[1*], Patrick Berka Njobeh[1], Janet Adeyinka Adebiyi[1], Sefater Gbashi[1], Judith Zanele Phoku[2] and Eugenie Kayitesi[1*]

*Address all correspondence to: oluwafemiadebo@gmail.com and eugeniek@uj.ac.za

1 Department of Biotechnology and Food Technology, University of Johannesburg, Johannesburg, Gauteng, South Africa

2 Water and Health Research Unit, University of Johannesburg, Johannesburg, Gauteng, South Africa

References

[1] FAO International Year of Pulses 2016 [Internet]. 2016. Available from: www.fao.org/pulses-2016/en/ [Accessed: December 20, 2016]

[2] Hu FB. Plant-based foods and prevention of cardiovascular disease: An overview. American Journal of Clinical Nutrition. 2003;**78**:544-551

[3] Jay JM, Loessner MJ, Golden DA. Modern Food Microbiology. 7th ed. India: Springer; 2005. DOI: 10.1007/b100840

[4] Galati A, Oguntotinbo FA, Moschetti G, Crescimanno M, Settanni L. The cereal market and the role of fermentation in cereal-based food production in Africa. Food Reviews International. 2014;**30**:317-337. DOI: 10.1080/87559129.2014.929143

[5] Adebiyi JA, Obadina AO, Adebo OA, Kayitesi E. Fermented and malted millet products in Africa: Expedition from traditional/ethnic foods to industrial value added products. Critical Reviews in Food Science and Nutrition. 2016. DOI: 10.1080/10408398.2016.1188056

[6] Wu H, Rui X, Li W, Chen X, Jiang M, Dong M. Mung bean (*Vigna radiata*) as probiotic food through fermentation with *Lactobacillus plantarum* B1-6. LWT-Food Science and Technology. 2015;**63**:445-451. DOI: 10.1016/j.lwt.2015.03.011

[7] Campos-Vega R, Loarca-Pina G, Oomah BD. Minor components of pulses and their potential impact on human health. Food Research International. 2010;**43**:461-482. DOI: 10.1016/j.foodres.2009.09.004

[8] Tsao R, Yu L, Shahidi F. Cereals and pulses—An overview. In: Yu L, Tsao R, Shadidi F. editors. Cereals and Pulses: Nutraceutical Properties and Health Benefits. London, UK: Wiley; 2012. pp. 1-6. DOI: 10.1002/9781118229415.ch13

[9] Tiwari BK, Gowen A, McKenna B. Pulses: What are they? In: Tiwari BK, Gowen A, McKenna B, editors. Pulse Foods: Processing, Quality and Nutraceutical Applications. London, UK: Academic Press; 2011. pp. 1-8.

[10] FAOSTAT Food and Agriculture Organization Statistics [Internet]. 2016. Available from: www.fao.org/faostat/en/#home [Accessed: December 20, 2016]

[11] Perez-Hidalgo MA, Guerra-Hernandez E, Garcia-Villanova B. Dietary fiber in three raw legumes and processing effect on chick peas by an enzymatic gravimetric method. Journal of Food Composition and Analysis. 1997;**10**:66-72. DOI: 10.1006/jfca.1997.0522

[12] Sathe SK. Dry bean protein functionality. Critical Reviews in Biotechnology. 2002;**22**:175-223. DOI: 10.1080/07388550290789487

[13] Kutos T, Golob T, Kac M, Plestenjak A. Dietary fiber of dry processed beans. Food Chemistry. 2002;**80**:231-235. DOI: 10.1016/S0308-8146(02)00258-3

[14] Osorio-Diaz P, Bello-Perez LA, Sayago-Ayerdi SG, del Pilar Benitez-Reyes M, Tovar J, Paredes-Lopez O. Effect of processing and storage time on in vitro digestibility and resistant starch content of two bean (*Phaseolus vulgaris*) varieties. Journal of the Science of Food and Agriculture. 2003;**83**:1283-1288. DOI: 10.1002/jsfa.1413

[15] Wang N, Toews N. Certain physicochemical and functional properties of fibre fractions from pulses. Food Research International. 2011;**44**:2515-2523. DOI: 10.1016/j.foodres.2011.03.012

[16] Frais J, Penas E, Martinez-Villaluenga C. Fermented pulses in nutrition and health promotion. In: Frias J, Martinez-Villaluenga C, Penas E, editors. Fermented Foods in Health and Disease Prevention. Academic Press; Rome, Italy: Food and Agriculture Organization of the United Nations. 2016. pp. 385-416. DOI: 10.1016/B978-0-12-802309-9.00016-9

[17] FAO Cereals, pulses, legumes and vegetable proteins. [Internet]. 2007. Available from: www. www.fao.org/3/a-a1392e.pdf [Accessed: February 1, 2017]

[18] Rehman ZU, Salariya AM, Zafar SI. Effect of processing on available carbohydrate content and starch digestibility of kidney beans (*Phaseolus vulgaris* L.). Food Chemistry. 2001;**73**:351-355. DOI: 10.1016/S0308-8146(00)00311-3

[19] Benouis K. Phytochemicals and bioactive compounds of pulses and their impact on health. Chemistry International. 2017;**3**:224-229.

[20] Singh J, Basu PS. Non-nutritive bioactive compounds in pulses and their impact on human health: An overview. Food and Nutrition Sciences. 2012;**3**:1664-1672. DOI: 10.4236/fns.2012.312218

[21] United States Department of Agriculture Food Composition Database. [Internet]. Available from www.nal.usda.gov/fnic/foodcomp/search/ [Accessed: January 19, 2017]

[22] Roy A, Moktan B, Sarkar PK. Traditional technology in preparing legume-based fermented foods of Orissa. Indian Journal of Traditional Knowledge. 2006;**6**:12-16.

[23] Mugendi JB, Njagi ENM, Kuria EN, Mwasaru MA, Mureithi JG, Apostolides Z. Effects of processing technique on the nutritional composition and anti-nutrient content of mucuna bean (*Mucuna pruiens* L.). African Journal of Food Science. 2010;**4**:156-166

[24] Satish KR, Kanmani P, Yuvaraj N, Paari KA, Pattukumar V, Arul V. Traditional Indian fermented foods: A rich source of lactic acid bacteria. International Journal of Food Science and Nutrition. 2013;**64**:415-428. DOI: 10.3109/09637486.2012.746288.

[25] Tamang JP, Sarkar PK, Hesseltine CW. Traditional fermented foods and beverages of Darjeeling and Sikkim—A review. Journal of the Science of Food and Agriculture. 1988;**44**:375-385. DOI: 10.1002/jsfa.2740440410

[26] Ross RP, Morgan S, Hill C. Preservation and fermentation: past, present and future. International Journal of Food Microbiology. 2002;**79**:3-16. DOI: 10.1016/S0168-1605(02)00174-5

[27] Aidoo KE, Nout NJR, Sarkar PK. Occurrence and function of yeasts in Asian indigenous fermented foods. FEMS Yeast Research. 2006;**1**:30-39. DOI: 10.1111/j.1567-1364.2005.00015.x

[28] Steinkraus KH. Handbook of Indigenous Fermented Foods. 2nd ed. New York: Marcel Dekker; 1996

[29] Thakur N, Savitri, Bhalla TC. Characterization of some traditional fermented foods and beverages of Himachal Pradesh. Indian Journal of Traditional Knowledge. 2004;**3**:325-335

[30] Rani DK, Soni SK. Applications and commercial uses of microorganisms. In: Soni SK, editor. Microbes: A Source of Energy for 21st Century. Tohtas Nagar, Shahdara, Delhi: Jai Bharat Printing Press; 2011:71-126

[31] Oboh G. Nutrient and antinutrient composition of condiments produced from some fermented underutilized legumes. Journal of Food Biochemistry. 2006;**30**:576-588. DOI: 10.1111/j.1745-4514.2006.00083.x

[32] Sha SP, Ghatani K, Tamang JP. *Dalbari*, a traditional pulses based fermented food of West Bangal. International Journal of Agriculture and Food Science and Technology. 2013;**4**:6-10

[33] Joshi N, Godbole SH, Kanekar P. Microbial and biochemical changes during *dhokla* fermentation with special reference to flavour compounds. Journal of Food Science and Technology. 1989;**26**:113-115

[34] Blandino A, Al-Aseeri ME, Pandiella SS, Contero D, Webb C. Cereal based fermented foods and beverages. Food Research International. 2003;**36**:527-543. DOI:10.1016/S0963-9969(03)00009-7

[35] Roy A, Moktan B, Sarkar PK. Survival and growth of foodborne bacterial pathogens in fermenting batter of *dhokla*. Journal of Food Science and Technology. 2009;**46**:132-135.

[36] Ramakrishnan CV. Studies on Indian fermented foods. Baroda Journal of Nutrition. 1979;**6**:1-54

[37] Roy A, Moktan B, Sarkar PK. Traditional technology in preparing legume based fermented foods of Orissa. Indian Journal of Traditional Knowledge. 2007;**1**:12-16

[38] Chettri R, Tamang JP. Microbiological evaluation of *maseura*, an ethnic fermented legume-based condiment of Sikkim. Journal of Hill Research. 2008;**21**:1-7

[39] Sharma N, Handa S, Gupta A. A comprehensive study of different traditional fermented foods/beverages of Himachal Pradesh to evaluate their nutrition impact on health and rich biodiversity of fermenting microorganisms. International Journal of Research in Applied, Natural and Social Sciences. 2013;**3**:19-28

[40] Dahal NR, Karki TB, Swamylingappa B, Li Q, Gu G. Traditional foods and beverages of Nepal—A review. Food Review international. 2005;**21**:1-25. DOI: 10.1081/FRI-200040579

[41] Dahal NR, Rao ER, Swamylingappa B. Biochemical and nutritional evaluation of *Masyaura*-a legume based traditional savory of Nepal. Journal of Food Science and Technology. 2003;**1**:17-22

[42] Nagai T, Tamang JP. Fermented soybeans and non-soybeans legume foods. In: Tamang JP, Kailasapathy K. editors. Fermented Foods and Beverages of the World. New York: CRC Press, Taylor and Francis Group; 2010. pp. 191-224

[43] Alexandraki V, Tsakalidou E, Papadimitriou K, Holzapfel WH. Status and Trends of the Conservation and Sustainable Use of Microorganisms in Food Processes. Commission on Genetic Resources for Food and Agriculture. FAO Background Study. London, UK 2013;65.

[44] Sandhu DK, Soni SK. Microflora associated with Indian Punjabi *warri* fermentation. Journal of Food Science and Technology. 1989;**26**:21-25

[45] Tewary HK, Muller HG. The fate of some oligosaccharides during the preparation of *wari*, an Indian fermented food. Food Chemistry. 1992;**2**:107-111. DOI: 10.1016/0308-8146(92)90222-N

[46] Kulkarni SG, Manan JK, Agarwal MD, Shukla IC. Studies on physico-chemical composition, packaging and storage of black gram and green gram Wari prepared in Uttar Pradesh. Journal of Food Science and Technology. 1997;**2**:119-122

[47] Bonar A. Functional Foods Market Is Expected to Grow 25% by 2017: Leatherhead [Internet]. 2014. Available from http://www.nutraingredients.com/Markets-and-Trends/Functional-foods-market-is-expected-to-grow-25-by-2017-Leatherhead [Accessed: January 31, 2017]

[48] Granito M, Frias J, Doblado R, Guerra M, Champ M, Vidal-Valverde C. Nutritional improvement of beans (*Phaseolus vulgaris*) by natural fermentation. European Food Research and Technology. 2002;**214**:226-231. DOI: 10.1007/s00217-001-0450-5

[49] Dueñas M, Hernández T, Estrella I. Occurrence of phenolic compounds in the seed coat and the cotyledon of peas (*Pisum sativum* L.). European Food Research and Technology. 2004;**219**:116-123. DOI: 10.1007/s00217-004-0938-x

[50] Pandey KB, Rizvi SI. Plant polyphenols as dietary antioxidants in human health and disease. Oxidative Medicine and Cellular Longevity. 2009;**5**:270-278. DOI: 10.4161/oxim.2.5.9498

[51] Manach C, Scalbert A, Morand C, Rémésy C, Jiménez L. Polyphenols: Food sources and bioavailability. The American Journal of Clinical Nutrition. 2004;**79**:727-747

[52] Ozcan T, Akpinar-Bayizit A, Yilmaz-Ersan L, Delikanli B. Phenolics in human health. International Journal of Chemical Engineering and Applications. 2014;**5**:393-396. DOI: 10.7763/ijcea.2014.v5.416

[53] Tharanathan RN, Mahadevamma S. Grain legumes a boon to human nutrition. Trends in Food Science and Technology. 2003;**14**:507-518. DOI: 10.1016/j.tifs.2003.07.002

[54] Xu BJ, Chang SKC. A comparative study on phenolic profiles and antioxidant activities of legumes as affected by extraction solvents. Journal of Food Science. 2007;**72**:159-166. DOI: 10.1111/j.1750-3841.2006.00260.x

[55] Oboh G, Ademiluyi AO, Akindahunsi AA. Changes in polyphenols distribution and antioxidant activity during fermentation of some underutilized legumes. Food Science and Technology International. 2009;**15**:41-46. DOI: 10.1177/1082013208101022

[56] Ademiluyi AO, Oboh G. Soybean phenolic-rich extracts inhibit key-enzymes linked to type 2 diabetes (α-amylase and α-glucosidase) and hypertension (angiotensin I converting enzyme) *in vitro*. Experimental and Toxicologic Pathology. 2013;**65**:305-309. DOI: 10.1016/j.etp.2011.09.005

[57] Ademiluyi AO, Oboh G, Boligon AA, Althayde ML. Dietary supplementation with fermented legumes modulate hyperglycemia and acetylcholinesterase activities in Streptozotocin-induced diabetes. Pathophysiology. 2015;**22**:195-201. DOI: 10.1016/j.pathophys.2015.08.003

[58] Megías C, Cortés-Giraldo I, Alaiz M, Vioque J, Girón-Calle J. Isoflavones in chickpea (*Cicer arietinum*) protein concentrates. Journal of Functional Foods. 2016;**21**:186-192. DOI: 10.1016/j.jff.2015.12.012

[59] Ho, CT. Phenolic compounds in food. In Ho C, Lee CY, & Huang M, editors. Phenolic Compounds in Food and their Effects on Health I. Analysis, Occurrence and Chemistry. Washington, DC: American Chemical Society. 1992. pp. 1-7

[60] Frassinetti S, Gabriele M, Caltavuturo L, Longo V, Pucci L. Antimutagenic and antioxidant activity of a selected lectin-free common bean (*Phaseolus vulgaris* L.) in two cell-based models. Plant Foods for Human Nutrition. 2015;**70**:35-41. DOI: 10.1007/s11130-014-0453-6

[61] Moktan B, Roy A, Sarkar PK. Antioxidant activities of cereal-legume mixed batters as influenced by process parameters during preparation of *dhokla* and *idli*, traditional steamed pancakes. International Journal of Food Sciences and Nutrition. 2011;**62**:360-369. DOI: 10.3109/09637486.2010.532116

[62] Yeap SK, Beh BK, Ho YW, Yusof HM, Mohamad NE, Ali NM, Jaganath IB, Alitheen NB, Koh SP, Long K. *In vivo* antioxidant and hypolipidemic effects of fermented mung bean on hypercholesterolemic mice. Evidence-Based Complementary and Alternative Medicine. 2015:1-6. DOI: 10.1155/2015/508029

[63] Scalbert A, Manach C, Morand C, Remesy C. Dietary polyphenols and the prevention of diseases. Critical Reviews in Food Science and Nutrition. 2005;**45**:287-306. DOI: 10.1080/1040869059096

[64] Arts ICW, Hollman PCH. Polyphenols and disease risk in epidemiologic studies. American Journal of Clinical Nutrition. 2005;**81**:317-325

[65] Anderson JW, Major AW. Pulses and lipaemia, short- and long-term effect: Potential in the prevention of cardiovascular disease. British Journal of Nutrition. 2002;**88**;263-271. DOI: 10.1079/BJN2002716

[66] McCrory MA, Hamaker BR, Lovejoy JC, Eichelsdoerfer PE. Pulse consumption, satiety, and weight management. Advances in Nutrition.2010;**1**:17-30. DOI: 10.3945/an.110.1006

[67] Agurs-Collins T, Smoot D, Afful J, Makambi K, Adams-Campbell LL. Legume intake and reduced colorectal adenoma risk in African-Americans. Journal of National Black Nurses Association. 2007;**17**(2):6-12

[68] Villegas R, Gao Y-T, Yang G, Li H-L, Elsay TA, Zheng W, Shu XO. Legume and soy food intake and the incidence of type 2 diabetes in the Shanghai Women's Health Study. American Journal of Clinical Nutrition. 2008;**87**:162-167

[69] Bartolomé B, Estrella I, Hernandez T. Changes in phenolic compounds in lentils (*Lens culinaris*) during germination and fermentation. Zeitschrift fur Lebensmitteluntersuchung und-Forschung A. 1997;**205**:290-294. DOI: 10.1007/s002170050167

[70] Oboh G, Rocha JBT. Polyphenols in red pepper (*Capsicum annum* var aviculare Tepin) and their protective effects on some pro-oxidant induced lipid peroxidation in brain and liver. European Food Research and Technology. 2007;**225**:239-247. DOI: 10.1007/s00217-006-0410-1

[71] Jiménez N, Esteban-Torres M, Mancheno JM, De Las eivas B, Munoz R. Tannin degradation by a novel tannase enzyme present in some *Lactobacillus plantarum* strains. Applied and Environmental Microbiology. 2014;**80**:2991-2997. DOI: 10.1128/AEM.00324-14

[72] Limon RI, Penas E, Torin MI, Martinez-Villaluenga C, Frias J. Fermentation enhances the content of bioactive compounds in kidney bean extracts. Food Chemistry. 2015;**172**:343-352. DOI: 10.1016/j.foodchem.2014.09.084

[73] Esteban-Torres M, Landete JM, Reveron I, Santamaria L, de las Rivas B, Munoz R. A *Lactobacillus plantarum e*sterase active on a board range of phenolic esters. Applied and Environmental Microbiology. 2015;**81**:3235-3242. DOI: 10.1128/AEM.00323-15

[74] Vasconcelos IK, Oliveria JTA. Antinutritional proprieties of plant lectins. Toxicon. 2004;**44**:385-403. DOI: 10.1016/j.toxicon.2004.05.005

[75] Boye J, Zare F, Pletch A. Pulse proteins: Processing, characterization, functional properties and applications in food and feed. Food Research International. 2010;**43**:414-431. DOI: 10.1016/j.foodres.2009.09.003

[76] Belitz HD, Weder JKP. Protein inhibitors of hydrolases in plants foodstuffs. Food Reviews International. 1990;**6**:151-211. DOI: 10.1080/87559129009540866

[77] Chen YW, Huang SC, Shiau SYL, Lin JK. Bowman –Birk inhibitor abates proteosome function and suppresses the proliferation of MCF7 breast cancer cells through accumulation of MAP kinase phosphatase-1. Carcinogenesis. 2005;**26**:1296-1306. DOI: 10.1093/carcin/bgi062

[78] Clemente A, Gee JM, Johnson IT, Mackenzie DA, Domoney C. Pea (*Pisum sativum*) protease inhibitors from the Bowman-Birk class influence the growth of human colorectal adenocarcinoma HT29 cells *in vitro*. Journal of Agriculture Food Chemistry. 2005;**53**:8979-8986. DOI: 10.1021/jf051528w

[79] Clemente A, Domoney C. Biological significance of polymorphism in legume protease inhibitors from the Bowman—Birk family. Current Protein and Peptide Science. 2006;**7**:201-216. DOI: 10.2174/138920306777452349

[80] Duranti M. Grain legume proteins and nutraceutical properties. Fitoterapia. 2006;**77**:67-82. DOI: 10.1016/j.fitote.2005.11.008

[81] Clemente A, Arques MDC. Bowman-Brik inhibitors from legumes as colorectal chemoprotective agents. World Journal of Gastroenterology. 2014;**20**:10305-10315. DOI: 10.3748/wjg.v20.i30.10305

[82] Cuadrado C, Hajos G, Burbano C, Pedrosa MM, Ayet G, Muzquiz M, Pusztai A, Gelencser E. Effect of natural fermentation on the lectin of lentils measured by immunological methods. Food and Agricultural Immunology. 2002;**14**:41-49. DOI: 10.1080/09540100220137655

[83] Klupšaitė D, Juodeikienė G. Legume: Composition, protein extraction and functional properties. A review. Chemical Technology. 2015;**66**:5-12. DOI: 10.5755/j01.ct.66.1.12355

[84] Hamid R, Masood A, Wani IH, Rafiq S. Lectins: Proteins with diverse applications. Journal of Applied Pharmaceutical Science. 2013;**3**:93-103. DOI: 10.7324/JAPS.2013.34.S18

[85] Khan FA, Khan A, Khattak MUR, Ullah N, Zahoor M, Shah SMM. Evaluation of inorganic profile and anti-nutritional values of *Cocculus hirsutus*. African Journal of Pharmacy and Pharmacology. 2012;**6**:144-147. DOI: 10.5897/ajpp11.410

[86] Hurrell RF, Reddy M, Cook JD. Inhibition of non-haem iron absorption in man by polyphenolic-containing beverages. British Journal of Nutrition. 1999;**81**:289-295

[87] Hanakahi LA, Bartlet-Jones M, Chappell C, Pappin D, West SC. Binding of inositol phosphate to DNA-PK and stimulation of double-strand break repair. Cell. 2000;**102**:721-729. DOI: 10.1016/s0092-8674(00)00061-1

[88] Shamsuddin AM. Anti-cancer function of phytic acid. International Journal of Food Science and Technology. 2002;**37**:769-782. DOI: 10.1046/j.1365-2621.2002.00620.x

[89] Minihane AM., Rimbach G. Iron absorption and the iron binding and anti-oxidant properties of phytic acid. International Journal of Food Science and Technology. 2002;**37**:741-748. DOI: 10.1046/j.1365-2621.2002.00619.x

[90] Kingsley O, Marshall AA. Medicinal potential of *Acalypha wilkesiana* leaves. Advances in Research. 2014;**2**:655-665. DOI: 10.9734/air/2014/9452

[91] Shears SB. Assessing the omnipotence of inositol hexakisphosphate. Cell Signalling. 2001;**13**:151-158. DOI: 10.1016/s0898-6568(01)00129-2

[92] Coda R, Melama L, Rizzello CG, Curiel JA, Sibakov J, Holopainen U, Pulkkinen M, Sozer N. Effect of air classification and fermentation by *Lactobacillus plantarum* VTT E-133328 on faba bean (*Vicia faba* L.) flour nutritional properties. International Journal of Food Microbiology. 2015;**193**:34-42. DOI: 10.1016/j.ijfoodmicro.2014.10.012

[93] Akpapunam MA, Achinewhu SC. Effects of cooking, germination and fermentation on the chemical composition of Nigerian cowpea (*Vigna unguicilata*). Plant Foods for Human Nutrition. 1985;**35**:353-358. DOI: 10.1007/BF01091780

[94] Khattab RY, Arntfield SD. Nutritional quality of legume seeds as affected by some physical treatments 2. Antinutritional factors. LWT-Food Science and Technology. 2009; **42**:1113-1118. DOI: 10.1016/j.lwt.2009.02.004

[95] Reddy NR, Pierson MD. Reduction in antinutritional and toxic components in plant foods by fermentation. Food Research International. 1994;**27**:281-290. DOI: 10.1016/0963-9969(94)90096-5

[96] Sandberg AS, Andlid T. Phytogenic and microbial phytases in human nutrition. International Journal of Food Science and Technology. 2002;**37**:823-833. DOI: 10.1046/j.1365-2621.2002.00641.x

[97] Tosh SM, Yada S. Dietary fibres in pulse seeds and fractions: Characterization, functional attributes, and applications. Food Research International. 2010;**43**:450-460. DOI: 10.1016/j.foodres.2009.09.005

[98] Van Horn L. Fibre, lipids, and coronary heart disease. A statement for healthcare professionals from the nutrition committee, American Heart Association. Circulation. 1997;**95**: 2701-2704. DOI: 10.1161/01.cir.95.12.2701

[99] Annor GA, Ma Z, Boye JI. Crops—legumes. In: Clark S, Jung S, Lamsal B, editors. Food Processing: Principles and Applications. 2nd ed. Chichester: John Wiley and Sons Ltd; 2014. pp. 305-337

[100] Fardet A. New hypotheses for the health-protective mechanisms of whole-grain cereals: What is beyond fibre? Nutrition Research Reviews. 2010;**23**:65-134. DOI: 10.1017/S0954422410000041

[101] Marlett JA, McBurney MI, Slavin JL. Position of the American Dietetic Association: Health implications of dietary fiber. Journal of the American Dietetic Association. 2002;**102**:993-1000. DOI: 10.1016/s0002-8223(02)90228-2

[102] Glore SR, Van Treeck D, Knehans AW, Guild M. Soluble fiber and serum lipids: A literature review. Journal of the American Dietetic Association. 1994;**94**:425-436. DOI: 10.1016/0002-8223(94)90099-x

[103] Hoover R, Hughes T, Chung HJ, Liu Q. Composition, molecular structure, properties, and modification of pulse starches: A review. 2010;**43**:399-413. DOI: 10.7324/JAPS.2013.34.S18

[104] Lehmann U, Robin F. Slowly digestible starch- its structure and health implications: A review. Trends in Food Science and Technology. 2007;**18**:346-355. DOI: 10.1016/j.tifs.2009.12.001

[105] Veena A, Urooj A, Puttaraj S. Effect of processing on the composition of dietary fibre and starch in some legumes. Molecular Nutrition and Food Research. 1995;**39**:132-138. DOI: 10.1002/food.19950390206

[106] Kavita V, Verghese S, Chitra GR, Prakash J. Effects of processing time, storage temperature on the resistant starch of foods. Journal of Food Science and Technology. 1998;**35**:299-304

[107] Granito M, Alvarez G. Lactic acid fermentation of black beans (*Phaseolus vulgaris*): Microbiological and chemical characterization. Journal of the Science of Food and Agriculture. 2006;**86**:1164-1171. DOI: 10.1002/jsfa.2490

[108] Guillon F, Champ MM. Carbohydrate fractions of legumes: Uses in human nutrition and potential for health. British Journal of Nutrition. 2002;**88**:293-306. DOI: 10.1079/bjn2002720

[109] Dilawari J, Kumar V, Bhatnagar R, Dash R. Effect of legumes on blood sugar in diabetes mellitus. Indian Journal of Medical Research. 1987;**85**:184-187

[110] Jenkins DJA, Wolever T, Jenkins AL. Starchy foods and glycemic index. Diabetes Care. 1988;**11**:149-159. DOI: 10.2337/diacare.11.2.149

[111] Messina MJ. Legumes and soybeans: overview of their nutritional profiles and health effects. The American Journal of Clinical Nutrition. 1999;**70**:439-450

[112] Luo J, Yperselle MV, Rizkalla SW, Rossi F, Bornet FJR, Slama G. Chronic consumption of short-chain fructooligosaccharides does not affect basal hepatic glucose production or insulin resistance in type 2 diabetes. Journal of Nutrition. 2000;**130**:1572-1577

[113] Atkinson FS, Foster-Powell K, Brand-Miller JC. International tables of glycemic index and glycemic load values: 2008. Diabetes Care. 2008;**31**:2281-2283. DOI: 10.2337/dc08-1239

[114] Yeap SK, Ali NM, Yusof HM, Alitheen NB, Beh BK, Ho WY, Koh SP, Long K. Anti-hyperglycemic effects of fermented and nonfermented mung bean extracts on alloxan-induced-diabetic mice. Journal of Biomedicine and Biotechnology. 2012;1-7. DOI: 10.1155/2012/285430

[115] Liu S, Stampfer M, Hu F, Giovannucci E, Rimm E, Manson J, Hennekens C, Willett W. Whole-grain consumption and risk of coronary heart disease: results from the Nurses' Health Study. American Journal of Clinical Nutrition. 1999;**70**:412-419

[116] Patterson CA, Maskus H, Dupasquier C. Pulse crops for health. Cereal Foods World. 2009;**54**:108-112

[117] Vidal-Valverde C, Frias J, Prodanov M, Tabera J, Ruiz R, Bacon J. Effect of natural fermentation on carbohydrates, riboflavin and trypsin inhibitor activity of lentils. Zeitschrift fur Lebensmittel-Untersuchung und -Forschung. 1993;**197**:449-452. DOI: 10.1007/BF01202616

[118] Yadav S, Khetarpaul N. Indigenous fermentation: effect on some antinutrients and *in-vitro* digestibility of starch and protein. Food Chemistry. 1993;**50**:403-406. DOI: 10.1016/0308-8146(94)90213-5

[119] Bhandal A. Effect of fermentation on *in vitro* digestibilities and the level of antinutrients in moth bean [*Vigna aconitifolia* (Jacq.) Marechal]. International Journal of Food Science and Technology. 2008;**43**:2090-2094. DOI: 10.1111/j.1365-2621.2008.01827.x

[120] Egounlety M, Aworh OC. Effect of soaking, dehulling, cooking and fermentation with Rhizopus oligosporus on the oligosaccharides, trypsin inhibitor, phytic acid and tannins of soybean (*Glycine max* Merr.), cowpea (*Vigna unguiculata* L. Walp) and groundbean (*Macrotyloma geocarpa* Harms). Journal of Food Engineering. 2003;**56**:249-254. DOI: 10.1016/S0260-8774(02)00262-5

[121] Adewumi GA, Odunfa SA. Effect of controlled fermentation on the oligosaccharides content of two common Nigerian *Vigna unguiculata* beans (*drum* and *oloyin*). African Journal of Biotechnology. 2009;**8**:2626-2630

[122] Odunfa SA. Carbohydrate changes in fermenting locust bean (*Parkia filicoidea*) during iru preparation. Plant Foods for Human Nutrition. 1983;**32**:3-10. DOI: 10.1007/bf01093924

[123] Roy F, Boye JI, Simpson BK. Bioactive proteins and peptides in pulse crops: Pea, chickpea and lentil. Food Research International. 2010;**43**:432-442. DOI: 10.1016/j.foodres.2009.09.002

[124] Sotelo A, Adsule RN. Chickpea. In: Nwokolo E, Smartt J, editors. Food and Feed from Legumes and Oilseeds. London, UK: Chapman and Hall; 1996. pp. 82-89

[125] Duranti M, Gius C. Legume seeds: Protein content and nutritional value. Field Crops Research. 1997;**53**:31-45. DOI: 10.1016/s0378-4290(97)00021-x

[126] Farrokhi N, Whitelegge JP, Brusslan JA. Plant peptides and peptidomics. Plant Biotechnology Journal. 2008;**6**:105-134. DOI: 10.1111/j.1467-7652.2007.00315.x

[127] Malaguti M, Dinelli G, Leoncini E, Bregola V, Bosi S, Cicero AFG, Hrelia S. Bioactive peptides in cereals and legumes: Agronomical, biochemical and clinical aspects. International Journal of Molecular Sciences. 2014;**15**:21120-21135. DOI: 10.3390/ijms151121120

[128] Coda R, Rizzello CG, Pinto D, Gobbetti M. Selected lactic acid bacteria synthesize antioxidant peptides during sourdough fermentation of cereal flours. Applied and Environmental Microbiology. 2011;**78**:1087-1096. DOI: 10.1128/aem.06837-11

[129] Sanjeewa WGT, Wanasundara JPD, Pietrasik Z, Shand PJ. Characterization of chickpea (*Cicer arietinum* L.) flours and application in low-fat pork bologna as a model system. Food Research International. 2010;**43**:617-626. DOI: 10.1016/j.foodres.2009.07.024

[130] Cabuk B, Korber D, Tanaka T, Nickerson M. [Internet]. Fermenting Pulses for Improved Nutrition and Functionality 2016. Available from: http://canadianfoodbusiness.com/2016/06/20/fermenting-pulses-for-improved-nutrition-and-functionality/ [Accessed: January 24, 2017]

[131] Zambrowicz A, Timmer M, Polanowski A, Lubec G, Trziszka T. Manufacturing of peptides exhibiting biological activity. Amino Acids. 2012;**44**:315-320. DOI: 10.1007/s00726-012-1379-7

[132] Tavano OL. Protein hydrolysis using proteases: An important tool for food biotechnology. Journal of Molecular Catalysis B. 2013;**90**:1-11. DOI: 10.1016/j.molcatb.2013.01.011

[133] Xiao Y, Xing, G, Rui X, Li W, Chen X, Jiang M, Dong M. Effect of solid-state fermentation with *Cordyceps militaris* SN-18 on physicochemical and functional properties of chickpea (*Cicer arietinum* L.) flour. LWT-Food Science and Technology. 2015;**63**:1317-1324. DOI: 10.1016/j.lwt.2015.04.046

[134] Lee IH, Hung YH, Chou CC. Solid-state fermentation with fungi to enhance the antioxidative activity, total phenolic and anthocyanin contents of black bean. International Journal of Food Microbiology. 2008;**121**:150-156. DOI: 10.1016/j.ijfoodmicro.2007.09.008

[135] Jung S, Murphy PA, Johnson LA. Physicochemical and functional properties of soy protein substrates modified by low levels of protease hydrolysis. Journal of Food Science. 2005;**70**:180-187. DOI: 10.1111/j.1365-2621.2005.tb07080.x

[136] Reddy NR, Pierson MD, Sathe SK, Salunkhe DK. Legume-based fermented foods: Their preparation and nutritional quality. Critical Review in Food Science and Nutrition. 1982;**17**:335-370. DOI: 10.1080/10408398209527353

[137] Prasad SK, Singh MK. Horse gram- an underutilized nutraceutical pulse crop: A review. Journal of Food Science and Technology. 2015;**52**:2489-2499. DOI: 10.1007/s13197-014-1312-z

[138] Faris MAE, Attlee A. Lentils (*Lens culinaris* L.): a novel functional food. In: Shekhar HU, Howlader ZH, Kabir Y, editors. Exploring the Nutrition and Health Benefits of Functional Foods, USA: ICI Global; 2017. pp. 42-72. DOI: 10.4018/978-1-5225-0591-4.ch003

[139] Bazzano LA, He J, Ogden LG, Loria C, Vupputuri S, Myers L, Whelton PK. Dietary potassium intake and risk of stroke in us men and women: National health and nutrition examination survey I epidemiologic follow-up study. Stroke. 2001;**32**:1473-1480. DOI: 10.1161/01.str.32.7.1473

[140] Lee JW, Shin JG, Kim EH, Kang HE, Yim IB, Kim JY, Joo HG, Woo HJ. Immunomodulatory and antitumor effects in vivo by the cytoplasmic fraction of *Lactobacillus casei* and *Bifidobacterium longum*. Journal of Veterinary Science. 2004;**1**:41-48

[141] Flight I, Clifton P. Cereal grains and legumes in the prevention of coronary heart disease and stroke : A review of the literature. European Journal of Clinical Nutrition. 2006;**60**:1145-1159. DOI: 10.1038/sj.ejcn.1602435

[142] Kandasamy M, Bay BH, Lee YK. Lactobacilli secreting a tumor antigen and IL15 activates neutrophils and dentritic cells and generates cytotoxic T lymphocytes against cancer cells. Cell Immunology. 2011;**271**:89-96. DOI: 10.1016/j.cellimm.2011.06.004

Chapter 6

The Role of Legumes in Human Nutrition

Yvonne Maphosa and Victoria A. Jideani

Additional information is available at the end of the chapter

http://dx.doi.org/10.5772/intechopen.69127

Abstract

Legumes are valued worldwide as a sustainable and inexpensive meat alternative and are considered the second most important food source after cereals. Legumes are nutritionally valuable, providing proteins (20–45%) with essential amino acids, complex carbohydrates (±60%) and dietary fibre (5–37%). Legumes also have no cholesterol and are generally low in fat, with ±5% energy from fat, with the exception of peanuts (±45%), chickpeas (±15%) and soybeans (±47%) and provide essential minerals and vitamins. In addition to their nutritional superiority, legumes have also been ascribed economical, cultural, physiological and medicinal roles owing to their possession of beneficial bioactive compounds. Research has shown that most of the bioactive compounds in legumes possess antioxidant properties, which play a role in the prevention of some cancers, heart diseases, osteoporosis and other degenerative diseases. Because of their composition, legumes are attractive to health conscious consumers, celiac and diabetic patients as well as consumers concerned with weight management. The incorporation of legumes in diets, especially in developing countries, could play a major role in eradicating protein-energy malnutrition especially in developing Afro-Asian countries. Legumes could be a base for the development of many functional foods to promote human health.

Keywords: legumes, nutrition, bioactive compounds, food security, proteins, micronutrients, malnutrition

1. Introduction

Legumes are plants belonging to the family Leguminosae also called as Fabaceae that produce seeds within a pod [1, 2]. Leguminosae is a large family with over 18,000 species of climbers, herbs, shrubs and trees of which only a limited number is used as human food. Common legumes used for human consumption include peas, broad beans, lentils, soybeans, lupins, lotus, sprouts, mung bean, green beans and peanuts and are referred to as grain legumes or food legumes [3, 4]. A variety of legumes are shown in **Figure 1**.

Figure 1. A variety of legumes [5].

Food legumes are divided into two groups, namely oil seeds and pulses. The former being legumes with high oil content such as soybean and peanuts and the latter being all dry seeds of cultivated legumes used as traditional food [4]. The Food and Agriculture Organisation of the United Nations [5] recognises 11 primary leguminous classes (**Table 1**). Legumes are believed to be one of the first crops cultivated by mankind and have remained a staple food for many cultures all over the world [2]. These seeds are valued worldwide as an inexpensive meat alternative and are considered the second most important food source after cereals [2]. Legumes are nutritionally valuable, providing proteins with essential amino acids, complex carbohydrates, dietary fibre, unsaturated fats, vitamins and essential minerals for the human diet [6–8]. In addition to their nutritional superiority, legumes have also been ascribed economical, cultural, physiological and medicinal roles owing to their possession of beneficial bioactive compounds [9].

The consumption of legumes has also been reported to be associated with numerous beneficial health attributes [10] such as hypocholesterolemic, antiatherogenic, anticarcinogenic and hypoglycemic properties [11].

Legumes have proven to be a cheap source of nutrients as well as a potential source of income for subsistence farmers who cultivate legumes at household level. They are excellent crops for

	Class	Examples of legumes
1	Dry beans (mainly species of *Phaseolus* and some beans classified as *Vigna*)	Kidney, haricot bean *(Ph. vulgaris)*, lima, butter bean *(Ph. lunatus)*, adzuki bean *(Ph. angularis)*, mungo bean, golden, green gram *(Ph. aureus)*, black gram, urd *(Ph. mungo)*, scarlet runner bean *(Ph. coccineus)*, rice bean *(Ph. calcaratus)*, moth bean *(Ph. aconitifolius)*, tepary bean *(Ph. acutifolius)*
2	Dry broad beans (*Vicia faba*)	Horse-bean *(Vicia faba equina), broad bean (Vicia faba major), field bean (Vicia faba minor)*
3	Dry peas (*Pisum spp.*)	Garden pea *(Pisum sativum), field pea (P. arvense)*
4	Chickpeas	Chickpea, Bengal gram, garbanzos *(Cicer arietinum)*
5	Dry cow peas	Cowpea, blackeye pea/bean *(Vigna sinensis; Dolichos sinensis)*
6	Pigeon peas	Pigeon pea, cajan pea, Congo bean *(Cajanus cajan)*
7	Lentils	Lentils (*Lens culinaris)*
8	Bambara beans	Bambara groundnut (*Vigna subterranean* (L.) Verdc), earth pea *(Voandzeia subterranea)*
9	Vetches *(Vicia sativa)*	Spring/common vetch
10	Lupins *(Lupinus spp.)*	Bitter lupin, sweet lupin
11	Minor pulses (Legumes not identified separately due to their minor relevance at international level)	lablab or hyacinth bean *(Dolichos spp.)*, jack/sword bean *(Canavalia spp.)*, winged bean *(Psophocarpus tetragonolobus)*, guar bean *(Cyamopsis tetragonoloba)*, velvet bean *(Stizolobium spp.)*, yam bean *(Pachyrrhizus erosus)*

Table 1. Classification of legumes.

local farmers that do not afford expensive irrigation systems and fertilisers. This is because legumes thrive in poor soils and adverse weather conditions, are highly disease and pest resistant, are cover crops; therefore, reduce soil erosion and have a symbiotic relationship with the nitrogen-fixing rhizopus resident in their root nodules, thus making them excellent rotation crops [12, 13].

It is of utmost importance to increase the utilisation of legumes and to introduce new legume-based products that will be affordable to low-income groups as a way to reduce poverty and alleviate malnutrition. Protein-energy malnutrition (PEM) is a major nutritional syndrome affecting over 170 million preschool children and lactating women in developing African and Asian countries [1, 12, 14]. The prevalence of PEM can be attributed to many factors such as the high price of animal protein (eggs, meat and milk), the staple cereal-based diet and the ever increasing price of food commodities becoming unaffordable to the lower income groups. Although, high protein legumes such as soybean and cowpea are available to consumers, their consumption rate surpasses their production rate; thus, an ever increasing demand has been observed [12]

The nutritional demand of legumes is increasing worldwide because of increased consumer awareness of their nutritional and health benefits. Furthermore, recent years have seen more people substituting animal protein with vegetable protein; thus, further increasing the demand for legumes as they are the chief source of plant proteins. To meet this demand, there is a need to direct attention to the nutritional profiling of various legumes, increase the utilisation of underutilised legumes, produce cheap, innovative value-added products from legumes, educate consumers on the nutritional value of legumes as well as find new ways of encouraging the use of existing legumes. **Figure 2** shows a comparison of the proximate composition of five common cereal grains and five common legumes. From the graph, it is evident that legumes have higher amounts of protein and dietary fibre than cereals.

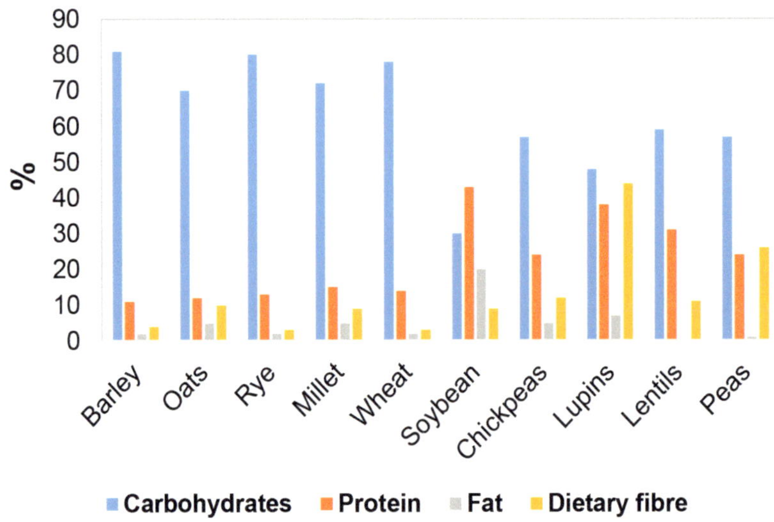

Figure 2. A comparison of the proximate composition of some common cereal grains and legumes [15, 16].

2. Protein content of legumes

Legumes are an excellent source of good quality protein with 20–45% protein that is generally rich in the essential amino acid lysine [9]. Peas and beans are on the lower side of the range with 17–20% proteins while lupins and soybeans are on the higher end of the range with 38–45% protein [2, 15]. Legumes have higher protein content than most plant foods with about twice the protein content of cereals (**Figure 2**) [2, 17, 18]. The high protein content of legumes can be attributed to their association with the activity of the nitrogen-fixing bacteria in their roots, which converts the unusable nitrogen gas into ammonium which the plant then incorporates into protein synthesis.

Leguminous proteins, except soy protein (**Table 2**), are however low in the essential sulphur-containing amino acids (SCAA), methionine, cystine and cysteine as well as in tryptophan (**Table 2**) and are therefore considered to be an incomplete source of protein [2]. The main fractions of leguminous protein are albumins and globulins which can be divided into two groups, namely vialin and legumin. Vialin is the major protein group in most legumes and is characterised by a low content of SCAA, thus explaining the low levels of SCAA in legumes [18]. The low level of SCAA in legumes is not completely a negative factor as it results in increased calcium retention. Hydrogen ions produced from the breakdown of SCAA cause the demineralisation of the bone and thus excretion of calcium in the urine. Therefore, leguminous

http://dx.doi.org/10.5772/intechopen.69127

Amino acid	BGN	CP	SB	AB	LP	LB	LT	CK	BB	KB
Arginine	4.0	1.6	7.2	1.3	3.9	2.2	2.2	1.8	0.7	1.5
Aspartic acid	5.0	2.8	11.7	2.4	3.9	2.9	3.1	2.3	0.8	2.9
Histidine	2.2	0.7	2.5	0.5	1.0	0.6	0.8	0.5	0.2	0.7
Serine	3.2	1.2	5.1	1.0	1.9	1.1	1.3	1.0	0.3	1.3
Glutamic acid	16.5	4.5	18.7	3.1	8.7	4.2	4.4	3.4	1.3	3.6
Proline	3.2	1.1	5.5	0.9	1.5	1.0	1.2	0.8	0.3	1.0
Glycine	3.3	1.0	4.2	0.8	1.5	1.1	1.1	0.8	0.3	0.9
Alanine	3.5	1.1	4.3	1.2	1.3	1.1	1.2	0.8	0.3	1.0
Lysine*	3.0	1.6	6.4	1.5	1.9	1.8	2.0	1.3	0.5	1.6
Threonine*	2.5	0.9	3.9	0.7	1.3	0.9	1.0	0.7	0.3	1.0
Valine*	3.8	1.1	4.8	1.0	1.5	1.2	1.4	0.8	0.3	1.2
Isoleucine*	3.8	1.0	4.5	0.8	1.6	1.0	1.2	0.8	0.3	1.0
Leucine*	6.8	1.8	7.8	1.7	2.7	1.8	2.0	1.4	0.6	1.9
Tyrosine*	3.2	0.8	3.1	0.6	1.4	0.7	0.8	0.5	0.2	0.7
Phenylalanine*	4.3	1.4	4.9	1.1	1.4	1.1	1.4	1.0	0.3	1.3
Tryptophan*	0.7	0.3	1.3	0.9	0.3	0.3	0.3	0.2	0.1	0.3
Cystine**	0.5	0.3	1.3	0.2	0.4	0.4	0.4	0.3	0.1	0.3
Methionine**	2.0	0.3	1.3	0.2	0.3	0.3	0.2	0.3	0.1	0.4

BGN: Bambara groundnut; CP: Cowpea; SB: soybean; AB: Adzuki bean; LP: Lupins; LB: Lima beans; LT: Lentils; CK: Chickpea; BB: Broad beans; KB: Kidney beans.
Essential amino acid.
Essential, sulphur-containing amino acid.

Table 2. Amino acid profiles of 10 legumes expressed as g/100 g protein [5, 17, 19, 20].

protein may improve calcium retention in comparison with high SCAA proteins of animal or cereal origin. Legume protein has also been reported to contribute to the reduction of low density lipoproteins, a known factor in the development of coronary heart diseases [9].

Legumes and cereals complement each other in terms of protein as cereals are high in SCAA (low in legumes) and have low in lysine (high in legumes) [1]. As such, protein quality is significantly improved when legumes are eaten in combination with cereals [18]. For nutritional balance, legumes and cereals are to be consumed in the ratio 35:65 [4]. Legumes are particularly important in vegetarian diets as they are the chief source protein and also provide vitamins and minerals [18]. For vegetarians to get a good balance of amino acids, their diets need to combine legumes with cereals. Common examples of such combinations are *dhal* with rice in India, beans with corn tortillas in Mexico, tofu with rice in Asia, peanut butter with bread in the USA and Australia [17], samp and beans (South Africa), Bambara groundnut and maize kernels (Zimbabwe), maize meal *pap* with beans (Southern Africa) and rice and beans (Southern Africa, Latin America). **Table 2** shows the amino acid profiles of several legumes.

3. Classification of carbohydrates in relation to legumes

Legumes are a source of complex, energy giving carbohydrates [17] with up to 60% carbohydrates (dry weight). Leguminous starch is digested slower than starch from cereals and tubers. As such, legumes have a low glycemic index (GI) rating for blood glucose control [9, 14] making them suitable for consumption by diabetic patients and those with an elevated risk of developing diabetes. Furthermore, legumes are gluten free, making them suitable for consumption by celiac disease patients or individuals sensitive to the proteins gliadin and glutenin [18]. Generally, legumes are important for individuals seeking a healthy, disease free lifestyle [8]. Legume starch isolates have been employed as thickeners in soups and gravies in the food industry [9].

Legumes are also a valuable source of dietary fibre (5–37%), containing significant amounts of both soluble and insoluble dietary fibre [2, 9, 17]. The monomers in legume dietary fibres include glucose, galactose, fucose, arabinose, rhamnose, xylose and mannose. Legumes also contain significant amounts of resistant starch and oligosaccharides, mainly raffinose, which have been reported to possess prebiotic properties [2]. These are fermented by probiotics to short chain fatty acids improving colonic health and reducing the risk of colon cancer. High dietary fibre diets are associated with many health benefits. These include the prevention and possible treatment of diseases and conditions like constipation, obesity, diabetes, heart complications, piles and some cancers [21–23]. In addition, dietary fibre, particularly soluble dietary fibre, has the ability to lower blood cholesterol, improve glucose tolerance and reduce glycaemic response by forming a protective gel lining along the intestinal walls thus reducing glucose and cholesterol assimilation into the bloodstream [22, 24, 25]. Insoluble dietary fibres are porous, have low densities, increase faecal bulk and promote normal laxation [26–28]. As such, legumes are an invaluable component of the human diet. Dietary fibre fractions from legumes have found use in the bakery, meat, extruded products and beverage industries as stabilisers, texturing agents, fortifiers, bulking agents, fat replacers and emulsion stabilisers [9, 10, 15, 17].

4. Fat and fatty acid composition of legumes

Legumes have no cholesterol and are generally low in fat, with ±5% energy from fat [10] with the exception of peanuts (±45%), chickpeas (±15%) and soybeans (±47%). The fat in legumes constitutes of significant amounts of mono- and polyunsaturated fatty acids (PUFA) and virtually no saturated fatty acids [2]. The highest amount of PUFA (71.1%) and monounsaturated fatty acids (34%) are reported in kidney beans and chickpeas, respectively [2]. The PUFAs present in some legumes include the essential omega-6 linoleic acid (C18:2, ω 6) and omega-3 alpha-linolenic acid (C18:3, ω-3). These PUFAs are essential for human health and since the human body cannot synthesise them, they must be included in the diet [18].

5. Clustering of legumes depending on their proximate composition

Using K-means cluster, 22 legumes were grouped into 3 cluster centres as shown in **Table 3**. Cluster 1 represented legumes that are high in carbohydrates (±63.8%), average in protein (±25.4%), low in fat (±2.6%) and low in dietary fibre (±9.3%). Cluster 2 represented legumes that are average in carbohydrates (±37.1%), high in protein (±36.1%), average in fat (±14.1%) and high in dietary fibre (±17.7%). Cluster 3 represented legumes that are low in carbohydrates (±19.3%), low in protein (±18.7%), high in fat (±55.0%) and average in dietary fibre (±13.3%).

Of the 22 legumes, 6% of the legumes fell into cluster 1, 18% into cluster 2 and 5% into cluster 3. Sword bean fell into clusters 1 and 2, hyacinth fell into clusters 1 and 3 and groundnut fell into clusters 2 and 3. It can be concluded that the majority of legumes are high in carbohydrates hence are high in energy, are a source of protein because even the cluster that is "low" in protein provides up to 19% protein which is significantly high and are low in fat with the exception of groundnut, hyacinth, lupins, soybean and sword bean.

	Cluster		
	1	**2**	**3**
Carbohydrate (%)	63.78	37.10	19.33
Protein (%)	25.44	36.09	18.73
Fat (%)	2.58	14.11	55.03
Dietary fibre (%)	9.32	17.72	13.28
Legumes	Adzuki bean, Green gram, Black gram, Pigeon pea, Cowpea, Lima bean, Broad bean, Kidney bean, Mung bean, African yam bean, Bambara groundnut, Lentil, Sword bean, Black velvet bean, White velvet bean, Pinto, Chickpea, Hyacinth	Sweet lupin, Bitter lupin, Soybean, Sword bean, Groundnut	Groundnut, Hyacinth

Table 3. Cluster centres for 22 legumes.

6. Micronutrients in legumes

Legumes are a good source of B-group vitamins such as folate, thiamin and riboflavin but are a poor source of fat soluble vitamins and vitamin C [2]. Folate is an essential nutrient and has also been reported to reduce the risk of neural tube defects like spina bifida in newly born babies [10, 18]. Legumes are also sources of the essential minerals zinc, iron, calcium, selenium, phosphorus, copper, potassium, magnesium and chromium [2, 29]. These micronutrients play important physiological roles such as bone health (calcium), enzyme activity and iron metabolism (copper), carbohydrate and lipid metabolism (chromium, zinc), haemoglobin synthesis (iron) as well as antioxidative activity, protein synthesis and plasma membrane stabilisation (zinc) [30]. Generally, legumes are low in sodium and this is desirable considering the recent trends encouraging sodium reduction [17, 31]. Although, legumes have high iron contents, the bioavailability of the iron is poor hence diminishing the value of legumes as a source of iron [10]. However, if legumes are consumed in combination with vitamin C rich foods, the absorption of iron is increased. In this manner, the high iron content would play a major role in the prevention of anaemia especially in women of reproductive age.

7. Bioactive compounds and non-nutrients in legumes

Legumes contain non-nutrient bioactive compounds such as phytochemicals and antioxidants [18]. These include isoflavones, lignans, protease inhibitors, trypsin and chymotrypsin inhibitors, saponins, alkaloids, phytoestrogens and phytates. Most of these chemicals are termed 'anti-nutrients' and although they are non-toxic, they generate adverse physiological effects and interfere with protein digestibility and the bioavailability of some minerals [32]. Most of these anti-nutrients are heat labile and since legumes are consumed after cooking, they do not pose a health hazard [32]. Legumes can also be detoxified by dehulling, soaking, boiling, steaming, sprouting, roasting and fermentation prior to processing [11].

Research has shown that most of these non-nutrients are phytochemicals with antioxidant properties which play a role in the prevention of some cancers, heart diseases, osteoporosis and other chronic degenerative diseases [8, 10]. The quantities of some non-nutrients present in legumes are given in **Table 4**. The antioxidant capacity of legumes allows them to inhibit or slow down oxidative processes which are largely responsible for degenerative diseases by interacting and scavenging free radicals and reactive oxygen species, chelating metal catalysts, activating antioxidant enzymes as well as inhibiting oxidases [22]. As such, the incorporation of legumes into human diets all over the world could offer protection against chronic diseases [33]. Therefore, legumes, especially underutilised legumes, should be explored for the development of innovative, value-added products (**Figure 3**).

Saponins and glycosides are another group of bioactive compounds present in legumes such as lentils, chickpeas, soy bean and peas. These compounds form insoluble complexes with 3-β-hydroxysteroids and form micelles with bile acid and cholesterol; thus, facilitating their

Legume	Polyphenols (%)	Phytic acid (%)	Tannins (%)	α-Galactosides (%)
Common bean (white)	0.3	1.0	0	3.1
Common bean (Brown)	1.0	1.1	0.5	3.0
Pea	0.2	0.9	0.1	5.9
Lentils	0.8	0.6	0.1	3.5
Faba bean	0.8	1.0	0.5	2.9
Chickpea	0.5	0.5	0	3.8
Soybean	0.4	1.0	0.1	4.0
Pigeon pea	0.2	0.1	0	0

Table 4. Some non-nutrients present in common legumes (% dry matter) [34, 35].

excretion from the human body. These compounds have also been reported to possess hypocholesterolemic and anticarcinogenic activity [2].

Other important bioactive compounds found in legumes include polyphenols and their derivatives such as flavanols, flavan-3-ols, anthocyanins/anthocyanidins, condensed tannins/proanthocyanidins and tocopherols [32]. The concentration of polyphenols such as glutathione and tocopherols in legumes ranges from 321 to 2404 μg/100 g. Although, tannins are generally considered undesirable because they render protein indigestible, recent studies have shown

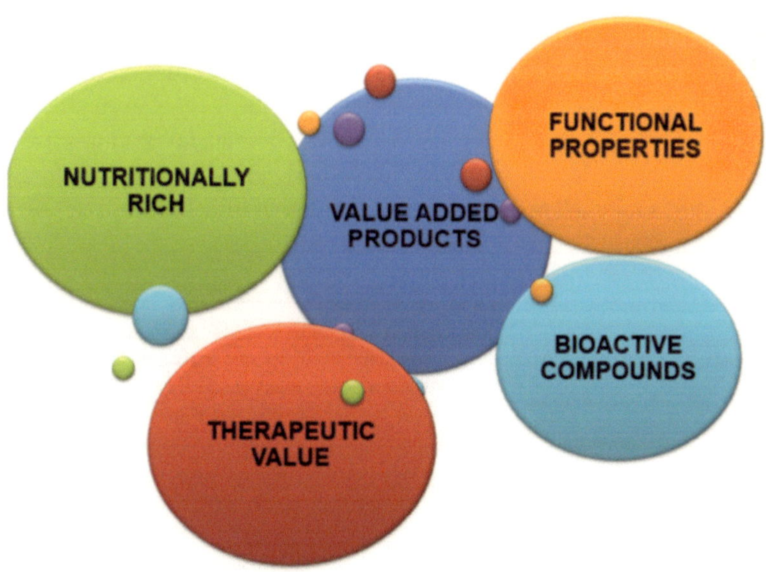

Figure 3. Potential of legumes in the production of value-added products.

their consumption to have an inverse correlation to the incidence of biological molecule (DNA, lipids and proteins) damage due to their reducing nature [11]. Legumes with coloured seed coats such as Bambara groundnut, black bean, red kidney bean and black gram, have long been associated with antioxidant and anticarcinogenic activity [2]. It is believed that the denser the colour of the seed coat, the higher the antioxidant activity.

7.1. Oligosaccharides

Most legumes contain up to 50 mg/g total oligosaccharides. Oligosaccharides are responsible for flatulence widely associated with the consumption of legumes. The absence of an α-galactosidase enzyme in the human gastrointestinal tract to cleave the α-1,6 galactose linkage in galactoside-containing oligosaccharides such as raffinose and stachyose means these oligosaccharides pass undigested to the colon where they are metabolised by bacteria forming large amounts of carbon dioxide, hydrogen and methane. These gases may cause bloating and gastric discomfort and are expelled from the body as flatulence. However, although the oligosaccharides in legumes are viewed negatively, their beneficial attributes outweigh their negative properties [10]. Oligosaccharides are prebiotic in nature and therefore, promote the growth of the probiotics, *Bifidobacteria* spp, which play a major role in the maintenance of a healthy colon. In Japan, soybean oligosaccharides have been suggested as a substitute for table sugar [10].

8. Legume consumption around the world

Legumes play an important role in many diets all over the world and are especially important in developing/third world countries in Africa, Latin America and Asia. Legumes have been labelled the 'poor man's meat' and this statement seems to hold some truth as observed in the consumption distribution in different regions, with an inverse relation between legume consumption and income being observed [10]. Emerging research is however changing the label of legumes to "health food", encouraging their inclusion in the diets of even affluent people [2]. Legumes have been used in the production of various commercial products such as textured vegetable protein (TVP), tofu, soy sauce, soy paste and curry. Some by-products of legumes include dietary fibre, single cell proteins, citric acid and enzymes. Legumes can be incorporated in various ways to increase their acceptance in balanced nutritious diets [8] as shown in **Table 5**.

Common name	Food uses
Soybean (*Glycine max*)	Asian dishes (tofu, natto miso), roasted snacks, milk, yoghurt, sprouted beans, curd, yuba, soy sauce, soy paste, TVP
Black gram (*Vigna mungo*)	Dhal, fermented products (idli, dosa, papad)
Lentils (*Lens culinaris*)	Dhal, papadums
Peas (*Pisum sativum*)	Soup, dhal
Peanut/Groundnut (*Arachis hypogaea*)	Peanut butter, peanut bar, flour, roasted/boiled snacks
Adzuki beans (*Vigna angularis*)	Japanese desserts and confections, soup ingredients for therapeutic purposes
Anasazi beans (*Phaseolus vulgaris*)	Boiled meal, snack, soup

http://dx.doi.org/10.5772/intechopen.69127

Common name	Food uses
Black-eyed peas (*Vigna unguiculata*)	Boiled snack/part of meal, fried cake *akara,* steamed pudding *moi moi* in West Africa
Chickpea (*Cicer arietinum*)	Middle Eastern and Mediterranean foods such as falafel and hummus, Boiled/fried/cooked/crushed snacks, dhal, curry, flour used in bread making, fermented food *(dhokla)*
Kidney beans (*Phaseolus vulgaris*)	Ingredient in Mexican chili; most-consumed legume in America
Lentils (*Lens culinaris*)	Soups and stews; most important legume in India
Lima beans (Phaseolus lunatus)	Cooked whole
Mung beans/Green gram (*Vigna radiate*)	Bean sprouts, cooked whole or with sugar into a dessert, soup, flour used for baking, transparent noodles, patties, sweets
Navy beans (*Phaseolus vulgaris*)	Baked beans
Black turtle beans (*Phaseolus vulgaris*)	Bean soup popular in latin American cuisine
Pinto beans (*Phaseolus vulgaris*)	Fried beans
Bambara groundnut (*Vigna subterranean* (L). Verdc)	Boiled whole or split, soups, milk, yoghurt, boiled/fried/cooked/crushed snacks, commercially canned in gravy, flour used in bread making
Yam bean (*Pachyrhizus* spp)	Tubers used as vegetables
Lupins (*Lupinus* spp)	High protein seeds
Rice bean (*Vigna umbellate*)	Boiled seeds, fodder
Winged bean (*Psophocarpus tetragonalobus*)	Boiled seeds
Faba bean (*Vicia faba*)	Whole food
Sword bean (*Canavalia gladiate*)	Mature beans and dried seeds used as food and for medicinal purposes
Hyacinth bean (*Lablab purpureus*)	Popular in south Asian dishes
Velvet bean (*Mucuna monosperma*)	Seeds used as food and for pharmaceutical application
African Yam bean (*Sphenostylis. stenocarpa*)	Bean seeds usually eaten alone or in combination with other foods
Tamarind (*Tamarindus indica*)	Pulp used for food and beverage preparation, flour used as soup thickener, remedy in diarrhoea and dysentery
Marama bean (*Tylosema esculentum*)	High nutritional value food

Table 5. Various ways in which legumes are eaten around the world [2, 19, 36–39].

9. Role of legumes in human health and food security

Many diseases of lifestyle are a result of a poor diet, high in animal products and low in plant matter. Legumes are high in dietary fibre, high in complex, low glycemic carbohydrates, high in bioactive compounds, low in saturated fat and no cholesterol (**Figure 4**). These dietary components promote health and longevity by decreasing insulin production and preventing chronic diseases such as diabetes, cancer, cardiovascular disease and obesity. As such, a legume-based diet can result in a longer, healthier life.

Although, legumes are the second most important crops after cereals, the inadequacy of the knowledge of their nutritional and functional benefits has resulted in them not being given enough attention. Therefore, future studies should look into harnessing the many desirable properties (**Figure 4**) of legumes in the development of inexpensive legume products that are available to all income groups [39]. Most legumes are cultivated by low-income groups at household level. The increased use of legumes would increase their demand and in turn would encourage local farmers to increase legume production, hence resulting in increased financial stability and food security. The functional properties (**Figure 4**) of legumes such as water binding, oil binding, emulsion stabilisation and gelling could be harnessed in the development of various food products. There is urgent need to educate communities worldwide about the nutritional value of legumes, methods of detoxifying legumes of anti-nutrients and various methods of making legumes more attractive to consumers. In addition, genetic modification could be explored in developing transgenic leguminous species that cook faster and have low levels of anti-nutrients.

Taking their nutritional superiority into consideration, it is expected that dieticians and nutritionists encourage the public through mass media such as television, press and radio, to increase their consumption of legumes.

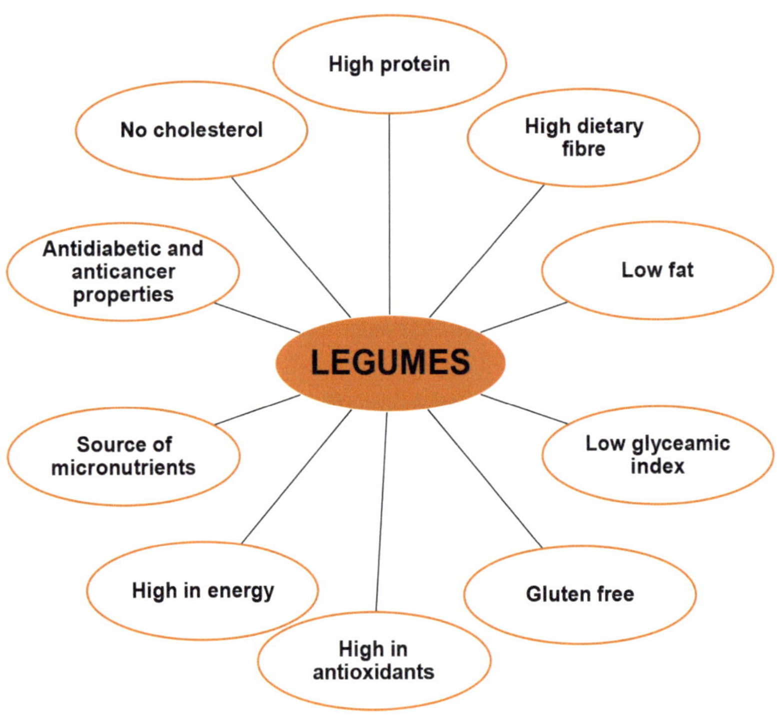

Figure 4. Desirable attributes of legumes.

10. Why underutilised legumes should be given more attention

Underutilised legumes also known as orphan crops, neglected crops or lesser crops such as Bambara groundnut, African locust bean, African yam bean, pigeon pea, kidney bean, lima bean and marama bean deserve to be given more attention [40]. Most of these underutilised legumes thrive in adverse conditions, are nutritionally superior and yield more than common legumes [40].

There is a pressing need in developing/poor countries such as those in sub Saharan Africa, for readily available, affordable, nutritional rich food supplements to cater for the ever increasing population. Underutilised legumes could be the answer to this demand. Most are cultivated only at household level as secondary crops. As such effort should be directed towards conducting extensive research to extend both technical and practical knowledge about these legumes so that their full potential may be achieved. These legumes' high nutritional could largely contribute to combating malnutrition [13]. It is envisaged that underutilised legumes could have an abundant amount of undiscovered bioactive compounds that could be employed in the production of therapeutic, affordable, functional foods. The increased use of underutilised legumes could reduce the overutilisation of common legumes such as soybean.

11. Constraints associated with the utilisation of legumes and possible solutions

Several factors contribute to the limited use of legumes. These include the presence of antinutrients, myths about legume consumption, their association with bloating and flatulence as well as their hard-to-cook phenomenon. There is a need to educate consumers about methods in which these negative properties of legumes can be reduced or removed completely. Processing methods such as soaking, germination, fermentation and cooking have been reported to detoxify the legume seed. Soaking prior to cooking also softens the seeds, significantly reducing cooking time.

Low yields, poor seed availability, lack of market, significant labour requirement at maturity, lack of awareness of indigenous legumes and the lack of convenient food applications also contribute to the low utilisation of some legumes [9]. The development of new legume products could lead to a higher demand of legumes hence prompting local farmers to increase the production of these legumes for commercial purposes [37]. To overcome the discomfort and embarrassment associated with bloating and flatulence caused by oligosaccharides, commercial digestive aids such as Beano (AkPharma Inc, Pleasantville, NJ) have been developed. These digestive aids contain the enzyme α-galactosidase, which breaks down the oligosaccharides, therefore avoiding gas production in the large intestines. Rinsing legumes and changing the boiling water several times also significantly reduces the amount of oligosaccharides in legumes. Several methods of overcoming constraints that limit the use of legumes are given in **Table 6**.

Constraint	Negative effect	Solution
Trypsin inhibitors and amylase inhibitors	Decreases protein digestibility and starch digestibility	Boiling dry beans generally reduces the content by 80–90% Fermentation
Phytate	Chelates with minerals resulting in poor mineral bioavailability	Dehulling, soaking, boiling, steaming, sprouting, roasting and fermentation, autoclaving, gamma irradiation
Lectins, saponins	Reduced bioavailability of nutrients	Most destroyed by cooking, soaking, boiling, sprouting, fermenting
Oligosaccharides	Flatulence and bloating	Digestive aids such as Beano, changing boiling water, soaking, cooking, germination
Hard-to-cook phenomenon	Energy and time consumption	Soak legumes before cooking them
Lack of convenient food applications	Boredom of eating the same food repeatedly	New product development of innovative legume products as well as increased utilisation of lesser legumes
Low levels of sulphur-containing amino acids	Incomplete protein source	Consumed in combination with cereals (high in sulphur-containing amino acids)
Lack of awareness, understanding and knowledge of nutritional value of legumes	Low intake of legumes	Increasing consumer awareness of the nutritional profile of legumes
Beliefs and taboos–for example, eating groundnuts can cause stomach upset	Low intake of legumes	Increasing consumer awareness of the nutritional profile of legumes and of methods to get rid of anti-nutrients and oligosaccharides
Reluctance to try a new kind of food or to change eating habits	Low intake of legumes	Development of innovative, attractive legume-based products to entice consumers
Low iron bioavailability	Poor source of iron	Consumed in combination with vitamin C rich foods, the absorption of iron would be increased

Table 6. Utilisation problem of legumes and possible solutions.

12. Role of legumes in weight management and satiety

Several studies have suggested that the consumption of legumes could aid in weight loss. This could be attributed to the low fat and high dietary fibre nature of legumes. The low GI nature of legume carbohydrates also aids in stabilising blood sugar and insulin levels resulting in the consumer feeling satiated for increased periods of time [18]. This in turn results in less and infrequent eating which is ideal for weight management. In a US National Health and Nutrition Examination Survey [41], it was concluded that eating legumes was associated with decreased body mass index (BMI), reduced waist circumference and reduced risk of obesity. More studies in Iran concluded that the risk of suffering from obesity was reduced in men who consumed at least 30 g of legumes a day [41]. More studies have reached the conclusion that the consumption of 3–5 cups of legumes as part of an energy-controlled diet results in the loss of 3.6–8.1 kg of body mass over 6–8 weeks [41].

13. Novel, healthy legume-based products

There are various products developed from legumes both at household level (**Table 5**) and commercially. Legumes provide high protein meat-substitutes for vegetarians, low fat substitutes for health conscious individuals and low cost products for low-income groups. One of the most utilised legumes is soybean [3]. Its high oil content makes it a suitable raw material for oil extraction [42]. From soybean, products such as milk, tofu, temper, soy sauce, yoghurt and cheese have been commercially produced (**Table 5**). Soymilk, cheese and yoghurt are excellent dairy substitutes for vegans and lactose intolerant individuals. Soycorn milk, a product produced from a mixture of soymilk and sweet corn is also available [42]. Blending sweet corn with soymilk helps in masking the beany flavour associated with legume milk as well as enhances its nutritional value [42]. Dairy substitutes have also been produced from Bambara groundnut. Bambara groundnut milk was patented by Ref. [38], these researchers also reported the production of yoghurt from Bambara groundnut milk.

Other leguminous products include texturised vegetable protein (TVP), canned beans, groundnuts/peanuts and flour. The term 'TVP' loosely refers to extruded defatted soy flour or concentrate with a meat-like chewy texture when cooked or hydrated [42]. This product is very popular amongst vegetarians. Canned legumes are a common sight in many supermarkets and small stores. Most legumes are canned in brine, sugar solution or tomato purees. Although, this technology preserves legumes allowing for their availability all year round, it increases their cost [42]. Groundnuts are another popular group of legumes. Commercially, they are used in the extraction of oil as well as in the manufacture of peanut butter or are sold as salted, boiled, roasted, shelled or unshelled (**Table 5**). Legumes are sometimes ground into flour for use as thickeners in soups, emulsion stabilisers or for baking [37]. Legume flour available in the food market includes that from cowpea, soybean, pigeon pea and African yam bean [42].

Research has begun exploring the technological function of leguminous ingredients in the formation of novel, healthier foods. Dietary fibres from legumes have high water binding, oil binding, swelling capabilities making them suitable for use as thickeners in soups, fat replacers in meat products, stabilisers in emulsions, texturisers in bread as well as in improving body and mouthfeel in products such a yoghurt [37]. In addition, dietary fibres extracted from legumes such as Bambara groundnut possess prebiotic properties and could be used in the production of prebiotic supplements [22]. Starch from legumes was reported to positively improve the stability and rheological properties of oil-in-water emulsions [43]. Soy protein finds use in protein shakes common amongst physically fit individuals [42].

14. Conclusions

Legumes are a sustainable and inexpensive source of protein, unsaturated fat, dietary fibre, complex carbohydrates, micronutrients and important bioactive phytochemicals, therefore their consumption could contribute to a healthier lifestyle. Their composition makes them attractive to health conscious consumers, celiac and diabetic patients as well as consumers concerned with weight management. To harness the nutritional benefits of legumes, they should be incorporated into children and infants' diets at home and through school feeding programs, especially in developing countries to reduce poverty and malnutrition. Furthermore, legumes could be a base for the development of many functional foods as well as a range of feed and raw material for industrial products.

Author details

Yvonne Maphosa* and Victoria A. Jideani

Address all correspondence to: yvonmaphosa@gmail.com

Department of Food Science and Technology, Cape Peninsula University of Technology, Bellville, South Africa

References

[1] Staniak M, Księżak J, Bojarszczuk J. Mixtures of legumes with cereals as a source of feed for animals. In: Pilipavicius V, editor. Organic Agriculture Towards Sustainability. InTech: Croatia, 2014. pp. 123–145. DOI: 10.5772/58358

[2] Kouris-Blazos A, Belski R. Health benefits of legumes and pulses with a focus on Australian sweet lupins. Asian Pacific Journal of Clinical Nutrition. 2016;**21**(1):1-17. DOI: 10.6133/apjcn.2016.25.1.23

[3] Yorgancilar M, Bilgicli N. Chemical and nutritional changes in bitter and sweet lupin seeds (*Lupinus albus* L.) during bulgur production. Journal of Food Science and Technology. 2014;**51**(7):1384-1389. DOI: 10.1007/s13197-012-0640-0

[4] Anonymous. Grain Composition. Lupin Food Australia. Perth: Australia. 2013. DOI: http://www.lupinfoods.com.au/grain-composition/

[5] FAO. International year of legumes: Nutritious seeds for a sustainable future. Food and Agriculture Organisation of the United Nations and World Health Organisation. Rome: FAO; 2016. DOI: www.fao.org/pulses-2016

[6] Rebello CJ, Greenway FL, Finley JW. A review of the nutritional value of legumes and their effects on obesity and its related co-morbidities. Obesity Reviews. 2014;**15**(5):392-407. DOI: 10.1111/obr.12144

[7] Annor GA, Zhen M, Boye JI. Crops–Legumes. In: Clark S, Jung S, Lamsal B, editors. Food Processing: Principles and Applications. Chichester: John Wiley & Sons, Ltd. 2014. pp. 305–337. DOI: 10.1002/9781118846315.ch14

[8] Bouchenak M, Lamri-Senhadji M. Nutritional quality of legumes, and their role in cardiometabolic risk prevention: A review. Journal of Medicinal Food. 2013;**16**(3):185-198. DOI: 10.1089/jmf.2011.0238

[9] Philips RD. Starchy legumes in human nutrition and culture. Plant Foods and Human Nutrition. 1993;**44**(3):195-211. DOI: 10.1007/BF01088314

[10] Messina MJ. Legumes and soybeans: Overview of their nutritional profiles and health effects. Asia Pacific Journal of Clinical Nutrition. 2016;**25**(1):1-17. DOI: 10.1.1.847.8636

[11] Ndidi US, Ndidi CU, Aimola IA, Bassa OY, Mankilik M, adamu Z. Effects of processing (Boiling and roasting) on the nutritional and antinutritional properties of Bambara groundnuts (*Vigna subterranean* [L.] Verdc.) from Southern Kaduna, Nigeria. Journal of Food Processing. 2014;**2014**:1-9. DOI: 10.1155/20172129

[12] Nedumaran S, Abinaya P, Jyosthnaa P, Shraavya B, Parthasarathy R, Bantilan C. Grain Legumes Production, Consumption and Trade Trends in Developing Countries. Working Paper Series No 60. International Crops Research Institute for the Semi-Arid Tropics (ICRISAT). Telangana: International Crops Research Institute for the Semi-Arid Tropics;2015. pp. 1–57. DOI: 10.7910/DVN/V61SNB

[13] Kalidass C, Mahapatra AK. Evaluation of the proximate and phytochemical compositions of an underexploited legume Mucuna pruriens var. utilis (Wall ex Wight) L.H. Bailey. International Food Research Journal. 2014;**21**(1):303-308

[14] Khalid II, Elharadallou SB. Functional properties of cowpea (*Vigna Ungiculata* L.Walp), and lupin (Lupinus Termis) flour and protein isolates. Journal of Nutrition and Food Science. 2013;**3**:234. DOI: 10.4172/2155-9600.1000234

[15] Mlyneková Z, Chrenková M, Formelová Z. Cereals and legumes in nutrition of people with celiac disease. International Journal of Celiac Disease. 2014;**2**(3):105-109. DOI: 10.12691/ijcd-2-3-3

[16] Haytowitz DB, Matthews RH. Composition of foods: Legumes and legume products. In: Agriculture Handbook 8-16. Washington DC: United States Department of Agriculture (USDA); 1986. pp. 1–156

[17] Leonard E. Cultivating good health. In: Grains and Legumes Nutrition Council. Adelaide: Cadillac Printing; 2012. pp. 3–18. ISSN 1039-6217

[18] FAO. Legumes can help fight climate change, hunger and obesity in Latin America and the Caribbean. Food and Agriculture Organisation of the United Nations and World Health Organisation. Santiago de Chile: FAO; 2016

[19] Olaleke AM, Olorunfemi O, Emmanuel AT. A comparative study on the chemical and amino acid composition of some Nigerian under-utilized legume flours. Pakistan Journal of Nutrition. 2006;**5**(1): 34-38. ISSN 1680-5194

[20] Yao DN, Kouassi KN, Erba D, Scazzina F, Pellegrini N, Casiraghi MC. Nutritive evaluation of the Bambara groundnut Ci12 landrace [*Vigna subterranea* (L.) Verdc. (Fabaceae)] Produced in Côte d'Ivoire. International Journal of Molecular Sciences. 2015;**16**:21428-21441. DOI: 10.3390/ijms160921428

[21] Anonymous. Grain Composition. Lupin Food Australia; 2013. DOI: http://www.lupinfoods.com.au/grain-composition/

[22] Maphosa Y, Jideani VA. Physicochemical characteristics of Bambara Groundnut dietary fibres extracted using wet milling. South African Journal of Science. 2016;**112**(1/2):1-8. DOI: 10.17159/sajs.2016/20150126

[23] Tamang JP, Shin DH, Jung SJ, Chae SW. Functional properties of microorganisms in fermented foods. Frontiers in Microbiology. 2016;**7**:578. DOI: 10.3389/fmicb.2016.00578

[24] Karner T. Effect of palatable soluble fibre-containing carbohydrate food on postprandial blood glucose response in healthy individuals. Masters Thesis; Aarhus University. Denmark; 2016

[25] Danish Whole Grain Partnership. Whole Grain Intake Sets New Record Facts about the Food Institute's Dietary Survey; 2014. pp. 1-3

[26] Bliss DZ, Savik K, Jung HG, Whitebird R, Lowry A, Sheng X. Dietary fibre supplementation for fecal incontinence: A randomized clinical trial. Research in Nursing and Health. 2014;**37**(5):367-378. DOI: 10.1002/nur.21616

[27] Myriam M, Grundy L, Edwards CH, Mackie AR, Gidley MJ, Butterwort PJ, Ellis PR. Re-evaluation of the mechanisms of dietary fibre and implications for macronutrient bioaccessibility, digestion and postprandial metabolism. British Journal of Nutrition. 2016;**116**(5):816-833. DOI: 10.1017/S0007114516002610

[28] Bliss DZ, Weimer PJ, Jung HG, Savik K. *In vitro* degradation and fermentation of three dietary fiber sources by human colonic bacteria. Journal of Agricultural and Food Chemistry. 2013;**61**:4614-4621. DOI: 10.1021/jf3054017

[29] Brigide P, Guidolin CS, Oliveira SM. Nutritional characteristics of biofortified common beans. Food Science and Technology (Campinas). 2014;**34**(3):493-500. DOI: 10.1590/1678-457x.6245

[30] Mogobe O, Mosepele K, Masa WRL. Essential mineral content of common fish species in Chanoga, Okavango Delta, Botswana. African Journal of Food Science. 2015;**9**(9): 480-486. DOI: 10.5897/AJFS2015.1307

[31] Foodstuffs, Cosmetics and Disinfectants Act. Regulations Governing the Labelling and Advertising of Foodstuffs, Regulation No. R146. In: Foodstuffs, Cosmetics and Disinfectants Act and Regulations, 54/1972. Updated 1 March 2010. Cape Town; 1972. Johannesburg: LexNexis Butterworths

[32] Sanchez-Chino X, Jomenez-Martinez C, Davila-Ortiz G, Alvarez-Gonzalez I, Madrigal-Bujaidar E. Nutrient and non-nutrient components of legumes and its chemopreventive activity: A review. Nutrition and Cancer. 2015;**67**(3):401-410. DOI: 10.1080/01635581.2015.100472

[33] Carbonaro, M. Chemico-physical and nutritional properties of traditional legumes (lentil, *Lens culinaris* L., and grass pea, *Lathyrus sativus* L.) from organic agriculture: an explorative study. Organic Agriculture. 2015;**5**(3):179-187. DOI: 10.1007/s13165-014-0086-y

[34] Gulewicz, P, Martinez-Villaluenga C, Kasprowicz-Potocka M, Frias J. Non-nutritive compounds in fabaceae family seeds and the improvement of their nutritional quality

by traditional processing–A review. Polish Journal of Food Nutrition and Science. 2014;**64**(2):75-89. DOI: 10.2478/v10222-012-0098-9

[35] Amarowicz R, Pegg RB. Legumes as a source of natural antioxidants. European Journal of Lipid Science and Technology. 2008;**110**:865-878. DOI: 10.1002/ejlt.200800114

[36] Levetin E, McMahon K. Plants and Society. 5th ed. New York: The McGraw Hill Company. ASIN: B008UB6X7K

[37] Maphosa Y, Jideani VA. Dietary fiber extraction for human nutrition–A review. Food Reviews International. 2015;**32**(1):98-115. DOI: 10.1080/87559129.2015.1057840

[38] Murevanhema YY, Jideani VA. Potential of bambara groundnut (*Vigna subterranea* (L.) Verdc) milk as a probiotic beverage–A review. Critical Reviews in Food Science and Nutrition. 2013;**53**(9):954-967. DOI: 10.1080/10408398.2011.574803

[39] Qayyum MMN, Butt MS, Anjum FM, Nawaz H. Composition analysis of some selected legumes for protein isolates recovery. The Journal of Animal & Plant Sciences. 2012; **22**(4):1156-1162. ISSN: 1018-7081

[40] Ebert AW. Potential of underutilized traditional vegetables and legume crops to contribute to food and nutritional security, income and more sustainable production systems. Sustainability. 2014;**6**:319-335. DOI: 10.3390/su6010319

[41] Polak R, Phillips EM, Campbell A. Legumes: Health benefits and culinary approaches to increase intake. Clinical Diabetes. 2015;**33**(4):198-205. DOI: 10.2337/diaclin.33.4.198

[42] Fasoyiro S, Widodo Y, Kehinde T. Processing and utilization of legumes in the tropics. In: Eissa AA, editor. Trends in Vital Food and Control Engineering. Croatia: InTech; 2012. pp. 71-84. ISBN: 978-953-51-0449-0

[43] Gabriel EG, Jideani VA, Ikhu-omoregbe DIO. Investigation of the emulsifying properties of Bambara groundnut flour and starch. International Journal of Food Science and Engineering. 2013;**7**:539-547

Chapter 7

Effect of Bioactive Nutriments in Health and Disease: The Role of Epigenetic Modifications

Pablo Bautista-García, Lorena González-López, Berenice González-Esparza and Camila Del Castillo-Rosas

Additional information is available at the end of the chapter

http://dx.doi.org/10.5772/intechopen.68789

Abstract

Recently, a list of clinical, physiopathological, and epidemiological studies has underlined the detrimental or beneficial role of nutritional factors in some chronic diseases such as obesity, type 2 diabetes, cardiovascular disease, and cancer. It has been described that lifestyle, environmental conditions, and nutritional compounds influence gene expression. In the last instance, it has been demonstrated that bioactive nutrimental components are important signal molecules that carry information from the external environment and could affect in biological terms, processes related to gene expression. Bioactive nutriments can work in different ways: regulating the chromatin structure or factors that directly regulate the activity of nuclear receptors. The relevance of the changes in the chromatin structure has been demonstrated by the fact that many chronic diseases and metabolic disorders are related with changes in DNA methylation patterns. For this reason, recently, the bioactive food nutriments have been investigated to characterize the molecular mechanism involved in changes of the chromatin structure, such as acetylation and methylation, and their potential benefit on chronic diseases. The dietary compounds intake involved in the regulation of epigenetic modifications can provide significant health effects and may prevent various pathological processes involved in the development of cancer and other serious diseases.

Keywords: bioactive nutriments, epigenomic changes, obesity, diabetes mellitus, carcinogenesis

1. Introduction

Bioactive food nutriments are constituents provided in food or dietary supplements; those have been characterized as biomolecules and have the capacity to regulate a myriad of metabolic processes in the body resulting in health benefits. In contrast, overload intake of bioactive nutriments can either be involved in the development of various stages of disease or may change the natural history of a disease. For this reason, the knowledge of these biological functional features can be applied in the treatment and prevention of human diseases.

Currently, advancements in biological and medical science have allowed a better understanding of physiopathological bases of disease, as well as identify the role of several bioactive components in food under metabolic processes. The development of new technologies have provided analytical and molecular tools for discerning the intricate relationship between a myriad of signaling pathways linked to pathological processes. The results have been useful to evaluate a vast numbers of food components and their role in disease prevention and health promotion.

Bioactive food nutriments can be provided in daily diet in many forms. Some of them can be found in conventional foods and others can be added to fortified foods, these kinds of supplements have been designed to reduce disease risk in special human groups with nutrimental deficiencies [1].

During decades, physicians and nutritionists have adopted nutritional guides, where they can find punctual information about nutritional recommendation for a large list of nutriments. However, the availability of nutritional guides for bioactive nutriments compounds is restricted because; these need more elements to evaluate dietary recommendation. One of the most important requirements to recommend a bioactive nutriment compound is based on the result obtained in clinical and experimental studies; this data must contain scientific evidence that shows a relationship between the bioactive nutriment compound and a beneficial health impact. In the same sense, other element that must be considered to choose a bioactive nutriment compound is whether the bioactive product exhibit side effects upon exposition.

For this purpose, researches must develop accurate biochemical markers to validate either the safety or hazardous effects of food intake for human, and finally physicians and nutritionist will decide the correct doses for each bioactive nutriment component, depending on many factors, such as sex, age, pregnancy, health, or pathological condition [2].

2. Bioactive nutriments in health: the role of epigenetic modifications

Human homeostasis is influenced by molecular signal pathways, exogenous factors, and diet habits. It has been demonstrated that bioactive nutriments have substantive impact on health and disease. A biological area that describes the molecular effect of certain nutriments on DNA expression is "Epigenomics," which can be defined as the study of the complete set of epigenetic modifications in a cell or in a tissue at a given time. The epigenome consists of chemical compounds that modify or mark the genome in such a way that can indicate how and when a specific set of genes are expressed in a cell or in a tissue, enhancing or inhibiting the production of a specific protein in a cell. These chemical modifications on DNA or histones have been characterized as "epigenetic marks" [3].

The epigenetic modifications are targeted to DNA or histones (DNA associated proteins), which induce modifications in chromatin without affecting the nucleotide sequence; these structural changes could modify the expression patterns of gene expression; however, these molecular modifications can be slow but progressive and potentially reversible. When epigenomic compounds attach to DNA and modify its structure and its transcriptional activity, they "marked" the genome. The biological transcendence of these marks is not to change the sequence of the DNA, conversely they change the way cells use the DNA's instructions. The marks are sometimes passed on from cell to cell as cells divide. They also can be passed down from one generation to the next.

The first type of mark, called DNA methylation, directly affects the DNA in a genome. In this process, a set of proteins chemically tag with methyl groups the DNA bases in specific places. The methyl groups can make DNA either more or less accessible to transcriptional apparatus, thus changing the expression patterns of specific genes.

The second kind of mark, characterized as histone modification, affects DNA indirectly because in this case DNA remains intact but the chemical modifications in histones affect the way in which DNA is wrapped around histone proteins, thus affecting the DNA structures and in consequence, the transcriptional activity of many proteins (**Figure 1**) [4].

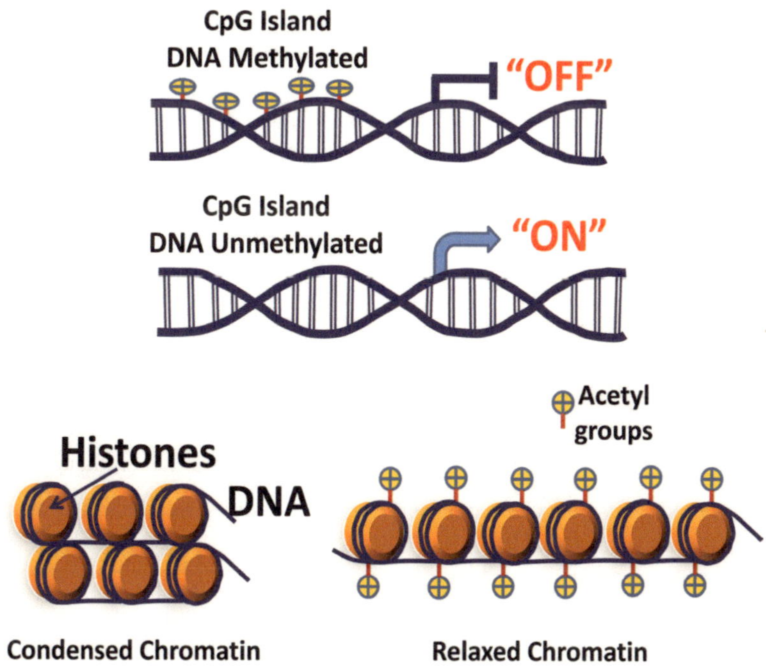

Figure 1. Activation/repression of DNA induced by epigenomic changes.

In the following paragraphs, we describe an increasing number of evidences that show how bioactive nutriments compounds can modify methylation patterns by interacting with enzymes that are able to place epigenetic marks on DNA or enhance the expression of specific proteins implicated in the formation of the epigenetic machinery.

2.1. Folates

Folate and folic acid are the forms of a water-soluble vitamin B; this can be obtained naturally in daily diet or in fortified foods and supplements. Sources include cereals, baked goods, spinach, broccoli, lettuce, asparagus, bananas, melons, lemons, legumes, yeast, mushrooms, beef liver, kidney, orange juice, and tomato juice. Folic acid supplements are effective for increasing folate levels in blood and decreasing symptoms associated with low folate levels. These kinds of supplements are prescribed for use in pregnancy women in order to prevent neural tube defects.

Folate is involved in DNA synthesis, repair, and methylation. After dietary ingestion, this compound undergoes many chemical reactions and is primary converted to tetrahydrofolate which is involved in the remethylation of homocysteine to methionine [5]. The relevance of this chemical reaction is that methionine is a precursor of SAM (S-adenosyl-L-methionine), the primary methyl donor group for most methylation reactions [6]. After transferring a methyl group, SAM is converted to S-adenosyl-L-homocysteine (SAH), an inhibitor of the methylation reactions.

This chemical event seems to be of particular relevance, because in the development of digestive neoplastic lesions related to folate deficiency may be involved in changes of the DNA methylation pattern in specific proto-oncogenes, such as c-myc, c-fos, and c-Ha-ras [7]. In all cases the malignant transformation was related to a significant decrease of SAM and global DNA hypomethylation, especially in DNA sequences where oncogenes are codified. In contrast, folic acid supplementation improved folate-related DNA biomarkers of cancer risk in colonic tissues adjacent to the former polyp site (**Figure 2**) [8].

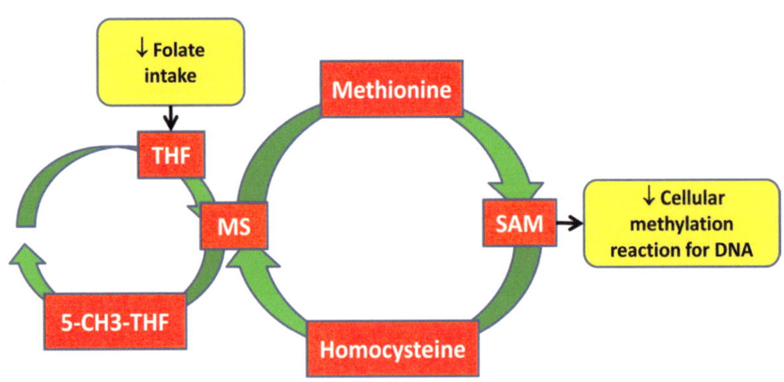

Figure 2. Association between folate deficiency and DNA methylation process. THF, tetrahydrofolate; MS, methionine synthase; 5-CH_3-THF, 5-methyltetrahydrofolate; SAM, S-adenosyl-L.

Paradoxically, hypermethylation was induced in DNA sequences coding for tumor suppressor genes. The changes in the methylation processes exerted by an increase in DNMTs (DNA methyltransferases) activity may explain the hypermethylation observed in these experimental models, whereas the stimulation of MBD2 and MBD4 (methyl-CpG-binding domain proteins) may explain the decrease on DNA methylation favoring the expression of oncogenes and prometastatic genes [9, 10].

The above mentioned data indicates that current nutritional recommendations of folate in daily diet must be considered more than a simple nutriment; it must be also considered an indispensable bioactive compound to avoid at least in some degree the aberrant expression of proto-oncogenes in many cellular contexts, thus decreasing the incidence of neoplastic process.

2.2. Vitamin A

Vitamin A is the name of a group of fat-soluble retinoids, including retinol, retinal, and retinyl esters. It is involved in many physiological functions, including: immune function, vision, reproduction, and cellular communication processes. Vitamin A also supports cell growth and differentiation, playing a critical role in the normal organogenesis and maintenance of heart, lungs, and kidneys functions. Preformed vitamin A is found in dark green and yellow vegetables, and yellow fruits, such as broccoli spinach, turnip greens, carrots, squash, sweet potatoes, pumpkin, cantaloupe, apricots, and food animal sources, including fish and meat. It must be metabolized intracellularly into retinal and retinoic acid, the active forms of vitamin A, to support the vitamin's physiological functions [11].

Once absorbed, these bioactive compounds are translocated to the nucleus where they bind to specific nuclear Retinoic Acid Receptors (RARs), which have been characterized as RARα, β, and γ that heterodimerize with Retinoid X Receptors (RXRs). The molecular complex binds to specific response elements and downregulates transcriptional activity of many genes, which includes AP-1 gene. This gene is involved in mediating transcriptional responses to many biological, pharmacological, or stress stimuli. Even more, AP-1 regulates the expression of several molecular mediators involved in oncogenic transformation and cellular proliferation and plays a regulatory role in S phase DNA replication and DNA damage repair [12].

Once p21 and AP-1 are activated by retinoids, the proteins encoded by these genes can interact with many proteins involved in DNA methylation changes, for example, p21 is able to compete with DNMT1 substrates for the same binding site on Proliferating Cell nuclear Antigen (PCNA), then affecting DNMT1 activity and its DNA methylation efficiency (see **Figure 4**) [13, 14]. Meanwhile, the mechanism for AP-1 involved its binding to the promoter of the DNMT1 regulatory region inducing the expression of DNMT1, favoring DNA methylation [15].

The biological transcendence exhibited by p21 and AP-1 expression induced by retinoids is the downregulation of enzymes that enhance DNA methylation events, which may contribute to increase the expression patterns of genes involved in antiproliferative, differentiating, and proapoptotic actions reducing the incidence of many types of cancers [16, 17].

Indeed, recently it has been demonstrated the antitumoral effect exerted by derivative of all trans retinoic acid in cellular cultures of human gastric cancer, which is related to the ability of these compounds to induce cycle cell arrest and cellular differentiation [18].

2.3. Vitamin D3

Vitamin D is found in many foods, including fish, eggs, fortified milk, and cod liver oil. However, Vitamin D can be also obtained by few minutes of sun exposition. There are several different forms of vitamin D. Two forms are important in humans: vitamin D2, which is made by plants, and vitamin D3, which is made by human skin when exposed to sunlight [19].

Although for VitD3, one of the most known physiological effects is the preservation of the calcium homeostasis. Currently, it has been explored other mechanisms not linked to calcium metabolism. In this sense, once VitD3 is converted into its active form (calcitriol), the biological actions of this vitamin share similar mechanisms to RA, because it must bind to specific vitamin D receptors (VDR), establishing homodimers, or heterodimers with RXR or RAR, and affect gene transcription through VDR responsive elements (VDRE) in target genes, such as p21 and PTE; this protein specifically catalyzes the dephosphorylation of the 3′ phosphate. This dephosphorylation is important because it results in inhibition of the AKT signaling pathway. Meanwhile, its weak protein phosphatase activity is also crucial for its role as a tumor suppressor, preventing cells from growing and dividing [20].

Bioactive nutriments	Natural sources	Antineoplastic effects	Epigenetic mechanisms of action
Folate,	spinach, asparagus, beans, peas, lentils, almonds	Anti-cancer, chemoprevention of malignant transformation	Regulation of SAM/SAH ratio, DNMT and MBD expression; regulation of tumor supressor miRNAs and oncogenic miRNAs
Retinoic acid	Mango, papaya, carrots, spinach, sweet potatoes	Anti-cancer, differentiating, pro-apoptotic	Regulation of DNMTs expression and activity, regulation of miRNAs targeting DNMTs; regulation of tumour suppressor miRNAs and oncogenic miRNAs; GNMT regulation; histone acetylation
Vitamin D3	Sun exposure, fish, fish liver oils	Anti-cancer, differentiating, pro-apoptotic	Regulation of DNMTs expression and enzyme activity; regulation of histone acetylation; regulation of oncogenic miRNAs
Resveratrol	Grapes, mulberries, apricots, pineapples, peanuts	Anti-cancer, antioxidant, anti-angiogenesis, pro-apoptotic	Regulation of DNMTs expression and enzyme activity; activation of deacetylase SIRT1 and p300 HAT; down-regulation of UHRF1; regulation of miRNAs
EGCG	Green tea	Anti-cancer, antioxidant, anti-angiogenesis, pro-apoptotic	Regulation of SAM/SAH ratio by COMT-mediated reactions; direct inhibition of DNMTs by binding to catalytic domain of the enzyme; regulation of tumour suppressor miRNAs
Curcumin	Spice turmeric	Anti-cancer, antioxidant, protects against heart failure	Direct inhibition of DNMTs by binding to catalytic domain of the enzyme; inhibition of HDACs and p300 HAT; regulation of tumour suppressor miRNAs and oncogenic miRNAs

Table 1. Epigenomic roles of bioactive nutriments.

In a similar way as for retinoid, the biological effect of VitD3 in cancer is linked to the ability of p21 to downregulate the activity of DNMT1 enzymes, which can modify the DNA methylation patterns of certain protective genes conferring an antitumoral role as was demonstrated for colon cancer [21] and more recently for metastatic castration-resistant prostate cancer patients (**Table 1**) [22].

One important fact is that in industrialized countries VitD3 intake is generally linked to calcium homeostasis but its role in the prevention of cancer development by epigenetic mechanisms is commonly unknown.

The date mentioned above can represent a new challenge for physicians and nutritionists to develop new strategies to raise awareness about the biological properties provided by many bioactive nutriments in the daily diet of the general population. This may contribute to reduce the incidence of most common types of cancer.

3. Nutrimentsv linked to disease: the role of epigenetic mechanism

3.1. High fat diet and induction of obesity

As mentioned before, we showed the beneficial effects of bioactive nutriments in health. In contrast, it has been demonstrated that the overfeeding of many of these nutriments can also participate in the evolution of several diseases. In this sense, there are lines of evidence that had proved the existence of obesity-genes. These genes are critical for energy balance and can be regulated by epigenetic mechanisms depending on nutritional environment conditions [23, 24]. For example, it has been proved that a long-term exposition to high fat diet in mice, MC4R promoter gene undergoes a reduced methylation in the brain of mice, promoting the fat storage and obesity [25].

In addition, it has been demonstrated that other genes may potentiate the effect of MC4R. For example, it has been shown that under a high-fat diet the methylation state of the Proopiomelanocortin (POMC) promoter can be modified, thus changing the correct balance between energy taken from the food and energy spent by the body, favoring obesity. It has been shown that proopiomelanocortin (POMC) deficiency causes severe obesity that begins at an early age. Affected infants usually have a normal weight at birth, but they are constantly hungry. Affected individuals experience excessive hunger and remain obese for life. It is unclear if these individuals are prone to weight-related conditions like cardiovascular disease or type 2 diabetes. Thus, changing the correct balance between energy taken from the food and energy spent by the body, favoring obesity [26].

The methylation changes observed in gene promoters involved in energy balance induced by long-term exposition to a fat diet in western countries, may explain the high incidence of obesity and metabolic diseases, which may be potentially prevented by healthy diet habits and exercise. Thereby, the current treatment of obesity must also consider the epigenetic effects on obesity genes exerted by overfeeding, more than considering surgery as first-line treatment, which only avoids the absorption of overcharged nutriments, but do not have any effect on the intrinsic mechanism of obesity.

4. Epigenetic changes associated with diabetic complications

Diabetes is a group of metabolic diseases characterized by hyperglycemia resulting from defects in insulin secretion, insulin action, or both. The chronic hyperglycemia of diabetes is associated with long-term damage, dysfunction, and failure of different organs, especially the eyes, kidneys, nerves, heart, and blood vessels [27].

One of the most frequent diabetes complications is diabetic vasculopathy, which is characterized by a vascular inflammation process. Recent studies have proposed that hyperglycemia may produce epigenetic modifications of specific genes involved in vascular inflammation. One of them is the transcription factor, Nuclear Factor-kB (NF-kB), which regulates the expression of a large list of genes who participate in inflammatory diseases, such as atherosclerosis and diabetic complications.

It has been demonstrated *in vitro* experiments that hyperglycemia increases NF-kB activity in monocytes thus enhancing gene expression of inflammatory cytokines. This step is the result of molecular interaction between the transcription factor (NF-kB) and histone acetyltransferases (HATs), resulting in hyperacetylation of target genes including the tumor necrosis factor (TNF)-and cyclooxygenase-2 promoters [28].

The data may suggest that the uncontrolled hyperglycemia in diabetic patients may produce epigenetic changes in specific genomic region which control the expression of proinflammatory genes, and subsequently the development of vascular inflammation. However, the control of hyperglycemia in patient is not enough to reduce the risk of diabetic complication because in this patient the risk of diabetic vasculopathy was not modified. The mechanism involved is that persistent hyperglycemia may induce "epigenetic marks" in proinflammatory promoters, thus enhancing the persistent expression of proinflammatory genes despite diabetic control. This finding suggests that the epigenetic modifications induced by long-term hyperglycemia may persist for a long time. These data must be considered in the future for the design of new strategies to decrease persistent hyperglycemia in diabetic patients and avoid the appearance of "epigenetic marks," which are associated with the development of diabetic complication [29, 30, 31].

5. Epigenetic effects of bioactive compounds in evolution of cancer

Cancer is the result to prolonged exposure to many carcinogenic factors, such as radiations, chemical substances, and prolonged exposure to sun. In industrialized countries the higher prevalence of cancer diseases is major in elderly people [32].

http://dx.doi.org/10.5772/intechopen.68789

It has been shown that cancer cells do not belong to a unique cellular lineage because into a malignant tumor or among the circulating cancerous cells, there can be a diversity of types of cells. Recently, it has been described a stem cell theory of cancer that proposes that among all cancerous cells, a few act as stem cells that reproduce themselves and sustain the cancer, much like normal stem cells that normally renew and sustain our organs and tissues. The idea that cancer is primarily driven by a smaller population of stem cells has important biological and clinical implications. Currently, many new anticancer therapies are evaluated based on their ability to decrease or eliminate tumors, but if the therapies are not killing the cancer stem cells, the tumor will soon grow back as well as the clinical symptoms. Therefore, if the special subpopulations of tumor cells characterized as "cancer stem cells" are destroyed, a full recovery is possible. Consequently, the new cancer therapy will be target to abolish or decrease the self-renewal capabilities of this subpopulation of cancer cells [33, 34].

The Wnt/β-catenin signaling pathway is one of the most conserved intercellular signaling cascades. Its pathway begins when a Wnt protein binds to the *N*-terminal extracellular cysteine-rich domain of a frizzled (Fz) family receptor. However, to facilitate Wnt signaling, coreceptors such as lipoprotein receptor-related protein-5/6 (LRP)-5/6) may be required alongside the interaction between the Wnt protein and Fz receptor. Upon activation of the receptor, a signal is conducted to the phosphoprotein disheveled (Dsh), which is located in the cytoplasm. Cytoplasmic β-catenin levels are normally kept low through continuous proteasome complex-mediated degradation (adenomatous polyposis coli (APC)/glycogen synthase kinase-3β (GSK-3β); however, when cells receive Wnt signals, the degradation of β-catenin is inhibited and levels of β-catenin build up in the cytoplasm and nucleus. Then, nuclear β-catenin interacts with transcription factors, such as T-cell factor/lymphoid enhancer-binding factor (Tcf/Lef) which is a transcription regulator for several genes that, in part, regulates tumor progression [35].

However, Wnt signaling is not only restricted to control self-renewal of stem cells in normal microenvironment, also this pathway is particularly active in a limited subpopulation of cells that display cancer stem properties.

The mechanism proposed for such effect is that once nuclear β-catenins are activated they could interact with transcription factors, such as T-cell factor/lymphoid enhancer-binding factor (Tcf/Lef) and increase the transcriptional activity of several genes involved in tumoral progression [36]. The biological significance of Wnt/β-catenin pathway in cancer was evidenced by the fact that in many neoplastic diseases (prostate, colon, and skin cancer) mutations have been detected in some Wnt-downstream effectors [37–39].

Currently, it has been explored many bioactive nutriments in cancer treatment due to their less toxic effects, as well as their property to exhibit less adverse effects compared to conventional antineoplastic drugs. The biological transcendence of many bioactive nutriments such as flavonoids, curcumin, green tea polyphenols, resveratrol, and lupeol lies in the fact that these compounds are able to disrupt β-catenin–mediated Wnt signaling (**Figure 3**). Their biological properties are mentioned below:

The flavonoids comprise a large class of low-molecular weight natural products of plant origin ubiquitously distributed in foods; many studies have demonstrated that these compounds upregulate the expression and activity of GSK-3β, an essential component of the Wnt/β-catenin pathway. GSK-3β is a kinase that phosphorylates β-catenin for its eventual degradation in cytoplasm, thus inhibiting the signaling linked to Wnt receptor activation. Thereby, the use of flavonoids in cancer may potentially inactivate Wnt/β-catenin signaling reducing the proliferation index in prostate cancer cells [40]. Unlike flavonois, curcumin (major component of turmeric and a member of the ginger curcuma longa) exhibits a different mechanism, which is based on constrain the transcriptional activity of Wnt target genes, such as c-myc, c-fos, c-jun, and iNOS, inhibiting cell proliferation and inducing apoptosis in human breast cancer cells [41].

Other important effect linked to curcumin administration is a dose-dependent decrease in expression of the nuclear p300 coactivator; p300 and CBP are thought to increase gene expression by relaxing the chromatin structure at the gene promoter through their intrinsic histone acetyltransferase (HAT) activity, recruiting the basal transcriptional machinery including RNA polymerase II to the promoter acting as adaptor molecules [42]. This is especially significant since nuclear beta-catenin forms a complex between Tcf4 and an important histone acetyltransferase mediator (p300 coactivator). This molecular interaction may change the DNA methylation patterns in Wnt target genes decreasing its transcriptional activity and consequently decreasing the tumoral progression [43].

Green tea polyphenols intake is linked to beneficial effects in many cancer cells, the mechanism involved is the ability of the polyphenols to downregulate β-catenin expression and consequently β-catenin/Tcf target genes (c-jun and cyclin D1). The clinical transcendence of these findings is that green tea intake may change the aberrant progression of many neoplastic during its early stages and consequently modify the clinical prognosis of the

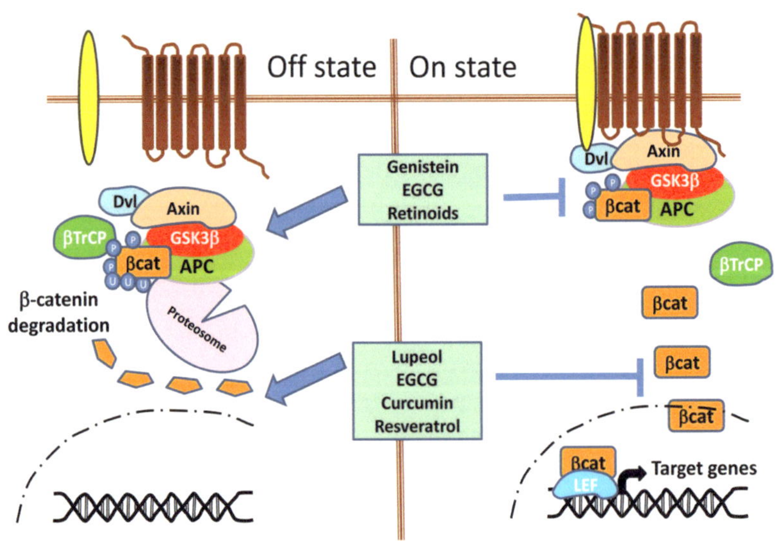

Figure 3. Effects of bioactive nutriments on Wnt/β-catenin pathway.

http://dx.doi.org/10.5772/intechopen.68789

disease [44]. However, the beneficial effects of green tea are not restricted to bowel cancer; it has been shown that Wnt/β-catenin signaling can be inhibited by polyphenols in a dose-dependent manner in breast cancer cells [45]. The beneficial role of green tea compounds make them excellent candidates to bioactive antineoplastic drugs in many tumor contexts without the adverse effects exhibited by conventional drugs (**Figure 4**).

Resveratrol, a dietary polyphenol can be provided by roots of hellebore, grapes, mulberries, apricots, pineapples, and peanuts. Its role as antineoplastic agent is associated with the reduced expression of a long noncoding metastasis associated lung adenocarcinoma transcript 1 (RNA-MALAT1), thus decreasing the amount and proportion of β-catenin in the nucleus in colon cancer cells [46]. In addition, the role of resveratrol is not only restricted to solid tumor, but it can also inhibit proliferation and induce cell cycle arrest and apoptosis in Waldenstrom's macroglobulinemia cells. These effects of resveratrol were found to be mediated via the downregulation of Akt, mitogen-activated protein kinase (MAPK), and Wnt signaling pathways [47]. Meanwhile, lupeol, a well-studied dietary triterpene found in several fruits (olives, figs, mangoes, strawberries, and grapes) and vegetables (green peppers, white cabbage, and tomato) has shown a significant growth inhibition role on melanoma cells that exhibit constitutive Wnt/β-catenin signaling decreasing its neoplastic potential [48].

Further, Fan et al. demonstrated that lupeol inhibits the localization of β-catenin into the nucleus and decreases the phosphorylation status of β-catenin at important serine sites (ser 552 and ser 675), which are the signals for their translocation into the nucleus and induce transcription of various downstream targets linked to neoplastic processes [49].

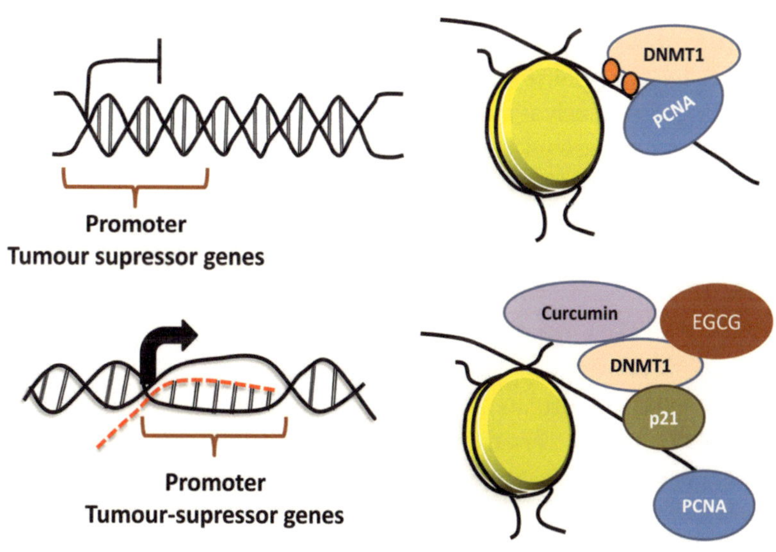

Figure 4. Antineoplastic effects of bioactive nutriments.

6. Novelty technologies for obtaining and delivering bioactive compounds for health and medical therapy

After identifying a potential new bioactive component, it is necessary to evaluate many factors for its availability, such as the efficacy and safety of the product, select the appropriate food vehicle, ensure the bioavailability, and accuracy of health claims, and finally ensure that during the process of synthesis, stabilization, and processing the bioactive product does not lose its biological properties.

Recently, a great effort has been performed to develop novel procedures to synthetize new bioactive formulations that can overcome poor bioavailability, stability limitations, and rapid metabolism of bioactive compounds.

In this sense, novel technologies have been developed to improve the process to obtain biological compounds at low cost, as well as new procedures to deliver these bioactive products in tissues to enhance their biological effects.

One of these novel procedures is the delivery of bioactive compounds by microorganisms; these procedures take the ability of microbiota to deliver bioactive compounds contained in dairy diet, which cannot be processed by digestive human enzymes. These microorganisms produce a set of digestive enzymes that overcome the human ability to entirely digest the biocomponents encrypted in diverse food matrixes. Lactic acid bacteria (LAB) have been chosen due to their property to release, almost completely, the bioactive compounds from food matrix.

LABs are ancient microorganisms adapted to anoxic conditions, but their functional capabilities to synthetize micronutriments are almost absent. Therefore, LAB evolved a very efficient proteolytic system, which allows them to release encrypted biomolecules present in different food matrices (alpha- and beta-caseins, albumin, and globulin from milk, rubisco from spinach, beta-conglycinin from soy, and gluten from cereals), which are linked to a myriad of physiological functions, such as mineral absorption, adaptive response to oxidative stress, hypoglycemic actions, cholesterol lowering, cardiovascular functions, and a highlight effect related to the control of food intake [50]. The bioenzymatic properties exhibited by LAB rise them as excellent candidates to be added to processed food to ensure the delivery of bioactive molecules encrypted in food matrix, which in normal conditions are not accessible to human proteolytic enzymes.

On the other hand, the potential benefits of nano-technology have been recognized by food industry sectors by its potential application, which include the development of nano-sensors, smart packaging, nano-encapsulation, and delivery of food compounds. However, nanotechnology can also be used to encapsulate in nano-emulsions many bioactive compounds to increase their bioavailability, stability, and reduce their biodegradation. Examples of ingredients encapsulated in nano-emulsion are: minerals, vitamins, enzymes, and bioactive ingredients. In this sense, currently it has been explored the use of an ROS-responsive polymeric nano-particles for efficient Cur delivery into cancer cells. This nano-system improves

Cur stability at physiological environment and enhances the Cur release in response to hydrogen peroxide. Both mechanisms displayed an antitumoral effect in a cellular culture of lung cancer. Thereby, the use of nano-technology to deliver bioactive compounds may have a potential application in medicine to improve the cancer treatment without the adverse effect observed in conventional drugs currently available [51].

Quercetin is a major constituent of various dietary products and recently its anticancer potential has been extensively explored, revealing its antiproliferative effect on different cancer cell lines. However, its medical applications are limited due to its low oral bioavailability, rapid clearance from body, high metabolic rate, and poor aqueous solubility.

Therefore, to overcome these biological disadvantages, novel quercetin-based nano-formulations are being developed due to their properties of bioavailability, gut absorption, and their capability to increase quercetin biological half-time in serum. The pharmacological effect of quercetin loaded/conjugated nano-particles majorly depends on the drug carriers used and the physicochemical properties of the nano-particulate system. These characteristics can increase the stability of quercetin, its bioavailability, and target specificity [52, 53]. However, the medical application of quercetin nano-particles is still under investigation likely due to the necessity of more stable and target-specific nano-particles.

Indeed, it has been explored other delivery system based on different matrix where bioactive compounds are encapsulated within PLGA (poly lactic-co-glycolic acid) and PLA (poly D,L-lactic acid) nano-particles. For example, it was observed a significant cytotoxic effect of quercetin encapsulated PLGA nano-particles in combination with ectopside-loaded PLGA nano-particles in a human lung adenocarcinoma epithelial cell line. Similarly, a significant reduction of breast cancer cells upon treatment with PLA-quercetin was shown, which support the clinical use of these novel technologies for cancer treatment [54]. Thereby, the nanotechnology can be used as a powerful tool to overcome the biochemical and physiological limitations of bioactive compounds, improve many pharmacodynamics parameters, and potentiate the pharmacological and functional effects exhibited by these compounds.

7. Conclusions

Although bioactive nutriments compounds have shown potential health benefits, currently, there are no nutritional guidelines to recommend intake levels as there are for other nutriments. The challenge for the future will be the establishment of nutritional recommendations for each bioactive food components. These kinds of products should not be only taken as vital nutriments, but also, they should be considered important molecules that depending on the nutritional conditions or cellular environments can modify the DNA methylation patterns and change the way that DNA is transcribed. The data provided show that the intake of bioactive products, provided in daily diet, may have a dual role depending on the load ingest, maintaining homeostasis at recommended levels, or induce the appearance of disease when are taken to overdoses. However, the most important role exhibited by bioactive nutriments is their antineoplastic effect, which depends on their molecular interactions with enzymes that modify the DNA structure or as methyl or acetyl donors to DNA or histones.

The clinical transcendence of the molecular effects displayed by bioactive nutriments is that they are exempt of the adverse effects of conventional antineoplastic drugs, which make them excellent candidates for cancer treatment in the future. Thereby, food companies must direct their effort to the development of novel and low cost processes to ensure an adequate bioactive concentration in the diverse food presentations of industrialized food, or these food preparations may be added with a special set of microorganisms to enhance the delivery of bioactive compounds encrypted in food matrixes.

Currently, novel nano-particles systems to carry bioactive compounds have been evaluated. These systems were designed to avoid the accelerated metabolism that bio compounds undergo along the gastrointestinal tract. Thereby, pharmaceutical industries must direct their efforts to design novel technologies to ensure the bioavailability of many bioactive compounds linked to antitumoral activity, preserving their biological activity in the affected tissue. Both strategies may have an enormous economic impact on pharmaceutical and food industries, lowering the production cost of antineoplastic drugs, and decrease the cytotoxic effect displayed by the actual antineoplastic conventional drugs. These novel technologies may also be useful to prevent the high incidence of cancer in general population providing an accurate concentration of bioactive compounds in industrialized food with the corresponding impact on social medical security, decreasing the economical inversion to cancer treatment in affected population.

Acknowledgements

This review was supported by the Centro de Investigación en Ciencias de la Salud (CICSA), Facultad de Ciencias de la Salud, Universidad Anáhuac México Campus Norte, Hiuxquilucan, Edo. de México, Mexico.

Author details

Pablo Bautista-García[1*], Lorena González-López[2], Berenice González-Esparza[3] and Camila Del Castillo-Rosas[1]

*Address all correspondence to: pablobautista1973@hotmail.com

1 Center for Research in Health Science (CICSA), Faculty of Science Health, Anáhuac University North Campus, Huixquilucan, Mexico

2 Center for Research and Advanced Studies (CINESTAV), Mexico City, Mexico

3 Technologic University of Mexico, Mexico City, Mexico

References

[1] Abuajah CI, Ogbonna AC, Osuji CM. Functional components and medicinal properties of food: A review. Journal of Food Science and Technology. 2015;**52**(5)2522-2529. DOI: 10.1007/s13197-014-1396-5

[2] Balentine DA, Erdman J, Courtney GP, Dwyer JT, Ellwood KC, Hu FB, Rusell RM. Are dietary bioactive ready for recommended intakes? Advances in Nutrition. 2013;**4**:539-541. DOI: 10.3945/an.113.004226

[3] Mazzio EA, Soliman KF. Epigenetics and nutritional environmental signals. Integrative and Comparative Biology. 2014;**54**(1):21-30. DOI: 10.1093/icb/icu049

[4] Ludwig AK, Zhang P, Cardoso MC. Modifiers and readers of DNA modifications and their impact on genome structure, expression, and stability in disease. Frontiers in Genetics. 2016;**7**:115. DOI: 10.3389/fgene.2016.00115

[5] Selhub J. Folate, vitamin B12 and vitamin B6 and one carbon metabolism. The Journal of Nutrition Health and Aging. 2002;**6**:39-42. DOI: PMID 11813080

[6] Shorter KR, Felder MR, Vrana PB. Consequences of dietary methyl donor supplements: Is more always better?. Progress in Biophysics & Molecular Biology. 2015;**118**(1-2):14-20. DOI: 10.1016/j.pbiomolbio.2015.03.007

[7] Duthie SJ, Narayanan S, Sharp L, Little J, Basten G, Power H. Folate, DNA stability and colo-rectal neoplasia. Proceedings of the Nutrition Society. 2004;**63**:571-578. DOI: 10.1079/PNS2004387

[8] O'Reilly SL, McGlynn AP, McNulty H, Reynolds J, Wasson GR, Molloy AM, Strain JJ, Weir DG, Ward M, McKerr G, Scott JM, Downes CS. Folic acid supplementation in post-polypectomy patients in a randomized controlled trial increases tissue folate concentrations and reduces aberrant DNA biomarkers in colonic tissues adjacent to the former polyp site. Journal of Nutrition. 2016;**146**(5):933-939. DOI: 10.3945/jn.115.222547

[9] Wu X, Xu W, Zhou T, Cao N, Ni J, Zou T, Liang Z, Wang X, Fenech M. The role of genetic polymorphisms as related to one-carbon metabolism, Vitamin B6, and gene-nutrient interactions in maintaining genomic stability and cell viability in Chinese breast cancer patients. International Journal of Molecular Sciences. 2016;**17**(7):1-24. pii: E1003. DOI: 10.3390/ijms17071003

[10] Detich N, Theberge J, Szyf M. Promoter-specific activation and demethylation by MBD2/demethylase. Journal of Biological Chemistry. 2002;**277**:35791-35794. DOI: 10.1074/jbc.C200408200

[11] Ball GFM. Vitamin A: Retinoids and carotenoids. In: Blackwell Publishing Ltd, editor. Vitamins: Their Role in the Human Body. 1st ed. UK: Oxford; 2004:133-187. DOI: ISBN 0-632-06478-1

[12] Benkoussa M, Braqnd C, Delmontte MH, Formstecher P. Retinoic acid receptors inhibit AP1 activation by regulating extracellular signal-regulated kinase and CBP recruitment to an AP1-responsive promoter. Molecular and Cellular Biology. 2002;**22**:4522-4534. DOI: 10.1128/MCB.22.13.4522-4534.2002

[13] Milutinovic S, Knox JD, Szyf M. DNA methytransferase inhibition induces the transcription of tumour suppressor p21 (WAF1/CIP1/sdi1). Journal of Biological Chemistry. 2000;**275**:6353-6359. DOI: 10.1074/jbc.275.9.6353

[14] Iida T, Suetake I, tajima S, Morioka H, Ohta S, Obuse C. PCNA clamp facilitates action of DNA cytosine methyltransferase 1 on hemimethylade DNA. Genes to Cells. 2002;**7**:997-1007. DOI: 10.1046/j.1365-2443.2002.00584.x

[15] Bigey P, Ramchandani S, Theberge J, Araujo FD, Szyf M. Transcriptional regulation of the human DNA methyltransferase (dnmt1) gene. Gene. 2000;**242**:407-418. DOI: 10.1016/S0378-1119(99)00501-6

[16] Pitha-Rowe I, Petty WJ, Kitareewan S, Dmitrovsky E. Retinoid target gen in acute promyelocytic leukemia. Leukemia. 2003;**17**:1723-1730. DOI: 10.108/sj.leu.2403065

[17] Kolb EA, Meshinchi S. Acute myeloid leukaemia in children and adolescents: Identification of new molecular targets brings promise of new therapies. Haematology American Society of Hematology Education Program. 2015;**2015**:507-513. DOI: 10.1182/asheducation-2015.1.507

[18] Xia Q, Zhao Y, Wang J, Qiao W, Zhang D, Yin H, Xu D, Chen F. Proteomic analysis of cell cycle arrest and differentiation induction caused by ATPR, a derivative of all-trans retinoic acid, in human gastric cancer SGC-7901 cells. Prot. Clin. Appl. 2017, 1600099

[19] Bouillon R, Suda T. Vitamin D: Calcium and bone homeostasis during evolution. Bonekey Report. 2014;**3**:480. DOI: 10.1038/bonekey.2013.214

[20] Saramaki A, Diermeier S, Kellner R, Laitinen H, Vaisanen S, Calberg C. Cyclical chromatin looping and transcription factor association on the regulatory regions of p21 (CDKN1A) gene in response to 1 alpha,25-dihydroxyvitamin D3. Journal of Biological Chemistry. 2009;**284**:8073-8082. DOI: 10.1074/jbc.M808090200

[21] Picotto G, Liaudat AC, Bohl L, Tolosa de Talamoni N. Molecular aspects of Vitamin D anticancer activity. Cancer Investigation. 2012;**30**(8):604-614. DOI: 10.3109/0735907.2012.721039

[22] Ben-Eltriki M, Deb S, Adomat H, Tomlinson Guns ES. Calcitriol and 20(S)-protopanaxadiol synergistically inhibit growth and induce apoptosis in human prostate cancer cells. The Journal of Steroid Biochemistry and Molecular Biology. 2016;**158**:207-219. DOI: 10.1016/j.jsbmb.2015.12.002

[23] Gerken T, Girard CA, Tung YC, Webby CJ, Saudek V, Hewitson KS. The obesity-associated FTO gene encodes a 2-oxoglutarate-dependent nucleic acid demethylase. Science. 2007;**318**:1469-1472. DOI: 10.1126/science.1151710

[24] Widiker S, Karst S, Wagener A, Brockmann GA. High-fat diet leads to a decreased methylation of the MC4R gene in the obese BFMI and lean B6 mouse lines. Journal of Applied Genetics. 2010;**51**:193-197. DOI: 10.1007/BF03195727

[25] Palou M, Pico C, McKay JA, Sánchez J, Priego T, Mathers JC, Palou A. Protective effects of leptin during the sucking period against later obesity may be associated with changes in promoter methylation of the hypothalamic pro-opiomelanocortin gene. British Journal of Nutrition. 2011;**106**:769-778. DOI: 10.1017/S000714511000973

[26] Milagro FI, Campion J, García-Dïaz DF, Goyenechea E, Paternain L, Martínez JA. High-fat diet-induced obesity modifies the methylation pattern of leptin promoter in rats. Journal of Physiology and Biochemistry. 2009;**65**:1-9. DOI: 10.1007/BF03165964

[27] American Diabetes Association. Standards of medical care in Diabetes-2010. Diabetes Care. 2010;**33**(Suppl. 1):S1-S61. DOI: 10.2337/dc10.S011

[28] Reddy MA, Natarajan R. Role of epigenetic mechanisms in the vascular complications of diabetes. Subcellular Biochemistry. 2013;**61**:435-454. DOI: 10.1007/978-94-007-4525-4_19

[29] Haberland M, Montgomery RL, Olson EN. The many roles of histonedeacethylases in development and physiology: Implications for disease and therapy. Nature Reviews Genetics. 2009;**10**:32-34. DOI: 101038/nrg2485

[30] Szyf M. Epigenetics, DNA methylation and chromatin modifying drugs. Annual Review of Pharmacology and Toxicology. 2009;**49**:243-263. DOI: 10.1146/annrev-pharmtox-061008-103102

[31] Bieliauskas AV, Pflum MK. Isoforms-selective histonedeacethylases inhibition. Chemical Society Reviews. 2008;**37**:1402-1413. DOI: 10.1039/b703830p

[32] Gloeckler LA, Reichman ME, Lewis DR, Hankey BF, Edwards BK. Cancer survival and incidence from the Surveillance. Epidemiology, and End Results (SEER) program. Oncologist. 2003;**8**:541-552. DOI: 10.1634/theoncologist.8-6-541

[33] Peitzsch C, Tyutyunnykova A, Pantel K, Dubrovska A. Cancer stem cells: The root of tumor recurrence and metastasis. Seminars in Cancer Biology. 2017. pii: S1044-579X(17)30031-7. DOI: 10.1016/j.semcancer.2017.02.011. [Epub ahead of print]

[34] Kaur S, Singh G, Kaur K. Cancer stem cells: An insight and future perspective. Journal of Cancer Research and Therapeutics. 2014;**10**(4):846-852. DOI: 10.4103/0973-1482.139264

[35] Zhan T, Rindtorff N, Boutros M. Wnt signaling in cancer. Oncogene 2017: 36, 1461-1473; doi:10.1038/onc.2016.304

[36] Zhao C, Blum J, Chen A, Kwon HY, Jung SH, Cook JM, Lo A, Reya T. Loss of beta-catenin impairs then renewal of normal and CML stem cells in vivo. Cancer Cell. 2007;**12:**528-541. DOI: 10.1016/j.ccr.2007.11.003

[37] Gerstein AV. APC/CTNNB1 (beta-catenin) pathway alteration in human prostate cancer. Genes Chromosomes Cancer. 2002;**34**:9-16. DOI: 10.1002/gcc.10037

[38] Mir R, Pradhan SJ, Patil P, Mulherkar R, Galande S. Wnt/β-catenin signaling regulated SATB1 promotes colorectal cancer tumorigenesis and progression. Oncogene. 2016;**35**:1679-1691. DOI: 10.1038/onc.2015.232

[39] Damsky WE, Curley DP, Santhanakrishnan M, rosenbaum LE, Platt JT, Gould BE, Taketo MM, Dankort D, Rimm DL, McMahon M, Boseberg M. Beta-catenin signaling controls metastasis in Braf-activated Pten-deficient melanomas. Cancer Cell. 2011;**20**(6):741-754. DOI: 10.10167j.ccr.2011.10.030

[40] Li Y, Wang Z, Kong D, Li R, Sarkar SH, Sarkar FH. Regulation of Akt/FOXO3a/GSK-3beta/AR signaling network by isoflavone in prostate cancer cells. Journal of Biological Chemistry. 2008;**283**:27707-2776. DOI: 10.1074/jbcM802759200

[41] Prasad CP, Rath G, Mathur S, Bhatnagar D, Ralhan R. Potent growth suppressive activity of curcumin in human breast cancer cells: Modulation of Wnt/beta-catenin signaling. Chemico-Biological Interactions. 2009;**181**:263-271. DOI: 10.1016/j.cbi.2009.06.012

[42] Ryu MJ, Cho M, Song JY, Yun YS, Choi IW, Kim DE, Park BS, Oh S. Natural derivaties of curcumin attenuate the Wnt/beta-catenin pathway through down-regulation of the transcriptional coactivator p300. Biochemical and Biophysical Research Communications. 2008;**377**:1304-1308. DOI: 10.1016/j.bbr.2008.10.171

[43] Gao Y, Teschendorff AE. Epigenetic and genetic deregulation in cancer target distinct signaling pathway domains. Nucleic Acids Research. 2017;**45**(2):583-596. DOI: 10.1093/nar/gkw1100

[44] Niedzwiecki A, Roomi MW, Kalinovsky T, Rath M. Anticancer efficacy of polyphenols and their combinations. Nutrients. 2016;**8**(9). pii: E552. DOI: 10.3390/nu8090552

[45] Kim J, Zhang X, Rieger-Christ KM, Summerhayes IC, Wazer DE, Paulson E, Yee AS. Suppression of Wnt signaling by the green tea compound epigallocatechin 3-gallate (EGCG) in invasive breast cancer cells. Requirements of transcriptional repressor HBP1. Journal of Biological Chemistry. 2006;**281**:10865. DOI: 10.1074/jbc.M513378200

[46] Ji Q, Liu X, Fu X, Zhang L, Sui H, Zhou L, Sun J, Cai J, Qin J, Ren J, Li Q. Resveratrol inhibits invasion and metastasis of colorectal cancer cells via MALAT1 mediated Wnt/β-catenin signal pathway. PLoS One. 2013;**8**(11):e78700. DOI: 10.1371/journal.pone.0078700

[47] Roccaro AM, Leleu X, Sacco A, Moreau AS, Hatjiharissi E, Jia X, Xu L, Ciccarelli B, Patterson CJ, Ngo T, Russo D, Vacca A, Dammaco F, Anderson KC, Ghobrial IM, Treon SP. Resveratrol exerts antiproliferative activity and induces apoptosis in Waldenstrom's macroglobulinemia. Clinical Cancer Research. 2008;**14**:1849-1858. DOI: 10.1158/1078-0432.CCR-07-1750

[48] Nitta M, Azuma K, Hata K, Takahashi S, Ogiwara K, Tsuka T, Imagawa T, Yokoe I, Osaki T, Minami S, Okamoto Y. Systemic and local injections of lupeol inhibit tumor growth in

a melanoma-bearing mouse model. Biomedical Reports. 2013;**1**(4):641-645. DOI: 10.3892/br.2013.116

[49] Fang D, Hawke D, Zheng Y, Xia Y, Meisenhelder J, Nika H, Mills GB, Kobayashi R, Hunter T, Lu Z. Phosphorylation of beta-catenin by AKT promotes beta-catenin transcriptional activity. Journal of Biological Chemistry. 2007;**282**:11221-11229. DOI: 10.1074/jbcM611871200

[50] Pessione E, Cirrincione S. Bioactive molecules released in food by lactic acid bacteria: Encrypted peptides and biogenic amines. Frontiers in Microbiology. 2016;**7**:876. DOI: 10.3389/fmicb.2016.00876

[51] Luo C-Q, Xing L, Cui P-F, Qiao J-B, He Y-J, Chen B-A, Jin L, Jiang H-L. Curcumin-coordinated nanoparticles with improved stability for reactive oxygen species-responsive drug delivery in lung cancer therapy. International Journal of Nanomedicine. 2017;**12**:855-869. DOI: 10.2147/IJN.S122678

[52] Nam J-S, Sharma AR, Nguyen LT, Chakraborty C, Sharma G, Lee S-S. Application of bioactive quercetin in oncotherapy: From nutrition to nanomedicine. Molecules. 2016;**21**:108. DOI: 10.3390/molecules21010108

[53] Pimple S, Manjappa AS, Ukawala M, Murthy RS. PLGA nanoparticles loaded with etoposide and quercetin dihydrate individually: In vitro cell line study to ensure advantage of combination therapy. Cancer Nanotechnology. 2012;**3**:25-36. DOI: 10.1007/s12645-012-0027-y

[54] Pandey SK, Patel DK, Thakur R, Mishra, DP, Maiti P, Haldar C. Anti-cancer evaluation of quercetin embedded PLA nanoparticles synthesized by emulsified nanoprecipitation. International Journal of Biological Macromolecules. 2015;**75**:521-529. DOI: 10.1016/j.ijbiomac.2015.02.011

Chapter 8

Functional and Biological Potential of Bioactive Compounds in Foods for the Dietary Treatment of Type 2 Diabetes Mellitus

Daniel Pelcastre Monjiote,
Edwin E. Martínez Leo and
Maira Rubi Segura Campos

Additional information is available at the end of the chapter

http://dx.doi.org/10.5772/intechopen.68788

Abstract

Type 2 diabetes mellitus (T2DM), or noninsulin-dependent diabetes, is a complex disease characterized by the alteration of oxidoreductive and proinflammatory mechanisms, which leads to disorders in the insulin receptor and consequent chronic hyperglycemia. The hypoglycemic, insulinomimetic, and lipid-lowering potential of food is a reality given the advances in understanding of the role of food in nutrition. Besides its nutritional content, food exerts a biological function in the organism, and this demonstrates the importance of redirecting therapeutic strategies as well as related prevention policies of T2DM. The present review evaluates the effect of food on T2DM treatment. Particular attention is paid to the consumption of nopal, soy, and oats for their hypoglycemic functions, as well as the consumption of omega-3 fatty acids, which are associated with the control of metabolic alterations of this disease.

Keywords: antioxidant, anti-inflammatory, functional foods, bioactive compounds, diabetes

1. Introduction

Type 2 diabetes mellitus (T2DM), also called noninsulin-dependent diabetes, is a complex and multifactorial disease. This review describes T2DM in the framework of oxidative stress and the inflammatory process, since its main etiological factor is obesity. These mechanisms can lead to various metabolic alterations, which have been proposed to be part of their chronicity and complexity [1].

According to the World Health Organization (WHO), there are 350 million people with diabetes worldwide, whereas the International Diabetes Federation (IDF) estimates that by 2013, 382 million people worldwide were diagnosed with some type of diabetes. This figure is expected to increase to 592 million by 2035 [2].

As a response to the increase in diseases related to the modern lifestyle, functional foods, such as soybean, nopal, oats, and foods with high antioxidant and omega-3 content, were developed in Japan in the 1980s, and these have become important alternatives for improving nutrition and public health. Hence, research into the benefits or effects of functional foods on T2DM is crucial and can determine whether these can be a true alternative for the prevention and control of this pathology, as well as for associated metabolic effects.

2. Physiopathogenesis of T2DM: oxidative stress and the inflammatory process

The alteration of some cellular biochemical processes is mainly caused by factors such as over-nutrition and decreased physical activity in the individual, as for glucose metabolism, specifically hyperglycemia, which in turn triggers:

- Cell overload of free fatty acids
- Endothelial dysfunction
- Insulin resistance in muscle
- Impaired insulin secretion in the beta cells of the pancreas.

T2DM includes several alterations in metabolism, including hyperglycemia, insulin resistance, dyslipidemia, and chronic low-grade inflammation, and these alterations arise from oxidative stress [3].

Oxidative stress is defined as the biochemical imbalance caused by the overproduction of reactive species (RS) and free radicals (FR) that cause oxidative damage to membrane lipids, carbohydrates, proteins, and DNA. In people with T2DM, free radicals are found in high concentrations, causing damage to various organs, such as the heart and blood vessels. This has been described as a risk factor for the development of complications in this disease [4].

As mentioned above, the excess of FR leads to the oxidation of macromolecules, which in turn leads to lesions at the cellular level; among them, the following effects are described:

- Lipids: During lipid peroxidation, unsaturated fatty acids react (in chains) with molecular oxygen and hydroperoxides are formed, which are degraded into various products, such as conjugated dienes, alkanes, aldehydes, and isoprostanes, among others. Damage from oxidation can affect both the lipids in cell membranes and those contained in plasma lipoproteins. In the first case, this would cause inadequate cellular functioning, which is presumed to be one of the causes of premature aging experienced by some individuals with diabetes [5].

In the case of plasma lipoproteins, damage to these in all known cases is derived from the oxidation of their lipids. Alterations of high density lipoproteins (HDL) and very low density lipoproteins (VLDL) can affect reverse cholesterol transport and clarification of plasma triglycerides, respectively [6].

On the other hand, the peroxidation of low-density lipoproteins (LDL) constitutes the major contribution of FR to the genesis and aggravation of atherosclerosis. Oxidative modifications of LDL confer greater atherogenic power on this macromolecule [6, 7].

It is also known that in diabetic patients with unacceptable metabolic control, there is greater susceptibility of LDL to oxidation and more oxidized LDL than in those with optimal control [6, 7].

- Protein: The mechanisms of damage in each radical-generating system may be different and may also vary depending on the affected protein. Oxidative modification of proteins increases their degradability and susceptibility to proteolysis, probably due to their increased hydrophobicity, which implies more rapid ubiquitination and degradation by the lysosomal pathway. Likewise, the alteration of free radical proteolysis is manifested both in intracellular protein catabolism and in extracellular protein systems, especially in proteins of the extracellular matrix [8].

  One protein that can undergo oxidative damage in people with T2DM is insulin. Oxidative damage causes chemical and structural changes in this hormone and, as a consequence, a loss of its biological function. It has been shown that human adipose tissue in the presence of oxidized insulin does not use glucose with the same efficiency as with native insulin [9].

  Also, carbonyl stress can also affect insulin receptors, and the molecules involved in the cellular response are appropriate to insulin stimulation [9].

- Deoxyribonucleic acid (DNA): There are many phenomena, associated with mutations and carcinogenesis, which are caused by damage to DNA. These include loss of expression or synthesis of a protein by damage to a specific gene, oxidative modifications of bases, fragmentations, stable interactions of DNA-proteins, chromosomal rearrangements, and demethylation of cytokines of the DNA that activates genes. The damage may be effected by such alterations; for example, via inactivation or loss of tumor suppressor genes, which may lead to the initiation, progression or both of carcinogenesis [10].

The above-described conditions are causes of metabolic alterations characteristic of T2DM. Also, oxidative stress present in people with T2DM is associated with the chronic hyperglycemia that characterizes this disease. Meanwhile, an excess of circulating glucose activates several metabolic pathways not very common in the organism, which leads to the generation of other metabolites, among which are oxygen FR [1, 4].

Regarding the sorbitol pathway, given the high circulating glucose levels in the blood, the metabolic pathway of the aldose reductase enzyme is followed: it is of low affinity to normal glucose concentrations, generates sorbitol from glucose and uses NADPH (nicotinamide adenine dinucleotide phosphate) as a cofactor. Because the antioxidant potential of glutathione depends on the NADPH supply (because glutathione requires it for regeneration), the flow of this cofactor by another route, such as that of sorbitol pathway, shifts the oxidant-antioxidant balance toward oxidative stress [11].

In turn, it has been shown that sorbitol affects the physiology of cells that do not use insulin-mediated transporters to take glucose (and which contain the enzyme aldose reductase), such as neurons, red blood cells, and the nephrons that undergo osmotic changes. In addition, the permeability of these cells may be altered due to the increase of sorbitol, leading to complications typical of T2DM [11].

Likewise, sorbitol has been linked to oxidative stress with low insulin levels in diabetic patients, since it has been shown that the beta cells of the pancreas are not immune to damage by FR. In this way, in patients who already have the disease, it is possible that symptoms worsen, since insulin secretion in the pancreas decreases because of interference of FR to the normal process of insulin production and secretion [1].

In addition to the increase in free radicals, there is also an increase in metabolic stress, which is the result of change in energy metabolism, in the level of mediators of inflammation and in the state of the antioxidant defense system. Therefore, the inflammatory process is also altered in patients with T2DM. Systemic inflammation is one of the most representative features of this type of diabetes, characterized by high systemic levels of pro-inflammatory cytokines damaging DNA and causing endothelial dysfunction, which causes microvascular and macrovascular complications in T2DM [1].

3. Mechanisms of antioxidant defense in T2DM

An antioxidant is a chemical entity that, at low concentrations and compared to the oxidant, retards or prevents the oxidation of a substrate, which includes lipids, proteins, carbohydrates, and DNA [12].

Antioxidants have been classified in different ways, of which the most used establish differences in chemical structure and biological function, dividing them into enzymatic and nonenzymatic [13].

- Exogenous: These come from the diet and include vitamin E, vitamin C, and carotenoids (beta carotenes, lycopenes, and xanthines). Vitamin C is the most abundant water-soluble antioxidant in the blood, whereas vitamin E is the major lipophilic antioxidant. Selenium, the most toxic mineral included in our diet, acts together with vitamin E as an antioxidant [13].

- Endogens: Antioxidant defenses consist of avoiding the univalent reduction of oxygen by enzymatic or nonenzymatic systems. A group of enzymes specialized in inactivating the reactive oxygen species (ROS) by different mechanisms has been described, such as catalase (CAT), glutathione peroxidase (GPX), and superoxide dismutase (SOD). Nonenzymatic antioxidants recognize amino acids, such as glycine, taurine, and the tripeptide glutathione [13].

In T2DM, a series of changes occur that indirectly indicate the existence of marked oxidative stress, due to the increase in formation of oxygen free radicals and the decrease of the plasma and intracellular levels of the antioxidants [4].

Carmeli et al. [14] confirmed that in people with T2DM, there is significantly decreased activity of the SOD enzyme as a consequence of high levels of hydrogen peroxide produced during the reaction, which inhibit the enzyme by negative feedback. Indeed, it was observed that an increase of SOD initially occurs in response to the high generation of the superoxide anion in the cell and its elimination by the enzyme. However, the intense production of this radical for a prolonged time exhausts the stimulation of enzymatic activity, since the product of the reaction can inhibit it.

With respect to the concentration of minerals (Cu^{2+} and Zn^{2+}), Devi et al. [15] found that patients with T2DM had significantly higher serum and erythrocyte copper levels. In addition, plasma copper levels have been reported to be higher in patients with complications. In this sense, it has been hypothesized that alterations in copper metabolism contribute to the progression of pathologies related to diabetes, because glycosylated proteins have a higher affinity for transition metals such as copper.

Nsonwu et al. [16] found that serum zinc levels were significantly lower in people with T2DM. This apparent hypozincemia may be the result of urinary loss, decreased intestinal absorption of this mineral or both conditions.

4. Inflammatory process and insulin resistance

Inflammation is a response of the body to exposure to infectious agents, antigenic stimuli or physical injury involving the nervous, vascular, and immune systems. Initially, it has a homeostatic function of protection or defense that is characterized by flushing, pain, swelling, edema, and lack of function in the affected area; however, if the process is inefficient and chronic, it becomes a pathophysiological process that favors the increase in FR and consequently oxidative stress [17].

In T2DM, there is a pathophysiological relationship with the chronic inflammatory process (CIP) by two mechanisms: one linked to obesity and the endocrine activity of adipose tissue, and the other involving the development of the immune response stimulated by generated AGEs because of the nonenzymatic glycosylation reaction of proteins [11].

The chronic inflammatory process is an alteration linked to obesity and T2DM, considering that adipose tissue, besides being an energy reserve, acts as a high activity endocrine gland, producing a wide variety of substances with effects at different levels in the body, including proinflammatory cytokines. In addition to secreting hormones, such as leptin, adiponectin, resistin and ghrelin, adipocytes synthesize and secrete cytokines associated with inflammation, such as IL-6 and TNF-α [18].

The mechanism by which the chronic process is linked to the development of diabetes mellitus occurs at the molecular level and implies insulin resistance. Briefly, the mechanism is as follows: when insulin binds to the extracellular alpha subunit of its receptor, it causes a conformational change that allows the binding of ATP to the intracellular beta subunit of the receptor. This promotes autophosphorylation of insulin and confers tyrosine kinase activity, which initiates tyrosine phosphorylation of intracellular proteins called insulin receptor substrate (IRS). IRS have a conserved region that, once activated, allows them to interact with other intracellular proteins, promoting the translocation of the glucose transporter (GLUT) to the cell membrane, with the subsequent entry of glucose [1, 19].

TNF-α causes an inhibition of the autophosphorylation of tyrosine residues of the insulin receptor and also causes the phosphorylation of a serine of the insulin receptor substrate (IRS). This in turn promotes the phosphorylation of a serine of the insulin receptor and inhibits the phosphorylation of tyrosine that is required to promote the cascade of signals for the capture of glucose; thus, this translates into insulin resistance. Also, it has been reported that IL-6 inhibits the signal of insulin transduction in the hepatocyte, which also causes insulin resistance [19].

Vozarova et al. [20] showed that markers of inflammation correlate with diabetes. The total leukocyte count is an indirect marker of inflammation and, specifically a higher neutrophil count than normal, is related to the insulin resistance characteristic of T2DM and cardiovascular diseases.

Inflammation of beta cells of the pancreas as a result of an autoimmune phenomenon has been recognized in type 1 diabetes mellitus and is increased in the pathogenesis of T2DM. Such inflammation is one of the pathways of the pathogenesis of T2DM and its complications [21, 22].

The main cell involved in the inflammatory process and in the insulin resistance of T2DM is the adipocyte, since insulin regulates glucose uptake and storage of triglycerides through these. Adipokines in turn also affect secretion and insulin resistance [23].

In particular, leptin, adiponectin, and resistin contribute to the dysfunction of the beta cells of the pancreas increasing insulin resistance. The adipose tissue also secretes dipeptidyl peptidase-4 (DPP-4) improving the degradation of glucagon in peptide 1 (GLP-1), which has an insulinotropic effect on beta cells [24].

On the other hand, the circulation of proinflammatory cytokines directly and indirectly affects the function of beta cells, increasing inflammation of the adipocyte. Cytokines such as TNF-α, beta-interleukin (IL-1β), and gamma interferon (IFN-γ) alter the regulation of intracellular calcium in beta cells and thus release insulin. In addition, TNF-α increases the expression of amyloid peptide (IAPP) in beta cells leading to accelerated death, which leads to insulin resistance [24].

Glucotoxicity, particularly lipotoxicity, increases fatty acids locally in the islets, and long chain fatty acids, especially palmitic acids, cause oxidative stress and the activation of N-terminal c-Jun kinases. These increase the secretion of adipokines, initiating a cycle that induces the dysfunction of the beta cells of the pancreas, which consequently increases inflammation [25].

5. New trends in the treatment of T2DM: functional foods and bioactive compounds

The World Health Organization (WHO) estimates that 50% of patients with T2DM do not comply with experts' recommendations regarding lifestyle and eating habits. In response to this problem, the science of nutrition faces a challenge: the search for new foods and/or food components that ensure health and reduce the risk of certain diseases. In addition, it could reduce future costs derived from the treatment of these diseases. At this point, the food industry plays a significant role, since it is the main producer and distributor of food [26, 27].

The concept of "functional food" was born as a convenient and economical solution for chronic health problems, being influential in many branches of science. Since 1984, the meaning of "functional food" has changed according to country and culture and has been defined and redefined over the past 30 years. A food may be considered "functional" if it has been satisfactorily demonstrated that, in addition to its nutritional effects, it beneficially affects one or more functions of the organism in a way that improves the state of health or well-being or reduces the risk of disease [27].

Therefore, in functional foods, two very important and different points are integrated. On the one hand, there is the science of nutrition, responsible for investigating and testing new compounds and/or foods that are being developed, and also, there is the industry, responsible for production and distribution of food that will eventually reach consumers [28].

In 1984, the Japanese government allocated funds for the study of functional foods or specific foods with therapeutic uses. Japan was the first country to use the definition of functional food as "fortified foods with special components that have beneficial physiological effects." To be considered as such, there was a legal category of food called FOSHU. In order of importance, the food had to meet three nutritional requirements:

1. It should be constituted by natural ingredients.
2. It should be consumed as part of a daily diet.
3. It should be a food that when consumed presents a particular function in the human body, such as:

- Improvement in biological defense mechanisms.
- Prevention or recovery of some specific diseases.
- Control of physical and mental conditions.
- Aging process delay [28, 29].

Subsequently, the term was adopted by Europe. In the United States, in 1994, the National Academy of Food Sciences and the Nutrition Board defined functional foods as "modified foods or ingredients that can improve health, beyond the nutrients they possess." In 2004, the American Dietetic Association (ADA) issued an institutional document on functional foods, where they were defined as foods that have potential beneficial effects on health when consumed as part of a varied diet, at effective levels. The definition covers whole, fortified, enriched, or improved (designed) foods [30].

In 2012, FFC (Functional Food Center) announced the new concept of functional food as: "natural or processed foods containing essential or nonessential biologically active compounds, which in specific amounts provide a clinically proven and documented health benefit for the prevention, management, or treatment of a chronic disease." This means that a functional food can be:

- Natural food.
- Food to which a component has been added.
- Foods to which a component has been removed.
- Foods to which the nature of one or more components has been changed.
- Food in which the bioavailability of one or more of its components has been modified.
- Any combination of the above possibilities [31].

At present, these foods are being greatly developed with emphasis on the following functions [31]:

- Regulation of basic metabolic processes: Foods that improve metabolic efficiency are sought. Metabolic efficiency includes glycemia optimization and foods that improve this would produce moderate glucose peaks. This involves developing new ingredients such as hydrogenated carbohydrates or trehalose.
- Defense against oxidative aggressions: The paradoxical relationship (i.e., respiration) is known, and certain toxic or harmful reactions occur, such as those occurring in the presence of reactive oxygen species (ROS) that act as powerful antioxidants. These possibly contribute to the appearance of aging processes, heart disease, cancer, cataracts and degenerative pathologies of the nervous system, such as those that occur in Parkinson's and Alzheimer's. The organic processes that defend against ROS can be complemented by several substances widespread in numerous foods, such as vitamin E, C, and carotenoids, as well as polyphenols of plant origin, which could reinforce the panoply of functional foods against oxidative aggression.
- Circulatory system: Functional foods may play a role in the different predisposing factors of cardiovascular diseases: arterial hypertension, vessel integrity, dyslipidemias, oxidized lipoproteins, elevated levels of homocysteine, increased blood coagulation, and low circulating vitamin K concentrations. Thus, blood lipids can be modified by the presence of certain fatty acids, fiber, and antioxidants, such as flavonoids in the diet. Vegetable components, such as phytosterols, may be able to lower LDL-cholesterol (LDL-C). The overall vascular integrity could also benefit from an increased concentration of folates, vitamin B6 and B12 in the diet, which will reduce plasma concentrations of homocysteine.

- Digestive system: The balance and variety of the microbial flora in the intestine are important factors in the maintenance of health. Prebiotics, probiotics, and symbiotics are considered as functional foods in this balance of the predominant flora in the intestine.

6. Potential functional foods and bioactive compounds with application in the treatment of T2DM

Currently, several foods with potential roles in the treatment of T2DM are associated. Mainly, the roles of nopal, soy, and oats are recognized because of their hypoglycemic, insulino-mimetic and lipid-lowering effects and of bioactive compounds such as antioxidants and omega-3 fatty acids. Oxidative stress and chronic inflammation are present in fresh fruits and vegetables, teas, and blue fish, respectively. The latter, in clinical studies, are treated as compounds characterized as nutraceuticals, given the low bioavailability they possess as part of a food matrix.

6.1. Nopal

The nopal belongs to the family of cactuses, which are fleshy, thickened, and spiny plants, and to the genus *Opuntia*, which is characterized by extended petals with an articulated stem. *Opuntia streptacantha* is the best studied of this genus and is more cultivated in arid and semi-arid zones of the Mexican territory [32].

Scientific evidence on nopal has shown a correlation between ethnomedical uses and experimental results, since people use this food as an alternative or combined treatment with T2DM drugs [32].

Pharmacological research of the nopal as a hypoglycemic agent began in 1964 and was continued in 1979 by the now-extinct Mexican Institute for the Study of Medicinal Plants (IMEPLAM). Researchers at this institute found that different preparations of liquefied raw nopal, administered by a nasogastric tube to rabbits with hyperglycemia induced by pancreatectomy or by administration of aloxane, produced a hypoglycemic effect. Four years later, Ibanez and Meckes (1983) showed that a semipurified fraction of fresh stem juice of *O. streptacantha* given to normoglycemic rabbits or with induced hyperglycemia produced a significant decrease in blood glucose and triglyceride levels [33].

Trejo-González et al. [34] performed a study in rats with streptozotocin-induced diabetes, who were given a simultaneous administration of *O. fuliginosa* (1 mg/kg) and insulin for 7 days. This induced decreased blood glucose and glycosylated hemoglobin to normal values. These values were maintained when insulin was withdrawn and only the cactus extract was administered.

Laurenz et al. [35] found that in pigs with chemically induced diabetes, oral administration of 250–500 mg/kg of *O. lindheimeri* extract maintained blood glucose at normal levels but did not modify the glycemia of nondiabetic pigs.

Frati-Munari et al. [36] administered 100 g of roasted cactus to both healthy and obese subjects with or without T2DM, 20 min before meals three times a day for 10 days, produced a significant decrease in total cholesterol, triglycerides, and total weight in nondiabetic obese subjects and type 2 diabetes obese subjects and in the glycemia of diabetic subjects. These results suggest that the effects observed with nopal are due to their fiber content. The fiber content is a mixture of lignin, cellulose, hemicellulose, pectin, mucilage and gums, which are capable of decreasing the gastrointestinal absorption of various nutrients and, consequently, decreasing blood levels of cholesterol, triglycerides, and glucose due to lack of absorption.

The group of Frati-Munari et al. [37] performed another study in patients with induced hyperglycemia and showed that the same dose as in the previous study of 100 g of roasted cactus, given to healthy volunteers, 20 min before starting the oral glucose tolerance test, prevented blood glucose elevation at 120 and 180 min and decreased blood insulin concentration. To explain this latter effect, a possible inhibitory action of the fiber on the gastric peptide was mentioned. This substance normally increases the sensitivity of the insulin receptor and induces the release of this hormone in the islets of Langerhans. Unfortunately, neither of these hypotheses have been experimentally studied.

In a subsequent study, it was reported that fresh nopal blotch, whose species was not identified, administered orally to healthy individuals, did not modify the basal glucose or blood insulin concentration. In contrast, an antihyperglycemic action was described in healthy individuals with orally, but not intravenously, induced hyperglycemia. These results suggest that liquefied cactus would only have an antihyperglycemic effect if it is ingested prior to food intake; this effect would prevent the complications of T2DM [37].

The same research group also showed that the decrease in blood glucose in individuals with type 2 diabetes is in direct proportion to the administered doses of roasted cactus. This effect which the authors called "acute hypoglycemia" is believed to be independent of that produced by the fiber at the level of the gastrointestinal tract [38].

This group also found that extracts of fresh crude nopal had virtually no "hypoglycemic" effect when given to type 2 diabetic patients under fasting conditions, whereas roasted cactus produced a "hypoglycemic" effect in the same type of patients but not in normoglycemic healthy subjects. These results call into question whether fresh nopal smoothies, which are consumed by much of the Mexican population, have any beneficial effect, especially if consumers are not diabetic [39].

In conclusion, nopal has different effects in the body. However, although it appears that this plant prevents glycemia elevation and has an insulinomimetic effect and lowers blood glucose levels below normal values, these effects only occur under certain conditions, such as the use of large doses (100–500 g) of roasted cactus.

Porrata et al. [40] emphasized the importance of a fiber-rich diet for the control of T2DM. In 6 months, 25 adults with T2DM treated with antihyperglycemic agents and a macrobiotic vegetarian diet with a majority of whole grains, vegetables, legumes, and green tea showed beneficial effects. These were evident in improved blood glucose control, decreased insulin requirements, slowed glucose absorption, increased peripheral tissue sensitivity to insulin, lowered cholesterol levels and triglycerides, controlled body weight and lowered blood pressure.

It was also observed that insulin has been shown to have a marked lipid-lowering effect in individuals with obesity and dyslipidemia. It has been recommended that 9 g/day of insulin for 4 weeks is sufficient to achieve a favorable effect on the lipid profile [40].

6.2. Soy

Soybean (*Glycine max*) is a species of the leguminous family (*Fabaceae*) cultivated for its seeds, which have medium oil and high protein content. Its composition is based on 40% protein and 20% oil. It is considered as the legume with the highest contribution of protein and its consumption produces hypoglycemic and hypolipidemic benefits, among others [41].

Céspedes et al. [42] conducted a study with 40 patients with T2DM to evaluate the effect of soy protein in this pathology. All patients received three servings of soy protein weekly as a nutritional contribution and performed physical exercises. The effect of the soy protein-enriched diet was highly significant for HDL cholesterol, suggesting that it could participate in the control of plasma concentrations of this lipoprotein by helping metabolic control of dyslipidemia, which is known to be a metabolic alteration characteristic of T2DM.

Garrido et al. [43] stated that soy consumption could confer benefits in the prevention of cardiovascular diseases, risk factors of which are T2DM, obesity, and corresponding dyslipidemias. In 2000, the state agency for the US Food and Drug Administration (FDA) allowed the use of a "health claim" for soy protein, associating consumption of this protein with a low saturated fat diet, with a decreased risk of cardiovascular disease. This measure was based on studies included in a meta-analysis of 38 controlled clinical studies using soy protein from the above, and it was concluded that the substitution of animal protein for soy protein significantly decreased total cholesterol, LDL-cholesterol and triglycerides without affecting HDL-cholesterol (HDL-C), and the effects were higher in subjects with higher basal cholesterol.

Each subject received six randomly tested foods: a standard glucose drink or a commercial low-carbohydrate soy drink (Ades Natural Light and Ades Chocolate Light), peanuts, a high-carbohydrate soy milk, or fiber drink. Before each session, the subjects were weighed and interviewed. Only water was allowed to be consumed during fasting, no caffeinated food was allowed. The subjects did not consume legumes and were not allowed to drink alcoholic beverages. The results showed that soy beverages should contain at least 6.25 g of protein per serving and that four servings per day should be consumed for a long time to see a possible beneficial effect on the blood lipid concentration. It is also recommended that soy products have a low concentration of maltodextrins and, if possible, contain soluble fiber to maintain low glycemic indexes and be usable in obese or diabetic patients. The consumption of soy protein (0.5 g/kg/day) in diabetic patients with renal impairment reduces the excretion of urinary albumin and increases HDL cholesterol, as well as improving glomerular filtration [44].

6.3. Oats

Oat is an annual herbaceous plant, belonging to the grass family. The most cultivated species are *Avena sativa* and *Avena byzantina*. It is rich in proteins of high biological value, fats and a large number of vitamins and minerals. It is the cereal with the greatest proportion of vegetable fat; 54% unsaturated fats and 46% linoleic acid. It also contains readily absorbed carbohydrates in addition to calcium, zinc, copper, phosphorus, iron, magnesium, potassium and sodium. In addition, it contains vitamins B1, B2, B3, B6 and E and contains a good amount of fiber, which is less important than nutrients, but contributes to good intestinal functioning [45].

Cabrera Llano and Cárdenas Ferrer [46] stated that in the past 30 years, multiple studies have shown that the administration of dietary fiber could reduce blood glucose levels in patients with both type 1 and type 2 diabetes.

The American Diabetes Association (ADA) continues to recommend a fiber intake between 20 and 35 g/day, both soluble and insoluble, to maintain better glycemic and insulin control, with the soluble fraction being the most effective in glycemic control [47].

The mechanisms proposed are delayed gastric emptying; decrease in glucose uptake by being trapped by fiber viscosity and thus less accessible to the action of pancreatic amylase and short chain fatty acid production; and propionate influences gluconeogenesis by reducing the hepatic production of glucose. Butyrate acts by reducing peripheral resistance to insulin by reducing the production of TNFα. Insulin resistance is one of the most important factors involved in the metabolic syndrome [48].

It is also important to take into account that insulin has, in addition to its metabolic action, an effect on vascular endothelium that facilitates the progression of atherogenesis. Therefore, it is proposed that oat hypoglycemic function is important in patients with T2DM and can be an alternative for the treatment of this. However, the hypolipidemic effect of oats is also noteworthy [48].

Regarding the lipid-lowering effect of oats, Kerckhoffs et al. [49] stated that daily consumption of approximately 3 g of soluble fiber can decrease total cholesterol by 0.13 mmol/L in normocholesterolemic and 0.41 mmol/L in hypercholesterolemic drugs, which would be a mechanism of prevention for one of the metabolic alterations of T2DM.

Ruiz et al. [50] carried out a study whose objective was to determine the effect of *Avena sativa* on the lipid profile of patients between 20 and 60 years old with diagnoses of dyslipidemias. Patients consumed 60 g of liquefied oats in water daily for 3 months, and total cholesterol, triglycerides, and LDL were measured at the beginning at 4 and 12 weeks. The results showed statistically significant decreases in total and LDL-C, without major changes in HDL-C and triglycerides.

Furthermore, a study performed by Raasmaja et al. [51] evaluated the effect of drink with symbiotic on the reduction of cholesterol, triglycerides, and glucose control by in vivo analysis with a model of 24 rats with genetic obesity exhibiting similar effects to the metabolic syndrome. These rats were randomly divided into three groups: group 1 control (water), group 2 (symbiotic), and group 3 (malted oats). Measurements of glucose, total cholesterol, and triglycerides in blood plasma were taken for 3 months on six occasions. The results showed that rats that consumed symbiotic beverages had decreased glucose, triglycerides, and weight. However, groups 1 and 3 showed a greater reduction of cholesterol in comparison with group 2. Therefore, it was concluded that the consumption of a symbiotic drink based on malted oats and *Lactobacillus casei* exerted a positive effect on the reduction of glucose and triglycerides in addition to showing a tendency for decreased weight. This type of drink may be a safe alternative for patients with T2DM since, in addition to glucose control, it exerts a lipid-lowering effect and a decrease in body weight.

6.4. Antioxidants

Dietary antioxidants play an important role in the defense against aging and chronic diseases such as T2DM, as these substances inactivate free radicals involved in oxidative stress and prevent its propagation. As previously described, T2DM is characterized by a chronic oxidative state. Therefore, the inclusion of antioxidants in the diet contributes to counteracting the effects of the oxidative state on the organism [52].

Supplementation of the diet with natural antioxidants may have a beneficial effect in improving the morbidity and mortality of diabetic patients, so that they could prevent and delay the development of chronic complications of T2DM [53].

Yusuf et al. [54] performed a study to evaluate the possible effects of antioxidants in the prevention and treatment of T2DM complications. In most studies, vitamin E was isolated or in combination. The doses of vitamin E used were 300–1800 IU/day, generally in the form of alpha-tocopherol. However, there were no significant data demonstrating a beneficial effect of vitamin E in the prevention of T2DM, but a beneficial role of vitamin E in endothelium-dependent vasodilation was observed in subjects with cardiovascular risk, such as diabetes. This directly associates improvement of function of endothelial activity with the reduction of oxidative stress, supporting that the benefit of vitamin E on endothelial function depends in part on its antioxidant effects.

Geohas et al. [55] evaluated metabolic effects of supplementation of chromium in different doses or chromium combined with biotin in a total of 216 type 2 diabetic patients. The study showed a reduction of glycosylated hemoglobin of up to 2%, postprandial glycemia, fructosamine, insulinemia, total cholesterol, HDL/LDL ratio, triglycerides, and atherogenicity index.

In addition, Lu et al. [56] found certain metabolic benefits for patients with T2DM by supplementing the diet with 3000 mg/day of vitamin C in a clinical trial. The metabolic benefits in the vitamin C group were manifested as a tendency to decrease glycosylated hemoglobin and total cholesterol, although there were no changes in the levels of interleukins, C-reactive protein, or in the oxidation of LDL-cholesterol particles.

Moreover, Porrata et al. [40] showed that the consumption of a large amount of green tea in the diet was related to the metabolic control of T2DM, due to the polyphenols it contains. These substances are considered as the main active ingredients in the protection against oxidative damage and in the anti-inflammatory activities of T2DM. They can also increase the activity of insulin, demonstrating an increase of insulin in vitro of more than 15 times. This potentiating activity is attributed to the epigallocatechin gallate contained in green tea.

This study described the benefits of tea on hypercholesterolemia and hypertriglyceridemia, which are metabolic alterations related to T2DM. This antilipemic effect of tea is due to the action of polyphenols leading to a decrease in the absorption of fats, as well as reduced fat storage in the liver and heart [40].

Likewise, Montano et al. [57] conducted a study of 22 patients (nine with T2DM), giving them 100 mg orally of coenzyme Q10 twice a day for 12 weeks. This resulted in a significant decrease in cholesterol and LDL levels, as well as glycosylated hemoglobin levels.

6.5. Omega-3 fatty acids

Long-chain polyunsaturated fatty acids (PUFAs) are dietary components that participate in multiple physiological processes, where they play a structural role in the phospholipids of cell membranes and are substrates for the synthesis of various physiological mediators. Within the PUFAs are two main groups: the omega-3 (ω-3) and omega-6 (ω-6) fatty acids. These are essential fatty acids (EFAs) for humans because the enzymatic machinery necessary to biosynthesize them is absent [58].

The first exponent of omega-3 fatty acids is α-linolenic acid which, via desaturases and elongases, can be transformed into eicosapentaenoic acid (EPA) and subsequently into docosahexaenoic acid (DHA) [59].

Food sources of α-linolenic acid are foods of plant origin, especially oils (soybean, flax, canola, among others) and nuts (almond, walnut, peanut, among others). The nutritional source of PUFAs derived from these is food of animal origin. Arachidonic acid (AA) is found in meats (beef, lamb, and pork). EPA and DHA are found in both marine animals and vegetables, particularly in fish with a high fat content, such as tuna, horse mackerel, and salmon, among others. AA, EPA, and DHA are important structural components of membrane phospholipids and are the substrates for the formation of a series of lipid derivatives called eicosanoids (derived from 20 carbon atoms in the case of AA and EPA) and docosanoids (derived from 22 carbon atoms, in the case of DHA), which exert important actions in cellular metabolism [60, 61].

Clinical and epidemiological evidence from multiple studies allows us to establish that ω-3 PUFAs are ideal therapeutic candidates for the prevention and/or treatment of a number of pathologies, especially those where inflammation plays a major role in its development as T2DM [62, 63].

Dietary supplementation with EPA and DHA can reduce the production of pro-inflammatory cytokines such as interleukin-1, interleukin-6, interleukin-8, and tumor necrosis factor-α (TNF-α), which are released when macrophages and monocytes are activated. Although these cytokines are potent activators of immune function, the excess activity of these substances contributes to pathological inflammation [64, 65].

Petrova et al. [66] obtained the first data that showed the cardioprotective effects of ω-3 PUFAs. This arose from studies performed in Eskimos (Inuits), who, despite having a high fat intake (more than 30% of energetic requirements), presented a very low incidence of cardiovascular diseases, identifying animals of marine origin (mammals and fish rich in these lipids) as the dietary source of these fats. These results were confirmed in studies carried out in populations with similar diets, which showed a low incidence of cardiovascular diseases.

Manerba et al. [67] conducted a study demonstrating that fish oils lowered plasma cholesterol and TG levels through the inhibition of very low-density lipoproteins (VLDL) and TG biosynthesis in the liver and unchanged biosynthesis of high density lipoproteins (HDL). They also indicated that ω-3 PUFAs have a number of potentially beneficial effects on smooth vascular muscles, by reducing intracellular calcium loss and decreasing smooth muscle cell proliferation (through the inhibition of growth) and increased production of nitric oxide. It is known that one of the main metabolic complications of a patient with T2DM is dyslipidemia, and ω-3 is considered as an alternative treatment for T2DM and, because of this, can be used to treat dyslipidemias.

Manerba et al. [67] also stated that the beneficial effects on cardiovascular health attributed to ω-3 PUFAs are the result of the following mechanisms: decreased plasma TG and LDL cholesterol, increased HDL cholesterol, decreased blood pressure, reduced platelet aggregation, and decreased incidence of arrhythmias.

Geleijnse et al. [68] noted that the type and form of fish preparation determine the cardioprotective effects of ω-3 PUFAs. The consumption of fish rich in ω-3 PUFAs (tuna, horse mackerel and salmon, among others) produced a significant decrease in the risk of presenting cardiac ischemia. This effect is observed when the fish is consumed roasted or baked, but not when consumed fried.

Nasiff-Hadad and Meriño [69] performed a review of the beneficial and detrimental effects of omega-3 fatty acids in subjects with T2DM, arterial hypertension and dyslipidemias, and their effects on hemostasis and other organs and systems. It was concluded that the ingestion of blue meat fish two or three times a week should be a dietary recommendation for the whole population and that the consumption of fish oils in moderate doses (up to 3 g/day) is beneficial for subjects with T2DM, hypertension and/or dyslipidemias as an adjuvant treatment. In these cases, this diet would also decrease platelet aggregation and reduce the synthesis of chemical mediators of inflammation. However, high doses of fish oils may be harmful to glycemic control, high blood pressure in susceptible persons and serum levels of LDLs and HDLs.

Table 1 shows a summary of the doses of the main foods or bioactive compounds used for the treatment of T2DM and which have updated evidence for their effects.

Food/bioactive compound	Dose	Effect	Reference
Nopal	300 g/day (roasted)	Significant decrease in total cholesterol, triglycerides, body weight, and glycemia	[39]
Insulin	9 g/day by 4 weeks	Improvement of the lipid profile	[40]
Soy protein	0.5 g/kg/day	Reduction of urinary albumin excretion, increase in HDL cholesterol and improve glomerular filtration	[44]
Soluble fiber	3 g/day	Total cholesterol reduction	[49]
	25–30 g/day	Delayed gastric emptying, decreased glucose uptake and short-chain fatty acid production	[47]
Liquefied oats with water	60 g/day	Significant decrease in total cholesterol and LDL	[50]
Vitamin E	300–1800 UI/day (α-tocoferol)	Improvement of endothelial function directly with the reduction of oxidative stress	[54]
Vitamin C	3000 mg/day	Decreased glycosylated hemoglobin and total cholesterol	[56]
Q10 coenzyme	100 mg/day (oral administration)	Significant decrease in the levels of cholesterol, LDL and glycosylated hemoglobin	[57]
Omega-3	3000 mg/day	Decreased platelet aggregation and reduced synthesis of chemical mediators of inflammation	[69]

Table 1. Food and bioactive compounds used in the treatment of T2DM.

7. Conclusion

T2DM is a complex disease with world prevalence, with important oxidative and proinflammatory components, in which lies its chronicity and complication. Nutrition based on the biological effects of food, beyond its nutritional component, is a dietary alternative that has repercussions on the health status and quality of life of patients with T2DM.

A diet based on the use of antioxidants, omega-3, or foods, such as soybean, nopal and oats, contributes to a better status of the metabolic imbalance produced in T2DM, as a product of carbohydrate metabolism, oxidative stress and inflammatory processes, with significant improvement in the biochemical and clinical markers that characterize this disease. In addition, the design of new policies and educational materials for this population should have a new direction, based on the functional potential of food, where studies have shown effective doses to counteract the chronicity and presence of complications.

Author details

Daniel Pelcastre Monjiote[1], Edwin E. Martínez Leo[1] and Maira Rubi Segura Campos[2*]

*Address all correspondence to: maira.segura@correo.uady.mx

1 Postgraduate and Research Unit, Latino University, Merida, Yucatan, Mexico

2 Faculty of Chemical Engineering, Autonomous University of Yucatan, Merida, Yucatan, Mexico

References

[1] Donath MY. Targeting inflammation in the treatment of type 2 diabetes. Diabetes, Obesity and Metabolism. 2013;**15**(3):193-196

[2] WHO. Obesity: preventing and managing the global epidemic, Report of a WHO Consultation. WHO Technical Report Series 894. Ginebra: Organización Mundial de la Salud; 2012

[3] Agrawal NK, Kant S. Targeting inflammation in diabetes: Newer therapeutic options, World Journal of Diabetes. 2014;**5**:697-710

[4] Muchová J, Országhová Z, Žitnanová I, Trebatický B, Breza J, Duracková Z. P63—The effect of natural polyphenols on the oxidative stress markers in patients with diabetic nephropathy. Free Radical Biology and Medicine. 2014;**75**:S42-S52

[5] Obregón O, Lares MC, Castro J, Garzazo G. Potencial de oxidación de las lipoproteínas de baja densidad en una población normal y en una población con diabetes mellitus tipo 2. Archivos Venezolanos de Farmacología y Terapéutica. 2004;**23**(1):1-12

[6] Yamageshi S, Nakamura K, Takeuchi M, Imaizumi T. Molecular mechanism for accelerated atherosclerosis in diabetes and its potential therapeutic intervention. International Journal of Clinical Pharmacology Research. 2004;**24**(4):129-134

[7] Karasik A. Glycaemic control is essential for effective cardiovascular risk reduction across the type 2 diabetes continuum. Annals of Medicine. 2005;**37**(4):250-258

[8] Vicedo A, Vicedo Y. Relaciones del estrés oxidativo con el catabolismo de proteínas. Revista Cubana de Investigaciones Biomédicas. 2000;**19**(3):206-212

[9] Olivares IM, Medina R, Torres YD, Montes DH. Daño a proteínas por estrés oxidativo: Lipoproteína de baja densidad e insulina. Revista de Endocrinologia y Nutrición. 2006;**14**(4):237-240

[10] Fraga CG, Shigenaga MK, Park JW, Degan P. Oxidative damage to DNA during aging. Proceedings of the National Academy of Science. 2000;**87**:4533-4537

[11] Singh VP, Bali A, Singh N, Jaggi AS. Advanced glycation end products and diabetic complications. Korean Journal of Physiology & Pharmacology. 2014;**18**(1):1-14

[12] Guerra J. Oxidative stress, diseases and antioxidant treatment. Annals of Internal Medicine. 2011;**18**:326-335

[13] Martínez E, Acevedo J, Segura M. Biopeptides with antioxidant and anti-inflammatory potential in the prevention and treatment of diabesity disease. Biomedicine and Pharmacotherapy. 2016;**83**:816-826

[14] Carmeli E, Coleman R, Berner Y. Activities of antioxidant scavenger enzymes (superoxide dismutase and glutathione peroxidase) in erythrocytes in adult women with and without type II diabetes. Experimental Diabesity Research. 2008;**5**:171-175

[15] Devi TR, Hijam D, Dubey A, Debnath S, Oinam P, et al. Study of serum zinc and copper levels in type 2 diabetes mellitus. International Journal of Contemporary Medical Research. 2016;**3**(4):1036-1040

[16] Nsonwu A, Usoro C, Etuko M, Usoro N. Glycemic control and serum and urine levels of zinc and magnesium in diabetics in Calar Nigeria. Pakistan Journal of Nutrition. 2006;**5**(1):75-78

[17] Wensveen F., Jelenčić V., Valentić S., Šestan M., Wensveen T., et al. NK cells link obesity-induced adipose stress to inflammation and insulin resistance. Nature Immunology. 2015;**16**(4): 376-85

[18] Rotter V, Nagaev I, Smith U. Interleukin-6 (IL-6) induces insulin resistance in 3T3–L1 adipocytes and is, like IL-8 and tumor necrosis factor-alpha, overexpressed in human fat cells from insulin-resistant subjects. Journal of Biological Chemistry. 2003;**278**:45777-45784

[19] Hotamisligil GS, Murray D, Choy L, Spiegelman B. Tumor necrosis factor a inhibits signaling from the insulin receptor. Proceedings of the National Academy of Sciences of the United States of America. 1994;**91**:4854-4858

[20] Vozarova B, Weyer C, Lindsay RS, Pratley RE, Bogardus C, Tataranni PA. High white blood cell count is associated with a worsening of insulin sensitivity and predicts the development of type 2 diabetes. Diabetes. 2002;**51**:455-461

[21] Larsen CM, Faulenbach M, Vaag A, Vølund A, Ehses JA, Seifert B, Mandrup-Poulsen T, Donath MY. Interleukin-1 receptor antagonist in type 2 diabetes mellitus. New England Journal of Medicine. 2007;**356**:1517-1526

[22] Goldfine AB, Fonseca V, Jablonski KA, Pyle L, Staten MA, Shoelson SE. The effects of salsalate on glycemic control in patients with type 2 diabetes: A randomized trial. Annals of Internal Medicine. 2010;**152**:346-357

[23] Dunmore SJ, Brown JE. The role of adipokines in β-cell failure of type 2 diabetes. Journal of Endocrinology. 2013;**216**:T37–T45

[24] Lamers D, Famulla S, Wronkowitz N, Hartwig S, Lehr S, Ouwens DM, et al. Dipeptidyl peptidase 4 is a novel adipokine potentially linking obesity to the metabolic syndrome. Diabetes. 2011;**60**:1917-1925

[25] Van Raalte DH, Diamant M. Glucolipotoxicity and beta cells in type 2 diabetes mellitus: target for durable therapy? Diabetes Research and Clinical Practice. 2011;**93**(1): S37–S46

[26] Roberfroid MB. Concepts and strategy of functional food science: The European perspective. American Journal of Clinical Nutrition. 2007;**71**(6):1669S–1664S

[27] Saito M. Role of FOSHU (Food for Specified Health Uses) for healthier life. Pharmaceutical Society of Japan. 2007;**127**:407-416

[28] Yamada K, Sato-Mito N, Nagata J, Umegaki K. Health claim evidence requirements in Japan. Journal of Nutrition. 2008;**138**:1192S–1198S

[29] Diplock AT, Aggett PJ, Ashwell M, Bornet F, Fern EB, Roberfroid MB, et al. Scientific concepts of functional foods in Europe: Consensus document. British Journal of Nutrition. 2000;**81**:S1–S27

[30] Hasler CM, Browm AC, American Dietetic Association. Position of the American Dietetic Association: Functional foods. Journal of the American Dietetic Association. 2009;**109**(4):735-746

[31] Martirosyan DM, Singh J. A new definition of functional food by FFC: What makes a new definition unique?. Journal of Functional Foods in Health and Disease. 2015;**5**(6):209-223

[32] Yeh GY, Eisenberg DM, Kaptchuk TJ, Phillips RS. Systematic review of herbs and dietary supplements for glycemic control in diabetes. Diabetes Care. 2003;**26**(4):1277-1294

[33] Ibanez-Camacho R, Meckes-Lozoya M. Effect of a semipurified product obtained from *Opuntia streptacantha* L. (a cactus) on glycemia and triglyceridemia of rabbit. Archivos De Investigacion Medica. 1993;**14**(4):437-443

[34] Trejo-González A, et al. A purified extract from prickly pear cactus (*Opuntia fuliginosa*) controls experimentally induced diabetes in rats. Journal of Ethnopharmacology. 2003;**55**:27-33

[35] Laurenz JC, Collier CC, Kuti JO. Hypoglycaemic effect of Opuntia lindheimeri Englem in a diabetic pig model. Phytotherapy Research. 2003;**17**:26-29

[36] Frati-Munari AC, Fernández-Harp JA, Banales-Ham M, ArizaAndraca CR. Decreased blood glucose and insulin by nopal (*Opuntia* sp.). Archivos De Investigacion Medica. 1993;**14**:269-274

[37] Frati-Munari AC, Yever-Garces A, Islas-Andrade S, Ariza-Andraca CR, Chavez-Negrete A. Studies on the mechanism of "hypoglycemic" effect of nopal (*Opuntia* sp.). Archivos De Investigacion Medica. 1997;**18**:7-12

[38] Frati-Munari AC, Gordillo BE, Altamirano P, Ariza CR. Hypoglycemic effect of Opuntia streptacantha Lemaire in NIDDM. Diabetes Care. 1998;**11**:63-66

[39] López P, Pichardo E, Avila A, Vázquez N, Tovar A, Pedraza J, et al. The effect of nopal (*Opuntia ficus indica*) on postprandial blood glucose, incretins, and antioxidant activity in Mexican patients with type 2 diabetes after consumption of two different composition breakfasts. Journal of the Academy of Nutrition and Dietetics. 2014;**114**:1811-1818

[40] Porrata C, Abuín A, Morales A, Vilá R, Hernández M, Menéndez J, et al. Efecto terapéutico de la dieta macrobiótica Ma-Pi 2 en 25 adultos con diabetes mellitus tipo 2. Revista Cubana de Investigaciones Biomédicas. 2007;**26**(2)

[41] Berk Z. Technology of production of edible flours and protein products from soybeans. Food and Agriculture Organization of the United Nations. 1ª ed. Haifa, Israel. 1992. ISBN 92-5-103118-5

[42] Céspedes EM, Riverón G, Alonso CA, Gordon L. Evolución metabólica de pacientes diabéticos tipo 2 sometidos a un tratamiento combinado de dieta y ejercicios yoga. Revista Cubana de Investigaciones Biomedicas. 2002;**21**(2):98-101

[43] Garrido GA, De la Maza P, Valladares BL. Fitoestrógenos dietarios y sus potenciales beneficios en la salud del adulto humano. Revista Médica de Chile. 2003;**131**(11):1321-1328

[44] Torres N, Palacios B, Noriega L, Tovar AR. Índice glicémico, índice insulinémico y carga glicémica de bebidas de soya con un contenido bajo y alto en hidratos de carbono. Revista de Investigación Clínica. 2006;**58**(5):487-497

[45] Sterna V, Zute S, Brunava L. Oat grain composition and its nutrition benefice. Agriculture and Agricultural Science Procedia. 2016;**8**:252-256

[46] Cabrera Llano JL, Cárdenas Ferrer M. Importancia de la fibra dietética para la nutrición humana. Rev. Cubana Med. Salud Pub [online]. 2006;**32**(4): [cited 2017-05-04], pp. 0-0. Disponible en: <http://scielo.sld.cu/scielo.php?script=sci_arttext&pid=S0864-3466200600 0400015&lng=es&nrm=iso>. ISSN 0864-3466

[47] American Diabetes Association, Bantle JP, Wylie-Rosett J, Albright AL, Apovian CM, Clark NG, et al. Nutrition recommendations and interventions for diabetes: a position statement of the American Diabetes Association. Diabetes Care. 2008;**31**(1): S61–S78

[48] Escudero E, González P. La fibra dietética. Nutrición Hospitalaria. 2007;**21**(2):61-72

[49] Kerckhoffs D, Hornstra G, Mensink R. Cholesterol lowering effect of b-glucan from oat bran in mildly hypercholesterolemic subjects may decrease when b-glucan is incorporated into brad and cookies. American Journal of Clinical Nutrition. 2006;**78**:221-227

[50] Ruiz E, Mejía O, Herrera A, Cortes J. Consumo de avena (*Avena sativa*) y prevención primaria de la dislipidemia en adultos sin restricción dietética. Atención Familiar. 2011;**18**(2):35-37

[51] Raasmaja A. et al. A water-alcohol extract of *Citrus grandis* whole fruits has beneficial metabolic effects in the obese Zucker rats fed with high fat/high cholesterol diet. Food Chemistry. 2013;**138**:1392-1399

[52] Brown AA, Hu FB. Dietary modulation on endothelial function: implications for cardiovascular disease. American Journal of Clinical Nutrition. 2007;**71**:673-686

[53] Ceriello A, Testa R. Antioxidant anti-inflammatory treatment in type 2 diabetes. Diabetes Care. 2009;**32**:S32–S36

[54] Yusuf S, Dagenais G, Pogue J, Bosch J, Sleight P. Vitamin E supplementation and cardiovascular events in high-risk patients. The Heart Outcomes Prevention Evaluation Study Investigators. New England Journal of Medicine. 2000;**342**:154-160

[55] Geohas J, Daly A, Juturu V, Finch M, Komorowski JR. Chromium picolinate and biotin combination reduces atherogenic index of plasma in patients with type 2 diabetes mellitus: A placebo-controlled, double-blinded, randomized clinical trial. American Journal of the Medical Sciences. 2007;**333**(3):145-153

[56] Lu Q, BjÖrkhem I, Wretlind B, Diczfalusy U, Henriksson P, Freyschuss A. Effect of ascorbic acid on microcirculation in patients with Type II diabetes: a randomized placebo-controlled cross-over study. Clinical Science (London). 2005;**108**(6):507-513

[57] Montano S, Grunler J, Nair D, Tekle M, Fernandes A, et al. Glutaredoxin mediated redox effects of coenzyme Q10 treatment in type 1 and type 2 diabetes patients. BBA Clinical. 2015;**4**:14-20

[58] Burr GO, Burr MM. On the nature and role of fatty acids essential in nutrition. Journal of Biological Chemistry. 2000;**86**:587-621

[59] Cunnane SC. Problems with essential fatty acids: Time for a new paradigm? Progress in Lipid Research. 2003;**42**:544-568

[60] Simopoulos AP. Genetic variants in the metabolism of omega-6 and omega-3 fatty acids: Their role in the determination of nutritional requirements and chronic disease risk. Experimental Biology and Medicine (Maywood, NJ). 2010;**235**:785-795

[61] Calder PC, Yaqoob P, Thies F, Wallace FA, Miles EA, et al. Fatty acids and lymphocyte functions. British Journal of Nutrition. 2007;**87**:S31–S48

[62] Serhan CN, Chiang N. Endogenous pro-resolving and anti-inflammatory lipid mediators: A new pharmacologic genus. British Journal of Pharmacology. 2008;**153**(Suppl 1):S200–S215

[63] Calder PC. n-3 Polyunsaturated fatty acid, inflammation, and inflammatory disease. American Journal of Clinical Nutrition. 2006;**83**(6):1505S–1519S

[64] Sampath H, Ntambi JM. Polyunsaturated fatty acid regulation of genes of lipid metabolism. Annual Review of Nutrition. 2005;**25**:317-340

[65] Xi S, Cohen D, Barve S, Chen LH. Fish oil suppressed cytokines and nuclear factor kappa B induced by murine AIDS virus infection. Nutrition Research. 2008;**21**:865-878

[66] Petrova S, Dimitrov P, Willett WC, Campos H. The global availability of n-3 fatty acids. Public Health Nutrition. 2011;**31**:1-8

[67] Manerba A, Vizzardi E, Metra M, Dei Cas L. n-3 PUFAs and cardiovascular disease prevention. Future Cardiology. 2010;**6**:343-350

[68] Geleijnse JM, de Goede J, Brouwer IA. Alpha-linolenic acid: Is it essential to cardiovascular health?. Current Atherosclerosis Reports. 2010;**12**:359-637

[69] Nasiff-Hadad A., Meriño E. Ácidos grasos omega 3: pescados de carne azul y concentrados de aceites de pescado. Lo bueno y lo malo. Rev. Cubana Med. 2006, 42(2): 49-55.

Section 2

Functional Foods to Help Prevent Diseases of a Society Increasingly Aware of Its Feeding

Chapter 9

Meat Product Reformulation: Nutritional Benefits and Effects on Human Health

Elisabeta Botez, Oana V. Nistor,
Doina G. Andronoiu, Gabriel D. Mocanu and
Ioana O. Ghinea

Additional information is available at the end of the chapter

http://dx.doi.org/10.5772/intechopen.69118

Abstract

This chapter aims to present the current state of the art in the field of meat product reformulation with respect to issues concerning the nutritional improvement and overall health benefits of such products. Our research team has recently finalised a national research project concerning this topic, and we feel that other food scientists could benefit from the theoretical and practical knowledge gathered during this time. The chapter will be divided into four subchapters. The first subsection will present the main targets of meat reformulation, such as lipid or protein profile modification, the use of bioactive compounds as additives, etc. The second subsection will discuss the bioavailability and bioaccessibility of carotenoids, phenolic compounds and other bioactive compounds, presenting these parameters from a nutraceutical perspective. The last subsections will include reported consumer attitudes. In this work, we will present data that could aid scientists in the field of food science to better grasp notions concerning consumer benefit, such as bioavailability, not only of a specific bioactive compound but also as part of a complex food matrix.

Keywords: reformulation, bioactive compounds, bioavailability, bioaccessibility, nutrition, human health

1. Introduction

Meat and meat products are a class of food products that are commonly included in human diet, due to the intake of good quality nutrients, diverse forms of presentation and highly appreciated sensorial characteristics. On the other hand, a number of studies have been published on the negative impact of meat consumption upon health. In 2007, a report of the World Cancer Research

Fund described a connection between the intake of processed red meat and the risk of colorectal cancer. Although this connection has not been fully clarified yet, it is presumed that cancer precursors could be excess fat, protein and iron, heat-processing compounds (heterocyclic amines) and various substances added during the technological process (sodium chloride and nitrates). The same report recommended the intake of less than 500 g cooked red meat per week [1].

Similarly, the intake of processed red meat was associated to an increased occurrence of cardiovascular disease and diabetes mellitus, but the triggering mechanisms of these conditions have yet to be fully understood. In order to meet the consumers' demands, as consumers have become increasingly concerned with the ingredients of the food products purchased, the present research in the field approaches the topic of *reformulating the meat products*, impacting upon obtaining functional products. The technological strategies used to reformulate meat products and obtain functional products are based on improving the fat content, incorporating proteins of vegetable origin, prebiotics and vegetable fibres: increasing the mineral content, including vitamins, antioxidants and vegetable compounds with a functional role [2], and reducing the exogenous compounds harmful to health.

2. Meat product reformulation

Reformulated meat products have been created to help consumers, who are constantly requiring nutritionally improved meat products, that is, with a lower content of fats, cholesterol, sodium chloride and nitrites, as well as a higher content of compounds beneficial to human health. The influence of the meat product composition on human health has long been well known, but the scientific foundation of the physiological role of bioactive compounds in modulating specific functions in the body is not yet fully understood.

Reformulating meat products may be achieved in the following manners [3]:

1. Increasing the concentration of a meat product (macronutrient or micronutrient) up to a desired level
2. Adding a component normally not existing in meat
3. Partial or total replacement of a macronutrient which may trigger nutritional deficiencies with a nutritionally beneficial component
4. Reducing the nutritionally harmful components
5. Improving component bioavailability or stability
6. Combinations of the above

2.1. Reduction of cholesterol, sodium, nitrite and phosphate contents

Depending on its concentration, circulation or accumulation in the human body, cholesterol may be desired or not in diet. Due to the association of cholesterol-rich diet with coronary heart disease, in the most situations, food containing the high level of cholesterol is avoided.

http://dx.doi.org/10.5772/intechopen.69118

Meat (especially red meat) and meat products are among this food. On the other hand, meat and meat products provide beneficial compound for the human body: high-quality proteins, high bioavailable iron, vitamin B12, zinc and selenium. In this context, there have been developed some possibilities to reduce the level of cholesterol in meat and meat products: lecithin treatment; short-path and path molecular distillation; supercritical carbon dioxide extraction; extraction by saponin, using cholesterol oxidase; etc. Some of these methods are costly, non-selective and not enough studied. The addition of cholesterol-lowering compounds, such as phytosterols and soy proteins, is more suitable for this purpose [4].

Sodium chloride (currently named *salt*) is widely used in meat products due to a series of technological benefits (increases the proteins' water-binding capacity, improves texture and shelf life). Because of the negative health impact of sodium consumption (high blood pressure) [5], several strategies for lowering salt content in meat products have been reported [6]: the use of salt substitutes (potassium chloride, magnesium chloride, calcium chloride, calcium ascorbate [7]), the use of flavour enhancers (monosodium glutamate or yeast extract) and the use of novel processing technologies (high-pressure processing and power ultrasound). These strategies have their limitations and may be combined.

Nitrite has numerous functions in meat products [8, 9]: prevents lipid oxidation, gives products the specific colour and provides antimicrobial activity. Their reduction implies the addition of other antioxidants (either natural or artificial), colourants or preservatives. In the manufacture of meat products, phosphates are used in order to increase the water-holding capacity, leading to a good texture and poor cooking loss. Due to their implication in setting of chronic diseases like diabetes, obesity or cardiovascular disease, phosphates are tending to be replaced by sodium citrate, carageenans or proteins of different origins (porcine blood, soybean and milk) [10, 11].

2.2. Enrichment of minerals and improvement of amino acid quality

Meat is known as an essential source of macro- and micro-nutrients indispensable to human diet as protein, fat, minerals and vitamins. While the minerals can be achieved only by exogenous sources in the body, the enrichment of meat products with minerals is important. Several studies demonstrated the beneficial cumulated effects of low fat or low salt and minerals (as potassium, calcium and magnesium) added in meat matrix on the plasma cholesterol in humans [12]. Triki et al. [13] had reformulated sausages by partially replacing the NaCl content by adding a mixture of KCl, $CaCl_2$ and $MgCl_2$. They have found that the product mineral profile was improved providing 10–15% of the recommended daily allowance (RDA) of potassium, 8–10% of the calcium RDA and 10–20% of the magnesium RDA. One of the most essential trace minerals is selenium being involved in regulating various physiological functions. In human metabolism, selenium deficiency is associated with decreased immune function resulting in increased susceptibility to some chronically diseases [14, 15]. The enrichment of meat with selenium could be reached by two ways: adding selenium in different meat matrices or by feeding the animals with fortified food [16]. Essential amino acids are integral part of meat and meat products. The umami taste could be intensified by the presence of sweet amino acids, such as glycine, alanine and serine [17]. A large increase in free amino acid quality occurs during long maturation and the curing of meat products. Other researchers have found that amounts of hydrophobic amino acids released during the fermentation or maturation process were significantly higher than other amino acids.

2.3. Incorporation of some healthy ingredients, reduction of fat content and improvement of fatty acid content

Within of the framework of the Nutritional Optimizing of Some Meat Products with Valorization of Plants Riched in Bioactive Compounds (OPTIMEAT) project, the P2 Partner ('Dunarea de Jos' University of Galati) has investigated two possibilities for reformulating meat products:

(a) Lipid reformulation by adding a vegetable ingredient made up of nuts and nut oil, sea buckthorn oil or sunflower seed oil

(b) Proteic reformulation by adding a vegetable ingredient made up of soy proteic isolate and juice of red beetroot or dry tomatoes

The main components used in the project are presented below.

2.3.1. Walnuts

Walnuts (*Juglans regia* L.) are common all over the world. Known under various names, such as Persian nut, white nut, English nut or common nut, it is used to be cultivated in the Eastern Balkans and the Western Himalayan range, but at present it can be found all over Europe. Worldwide, there are many types of nuts, such as almonds, peanuts, earth nuts, cashew nuts, macadamia nuts, pistachios and pecan nuts.

It may be said that adding walnuts has positive implications in the creamy consistency of frankfurters as compared to traditional products where pork fat has a tougher consistency. There are also alterations in the fat-protein-fibre interactions supporting the gel formation process, which is essential in frankfurter manufacture. Thus, adding walnuts increases product consistency and at the same time the nutritional value of the product, becoming a viable alternative for this product [18–23]. The nutritional profile of products in which animal fat was replaced by walnuts is by far healthier and richer than that of the traditional products. By adding walnuts to products, an increase in the nutritional value and the quantity of biologically active compounds beneficial to human health can be observed. It may be observed that the number of studies in the field is relatively low, and the existing ones mention the need for further research, more detailed and on other products, and also in comparison to other products available on the market. Also, it is recommendable to study the stability and shelf life of these new products. The results of the academic studies are very valuable and recommend the use of walnuts in optimising the nutritional characteristics in meat products.

2.3.2. Tomatoes

Several studies using tomatoes and their derivatives were reported in improving meat products. Deda et al. [24] analysed the influence of adding tomato paste in pork frankfurters, reaching the conclusion that it enhances the colour and attractiveness of the final product. Similar results were obtained by Eyiler and Oztan [25] for adding tomato powder. Calvo et al. [26] studied the implications of adding tomato skins to raw-dried sausages, while Savadkoohi et al. [27] added extracts from tomato skins and seeds to frankfurters and beef ham. All these studies evinced the improved colour of the meat products obtained, as well as the improved texture and water-bonding ability. These effects are due to the high content of lycopene and beta carotene, as well as soluble fibres contained in tomatoes.

The bioavailability of lycopene depends on the following factors: the components of the food matrix, the physical state of lycopene, the size of particles before and after chewing, the intensity of digestive processes and the presence of fibres [28, 29]. Red tomatoes contain 95% lycopene as a *trans*-isomer (the most stable form of lycopene) [30, 31].

In addition to the beneficial effects on human health, tomato-derived products may contribute to reducing the added synthetic colourings in meat products, such as hamburgers, fresh sausages, salami or frankfurters, at the same time improving the nutritional profile by the content of bioactive components [24, 25, 32–34]. Certain synthetic colourings are considered responsible for allergic reactions or harmful side effects, and that is why consumers associate the presence of natural colourings with healthy and qualitative food products.

2.3.3. Soy protein isolate

Proteins from plants are used in meat industry for technological reasons, such as cutting costs and nutritional reasons and lately their health-promoting properties [39]. Soy beans contain the average 40% of protein and 20% of fat. By removing fat at low temperatures, the soy protein isolate is obtained, which is highly used in food industry. The predominant proteins in soy protein isolate are β-conglycinin and glycine. Their structure was thoroughly investigated by various methods, leading to the conclusion that glycine contains a multitude of disulphide groups, which is why its ability of foaming and emulsification is slow, as compared to β-conglicynin [35].

Proteic ingredients are the main vegetable component used in manufacturing meat products, for technological purposes—cutting costs—as well as for nutritional benefits, reducing the cholesterol level, increasing the proteic components and improving the amino acid profile. In meat industry, soy proteins are used in obtaining meat pasta to increase emulsion stability by forming a protein matrix that includes water and fat droplets [36]. Specialised literature in the field shows that adding soy proteins in products containing meat pasta has beneficial effects: Matulis et al. [37] reported a less rubbery texture of frankfurters with a low-fat content and Rahardjo et al. [38] reported lower cooking losses and improved texture of pork sausages. Das et al. [39] analysed the effects of adding soy (as pasta or textured granules) on the quality and storability of the nugget-type products made of goat meat. The findings of the study were that adding soy improves the appearance, texture and water-retaining capacity while slowing down fat oxidation during frozen storage. The data published by Youssef and Barbut [40] show that using soy proteins in obtaining meat paste improves the water-bonding ability, emulsion stability, appearance and texture while decreasing thermal treatment losses. The authors mentioned above analysed the microstructure of the samples obtained, concluding that adding soy proteins lowers the aggregation degree of meat proteins and reduces the size of fat droplets.

Although the influence of the soy protein addition on meat products has been thoroughly studied, their use is limited by the negative influence on taste, smell and colour. Under these circumstances studies are needed regarding the percentage of soy protein isolate that may be added to meat products in order to improve their quality.

2.3.4. *Red beetroot juice*

Red beetroot juice contains important quantities of antioxidants [41] together with micronutrients such as potassium, magnesium, folic acid, iron, zinc, calcium, phosphorus, vitamin B_6, soluble fibres and pigments (betalains—compounds of betacyanins and betaxanthins). Specialists have been increasingly interested in red beetroot juice due to the content of phenolic compounds [42, 43]. Red beetroot juice mainly contains pigments called betalains, a class of compounds derived from betalamic acid, mainly composed of betacyanin and betaxanthin. In addition to these, red beetroot juice also contains small amounts of gallic, syringic and caffeic acid, as well as flavonoids [44]. Betalains are used in food industry as natural colourings, but a series of health benefits were also found, antioxidant and anti-inflammatory [45, 46], inhibiting lipid peroxidation [47] and increased resistance to lipoprotein oxidation in low density [48].

2.3.5. *Vegetable oils*

Vegetable oils play an important role in the human diet and are an important energy source. The main constituents of oils are fatty acids, classified as saturated fatty acids (SFA), monounsaturated fatty acids (MUFA) and polyunsaturated fatty acids (PUFA). Polyunsaturated fatty acids determine the regulation at an optimum level of lipids, mainly low density lipid (LDL) cholesterol in the human body [49–51]. **Table 1** shows the fatty acid percentage for oils expressed from sunflower seeds, soy, palm and walnuts.

The partial or total replacement of animal fat in meat products by vegetable oils may be seen as an efficient strategy of nutritional improvement and a means of increasing oxidative stability.

Fatty acids (%)	Sunflower seed oil	Soy oil	Palm oil	Walnut oil
Saturated fatty acids	8.51 ± 1.91	18.26 ± 0.67	46.34 ± 0.40	9.18 ± 1.09
Monounsaturated fatty acids	45.5 ± 16.89	23.28 ± 1.99	41.46 ± 0.56	23.22 ± 2.87
Polyunsaturated fatty acids	46.10 ± 14.92	57.86 ± 1.20	11.84 ± 0.92	63.45 ± 4.66

Table 1. The content of fatty acids for certain types of vegetable oils.

3. Bioavailability and bioaccessibility of bioactive compounds

In pharmacology, bioavailability is defined as the ratio between the amount of active substance and the speed at which it is yielded and absorbed into the body, then reaches its point of action and manifests its biological effect. By definition, if the medicine is intravenuously administered, its bioavailability is 100%.

As far as food supplements are concerned, since their administration is most often than not oral, the bioavailability is the ratio between the amount of ingested substance and the amount of the absorbed substance [52]. The nutrients existing in food are not absorbed and used by the body in their entirety. Among the factors responsible for this phenomenon, there are a number of nutrient-related factors (chemical formula, the presence of inhibitors or enhancers, the possibility of interacting with other components) and a number of factors related to the organism using that nutrient (duration of intake, volume of enzymatic secretion, activity of intestinal microflora, state of health, eating style, etc.) [53].

Fat-soluble vitamins (e.g. A, D and E) as well as the ω-3 fatty acids, carotenoids, conjugated linoleic acid (CLA) or curcumin are micronutrients with a hydrophobic behaviour which may play a potential functional role when included in the diet or a food product. Many studies showed that due to the hydrophobic behaviour of these micronutrients, bioavailability is slow or variable [54]. The factors contributing to decreasing bioavailability are grouped into three categories – bioaccessibility, absorption and transformation. Bioavailability refers to the low release into the food matrix, low solubilisation in the gastrointestinal fluids as well as interaction with other insoluble components. Deficient absorption is due to the transportation through the stomach membrane or inhibiting active transporters. Transformation refers to the multiple chemical or metabolic processes in which micronutrients may participate.

Bioactive compounds have various characteristics such as structure and molecular weight, polarity and physical state. They may be introduced directly in a food matrix or indirectly by means of a transportation system. The transportation system focuses on maintaining or improving the bioavailability of the bioactive components and has to possess the following characteristics: protection against chemical or biological spoilage (especially for oxidation and hydrolysis), control of the release of the bioactive component (depending on pH, temperature and other factors) as well as the compatibility between the bioactive component and other parameters of the food matrix [55]. The bioavailability of bioactive compounds is generally low and depends on the components of food matrix. Some processes like ingestion, diffusion, solubilisation, movement across intern membrane and enters in the lymphatic system and circulation affect the bioavailability of bioactive compounds.

4. Effects on human health

The bioactive compounds from the selected sources (described in Subchapter 1.3) have some benefits to human health:

- ω-3 Fatty acids (from walnuts)—anti-inflammatory activity, reduces the risk of cardiovascular disease [56].
- Sterols and stanols (from nuts and vegetable oils)—reduce the total cholesterol level, protection against certain types of cancer, anti-inflammatory activity and improve blood pressure [41, 57, 58].

- Lycopene (from tomatoes)—antioxidant, reduces the risk of cardiovascular disease and protection against certain types of cancer [59, 60].
- Isoflavones (from soy proteic concentrate)—reduce the risk of cardiovascular disease [61].

Nut consumption as a trend is on the increase, especially due to the major nutritional components (proteins, unsaturated fatty acids and fibres), as well as micronutrients (sterols, vitamins, minerals, fatty acids and phenolic compounds) [61–63] and antioxidants [64]. As expected, the consumption of nuts is on the increase owing to their antioxidant properties, mainly responsible for the lowering of LDL cholesterol and associated triglycerides, leading to better results than traditional low-calorie diets, in which the consumption of oils or carbohydrates is replaced by nuts. As the consumption of nuts by Mediterranean population is higher as compared to other areas, the mortality rate caused by heart disease or cancer is low [65].

Walnuts are well known for their nutritional value and the high content of bioactive compounds, such as antioxidants, vitamins, essential amino acids and minerals [66, 67]. It is a common knowledge that free radicals are the main factors causing human illnesses, with implications in the pathology of cancer, atherosclerosis or inflammatory disease [68], and that is why regular intake of nuts and thus of antioxidants is essential. *J. regia* Linn may be used in traditional medicine in preventing or treating helminths, diarrhoea, sinus ailments, gastritis, arthritis, asthma, eczema, dermatitis and the various endocrine diseases, such as diabetes, anorexia, thyroid problems, infectious diseases and cancer. Walnuts are also well known for their rich content of unsaturated fatty acids, vitamin E, fibres, magnesium and potassium [69]. As compared to other nut types (macadamia nuts, pistachios, almonds, cashew nuts, earth nuts, pecans, etc.), which mainly contain monounsaturated fatty acids (MUFA), walnuts are rich in omega-6 and omega-3 polyunsaturated fatty acids (PUFA), which play an essential role in daily diet [70].

These properties qualify walnuts as unique in each consumer's diet. Many studies showed that walnut intake may protect the human body against cardiovascular disease [71] and work as blood pressure regulator by their content of magnesium and potassium, respectively. Replacing saturated fats in daily diet with other mono- or polyunsaturated fatty acids (MUFA or PUFA) decreases the concentration of LDL cholesterol in the plasmatic liquid. The chemical and mineral components may differ according to the variety of genotype conditions, ecological, technical and cultural conditions, climate conditions.

Walnuts are tremendously beneficial to the human body because of their chemical composition; they are also a rich source of fatty acids (mainly the linoleic acid, followed by the oleic, linolenic, palmitic and stearic acids) [70, 72] and tocopherols [70, 73]. In addition, they contain other components beneficial to human health, such as proteins, vegetable fibres, sterols [70], melatonin [74], folates, tannins and polyphenols [75].

Walnuts were selected as potential functional component in reformulating meat products due to the composition of the lipid fractions, especially ω-3 and ω-6 acids and ϒ-tocopherol. Numerous studies [76–79] show that reformulating meat products by adding walnuts in various ratios leads to reducing the risk of cardiovascular disease. Although the action mechanism is not yet fully understood, this effect is due to the high content of lipids (62–68% of the dry substance) and the high ratio of monounsaturated (MUFA) and polyunsaturated fatty acids (PUFA).

Selecting tomatoes as a source of bioactive compounds was based on lycopene, the main pigment in the carotenoid class contained by tomatoes. This carotenoid was studied by many researchers, who found proof in favour of its antioxidant and cancer-preventing properties [80–84]. Together with lycopene, tomatoes are an important source of vitamins A and C, as well as a high content of carotenoids. The role of these antioxidants is to neutralise free radicals and to prevent the decay of cells and membranes, swelling and the occurrence of diseases like atherosclerosis, asthma, diabetes and cancer [85]. Tomatoes also contain high amounts of potassium, niacin, vitamin B_6, folates and riboflavin.

Soy protein isolate has a series of nutritional benefits due to the lower energy value and cholesterol content (when used as fat replacements), the higher protein content, the balanced amino acid profile and the incorporated bioactive compounds [86, 87]. Certain vegetable proteins (sunflower, walnuts) were used in meat systems to balance the lysine/arginine ratio [88]. Soy proteins have been focused on by meat specialists for numerous reasons, such as they ensure a balance in amino acid composition, contain beneficial bioactive components decreasing the cholesterol level in the bloodstream and reduce the risk of cardiovascular disease, and have excellent technological properties like jellification, emulsification and the ability to retain water and fats [35, 89]. Soy proteins are well known for their preventative and therapeutic effect in heart disease, cancer and osteoporosis [90]. Clinical studies on the bioavailability of the soy isoflavone forms (such as food supplements, additives or soy food products) were performed in various geographical areas [91, 92]. However, it may be stated that data are inconclusive for a definite conclusion because of the different dietary habits of the individuals included in the studies, the composition of isoflavones and the amount and quality of the meals under study.

The studies carried out by Wootton-Beard and Ryan [93] showed that red beetroot juice is an important source of antioxidants and polyphenols, which were quantified by various biochemical methods before and after in vitro digestion. McDougall and Stewart [94] proved that polyphenols inhibit α-glycosidase resulting in the stimulation of insulin secretion, thus reducing the absorption of glucose into the bloodstream. Polyphenols increase the glutathione level and the level of antioxidant enzymes (glutathione peroxidase, catalase and superoxide dismutase), being capable of reducing the oxidative stress which is the cause of dysfunctions in the case of cardiovascular disease, diabetes and autoimmune diseases. Being natural products, polyphenols may act on various paths in order to prevent chronic inflammation and are more efficient than synthetically obtained anti-inflammatory medication [95].

Many types of vegetable oils are considered as food products with multiple benefits to human health. Especially, cold-expressed oils are a great source of bioactive lipids, phenolic compounds with an antioxidant role, which may contribute to improving human health [96]. Antioxidants play an important role in maintaining the stability of vegetable oils and reduce oxidative stress in vivo.

5. Consumer attitudes to food reformulation

Meat and meat products many times are comprehended by the consumers like unhealthy. A chance for meat industry to change this perception may be represented by functional or reformulated meat products [1]. To answer the consumers' needs, the reformulated meat products have been developed. According to Jiménez-Colmenero et al. [88], the consumers may approve the reformulated products if they are promoted like 'healthier' products. To satisfy these needs, the meat industry is encouraged to make new meat products. However, it is a provocation to convince the consumers [97] (as well as the media, nutritionists and legislative authorities) that meat is a suitable carrier for functional ingredients [1, 98, 99]. It is significant to present to the consumer that reformulated meat products can be performed in a manner which will meet all the relevant qualities which consumers look for in traditional meat products [91].

6. Conclusions

Functional foods represent a good opportunity for the meat industry, in order to improve the quality of meat, and create meat products with health beneficial properties. Meat and meat products are excellent foods for delivering bioactive compounds without changing dietary habits. Some bioactive compounds from fruits and vegetables (walnuts, tomatoes, soy protein isolate, red beetroot juice and vegetable oils) appear to play an important role in the prevention of specific diseases like cardiovascular diseases, cancers and diabetes mellitus. These compounds are able to reduce the oxidative stress, which has been associated to the occurrence of chronic diseases, and maintain the health. Nowadays, the consumers demand natural and healthy food products, including meat products, with better nutritional properties. Promoting health through nutrition is an important objective of nutrition and public health programmes in a large number of European countries.

Acknowledgements

This work was supported by the CNCSIS-UEFISCDI Romania as National Project II No. 115/2012—Nutritional Optimizing of Some Meat Products with Valorization of Plants Riched in Bioactive Compounds (OPTIMEAT).

Author details

Elisabeta Botez[1]*, Oana V. Nistor[1], Doina G. Andronoiu[1], Gabriel D. Mocanu[1] and Ioana O. Ghinea[2]

*Address all correspondence to: Elisabeta.Botez@ugal.ro

1 Department of Food Science, Food Engineering and Applied Biotechnology, Faculty of Food Science and Engineering, "Dunarea de Jos" University of Galati, Galati, Romania

2 Department of Chemistry, Physics and Environment, Faculty of Science and Environment, "Dunarea de Jos" University of Galati, Galati, Romania

References

[1] Grasso S, Brunton NP, Lyng JG, Lalor F, Monahan FJ. Healthy processed meat products – Regulatory, reformulation and consumer challenges. Trends in Food Science & Technology. 2014;**39**(1):4-17. DOI: http://dx.doi.org/10.1016/j.tifs.2014.06.006

[2] Olmedilla-Alonso B, Jiménez-Colmenero F, Sánchez-Muniz FJ. Development and assessment of healthy properties of meat and meat products designed as functional foods. Meat Science. 2013;**95**(4):919-930. DOI: 10.1016/j.meatsci.2013.03.030

[3] Ashwell M. Concepts of Functional Foods. Brussels, Belgium: ILSI Press; 2002. 39 p

[4] Cohn JS, Kamili A, Wat E, Chung RWS, Tandy S. Reduction in intestinal cholesterol absorption by various food components: Mechanisms and implications. Atherosclerosis Supplements. 2010;**11**(1):45-48. DOI: http://dx.doi.org/10.1016/j.atherosclerosissup.2010.04.004

[5] Aburto NJ, Ziolkovska A, Hooper L, Elliott P, Cappuccio FP, Meerpohl JJ. Effect of lower sodium intake on health: Systematic review and meta-analyses. British Medical Journal. 2013;**346**:1-20. DOI: https://doi.org/10.1136/bmj.f1326

[6] Inguglia ES, Zhang Z, Tiwari BK, Kerry JP, Burgess CM. Salt reduction strategies in processed meat products – A review. Trends in Food Science & Technology. 2017;**59**:70-78. DOI: http://dx.doi.org/10.1016/j.tifs.2016.10.016

[7] García-Íñiguez de Ciriano M, Berasategi I, Navarro-Blasco Í, Astiasarán I, Ansorena D. Reduction of sodium and increment of calcium and u-3 polyunsaturated fatty acids in dry fermented sausages: Effects on the mineral content, lipid profile and sensory quality. Journal of the Science of Food and Agriculture. 2013;**93**(4):876-881. DOI: 10.1002/jsfa.5811

[8] de Oliveira TLC, de Carvalho SM, de Araújo Soares R, Andrade MA, das Graças Cardoso M, Ramos EM, Hilsdorf Piccoli R. Antioxidant effects of *Satureja montana* L. essential oil on TBARS and color of mortadella-type sausages formulated with different levels of sodium nitrite. LWT – Food Science and Technology. 2012;**45**(2):204-212. DOI: http://dx.doi.org/10.1016/j.lwt.2011.09.006

[9] Bedale W, Sindelar JJ, Milkowski AL. Dietary nitrate and nitrite: Benefits, risks, and evolving perceptions. Meat Science. 2016;**120**:85-92. DOI: http://dx.doi.org/10.1016/j.meatsci.2016.03.009

[10] Alvarado C, McKee S. Properties and safety of poultry meat. Journal of Applied Poultry Research. 2007;**16**(1):113-120. DOI: https://doi.org/10.1093/japr/16.1.113

[11] Hurtado S, Dagà I, Espigulé E, Parés D, Saguer E, Toldrà M, Carretero C. Use of porcine blood plasma in "phosphate-free frankfurters". Procedia Food Science. 2011;**1**:477-482. DOI: http://dx.doi.org/10.1016/j.profoo.2011.09.073

[12] Tapola NS, Lyyra ML, Karvonen HM, Uusitupa MI, Sarkkinen ES. The effect of meat products enriched with plant sterols and minerals on serum lipids and blood pressure.

International Journal of Food Sciences and Nutrition. 2004;**55**(5):389-397. DOI: http://dx.doi.org/10.1080/09637480400002842

[13] Triki M, Herrero AM, Jiménez-Colmenero F, Ruiz-Capillas C. Storage stability of low-fat sodium reduced fresh merguez sausage prepared with olive oil in konjac gel matrix. Meat Science. 2013;**94**(4):438-446. DOI: http://dx.doi.org/10.1016/j.meatsci.2013.03.019

[14] Gramadzińska J, Reszka E, Bruzelius K, Wasowicz W, Åkesson B. Selenium and cancer: Biomarkers of selenium status and molecular action of selenium supplements. European Journal of Nutrition. 2008;**47**(Suppl. 2):29-50. DOI: 10.1007/s00394-008-2005-z

[15] Müller AS, Mueller K, Wolf NM, Pallauf J. Selenium and diabetes: An enigma?. Free Radical Research. 2009;**43**(11):1029-1059. DOI: 10.1080/10715760903196925

[16] Fisinin VI, Papazyan TT, Surai PF. Producing selenium-enriched eggs and meat to improve the selenium status of the general population. Critical Reviews in Biotechnology. 2009;**29**(1):18-28. DOI: 10.1080/07388550802658030

[17] Kawai M, Okiyama A, Ueda Y. Taste enhancements between various amino acids and IMP. Chemical Senses. 2002;**27**(8):739-745. DOI: https://doi.org/10.1093/chemse/27.8.739

[18] Álvarez D, Xiong YL, Castillo M, Payne FA, Garrido MD. Textural and viscoelastic properties of pork frankfurters containing canola–olive oils, rice bran, and walnut. Meat Science. 2012;**92**(1):8-15. DOI: http://dx.doi.org/10.1016/j.meatsci.2012.03.012

[19] Ayo J, Carballo J, Serrano J, Olmedilla-Alonso B, Ruiz-Capillas C, Jiménez-Colmenero F. Effect of total replacement of pork backfat with walnut on the nutritional profile of frankfurters. Meat Science. 2007;**77**(2):173-181. DOI: http://dx.doi.org/10.1016/j.meatsci.2007.02.026

[20] Ayo J, Carballo J, Solas MT, Jiménez-Colmenero F. Physicochemical and sensory properties of healthier frankfurters as affected by walnut and fat content. Food Chemistry. 2008;**107**(4):1547-1552. DOI: http://dx.doi.org/10.1016/j.foodchem.2007.09.019

[21] Serrano A, Cofrades S, Ruiz-Capillas C, Olmedilla-Alonso B, Herrero-Barbudo C, Jiménez-Colmenero F. Nutritional profile of restructured beef steak with added walnuts. Meat Science. 2005;**70**(4):647-654. DOI: http://dx.doi.org/10.1016/j.meatsci.2005.02.014

[22] Botez E, Mocanu GD, Stoian I, Nistor OV, Andronoiu DG, Mihociu T, Şerban MA. Healthy lipid combination. Effect of thermal processing on the quality characteristics of meat products. Bulgarian Chemical Communications. 2014;**46**(Special Issue B):49-52

[23] Mocanu GD, Barbu M, Nistor OV, Andronoiu DG, Botez E. The effect of the partial substitution of pork back fat with vegetable oils and walnuts on the chemical composition, texture profile and sensorial properties of meatloaf. The Annals of the University Dunărea de Jos of Galaţi Fascicle VI – Food Technology. 2015;**39**(1):58-69

[24] Deda MS, Bloukas JG, Fista GA. Effect of tomato paste and nitrite level on processing and quality characteristics of frankfurters. Meat Science. 2007;**76**(3):501-508. DOI: http://dx.doi.org/10.1016/j.meatsci.2007.01.004

[25] Eyiler E, Oztan A. Production of frankfurters with tomato powder as a natural additive. LWT-Food Science and Technology. 2011;**44**(1):307-311. DOI: http://dx.doi.org/10.1016/j.lwt.2010.07.004

[26] Calvo MM, Garcia ML, Selgas MD. Dry fermented sausages enriched with lycopene from tomato peel. Meat Science. 2008;**80**(2):167-172. DOI: http://dx.doi.org/10.1016/j.meatsci.2007.11.016

[27] Savadkoohi S, Hoogenkamp H, Shamsi K, Farahnaky A. Color, sensory and textural attributes of beef frankfurter, beef ham and meat-free sausage containing tomato pomace. Meat Science. 2014;**97**(4):410-418. DOI: http://dx.doi.org/10.1016/j.meatsci.2014.03.017

[28] Johnson EJ. Human studies on bioavailability and plasma response of lycopene. Experimental Biology and Medicine. 1998;**218**(2):115-120. DOI: https://doi.org/10.3181/00379727-218-44284a

[29] Rodriguez-Amaya DB. Natural food pigments and colorants. Current Opinion in Food Science. 2016;**7**:20-26. DOI: http://dx.doi.org/10.1016/j.cofs.2015.08.004

[30] Sesso HD, Liu S, Gaziano JM, Buring JE. Dietary lycopene, tomato-based food products and cardiovascular disease in women. Journal of Nutrition. 2003;**133**(7):2336-2341

[31] Singh P, Goyal GK. Dietary lycopene: Its properties and anticarcinogenic effects. Comprehensive Rewiews in Food Science and Food Safety. 2008;**7**(3):255-270. DOI: 10.1111/j.1541-4337.2008.00044.x

[32] Doménech-Asensi G, Garcia-Alonso FJ, Martinez E, Santaella M, Martin-Pozuelo G, Bravo S, Periago MJ. Effect of the addition of tomato paste on the nutritional and sensory properties of mortadella. Meat Science. 2013;**93**(2):213-219. DOI: http://dx.doi.org/10.1016/j.meatsci.2012.08.021

[33] Garcia ML, Calvo MM, Selgas MD. Beef hamburgers enriched in lycopene using dry tomato peel as an ingredient. Meat Science. 2009;**83**(1):45-49. DOI: http://dx.doi.org/10.1016/j.meatsci.2009.03.009

[34] Mercadante AZ, Capitani CD, Decker EA, Castro IA. Effect of natural pigments on the oxidative stability of sausages stored under refrigeration. Meat Science. 2010;**84**(4):718-726. DOI: http://dx.doi.org/10.1016/j.meatsci.2009.10.031

[35] Nishinari K, Fang Y, Guo S, Philips GO. Soy proteins: A review on composition, aggregation and emulsification. Food Hydrocolloids. 2014;**39**:301-318. DOI: http://dx.doi.org/10.1016/j.foodhyd.2014.01.013

[36] McArdle R, Hamill R, Kerry JP. Utilisation of hydrocolloids in processed meat systems. In: Kerry JP, Kerry JF, editors. Processed Meats: Improving Safety, Nutrition and Quality. 1st ed. Cambridge, UK: Woodhead Publishing Limited; 2011. pp. 243-269. DOI: http://dx.doi.org/10.1016/B978-1-84569-466-1.50028-3

[37] Matulis RJ, McKeith FK, Sutherland JW, Brewer MS. Sensory characteristics of frankfurters as affected by salt, fat, soy protein, and carrageenan. Journal of Food Science. 1995;**60**(1):48-54. DOI: 10.1111/j.1365-2621.1995.tb05604.x

[38] Rahardjo R, Wilson LA, Sebranek JG. Spray dried soymilk used in reduced fat pork sausage patties. Journal of Food Science. 2006;**59**(6):1286-1290. DOI: 10.1111/j.1365-2621.1994.tb14697.x

[39] Das AK, Anjaneyulu ASR, Gadekar YP, Singh RP, Pragati H. Effect of full-fat soy paste and textured soy granules on quality and shelf-life of goat meat nuggets in frozen storage. Meat Science. 2008;**80**(3):607-614. DOI: 10.1016/j.meatsci.2008.02.011

[40] Youssef MK, Barbut S. Effects of two types of soy protein isolates, native and preheated whey protein isolates on emulsified meat batters prepared at different protein levels. Meat Science. 2011;**87**(1):54-60. DOI: http://dx.doi.org/10.1016/j.meatsci.2010.09.002

[41] Woyengo TA, Ramprasath VR, Jones PJH. Anticancer effects of phytosterols. European Journal of Clinical Nutrition. 2009;**63**(7):813-820. DOI: 10.1038/ejcn.2009.29

[42] Kaur C, Kapoor HC. Anti-oxidant activity and total phenolic content of some Asian vegetables. International Journal of Food Science and Technology. 2002;**37**(2):153-161. DOI: 10.1046/j.1365-2621.2002.00552.x

[43] Pitalua E, Jimenez M, Vernon-Carter EJ, Beristain CI. Antioxidative activity of microcapsules with beetroot juice using gum Arabic as wall material. Food and Bioproducts Processing. 2010;**88**(2-3):253-258. DOI: http://dx.doi.org/10.1016/j.fbp.2010.01.002

[44] Kazimierczak R, Hallmann E, Lipowski J, Drela N, Kowalik A, Püssa T, Matt D, Luik A, Gozdowski D, Rembia XKE. Beetroot (*Beta vulgaris* L.) and naturally fermented beetroot juices from organic and conventional production: Metabolomics, antioxidant levels and anti-cancer activity. Journal of the Science of Food and Agriculture. 2014;**98**(13):2618-2629. DOI: 10.1002/jsfa.6722

[45] Georgiev VG, Weber J, Kneschke EM, Denev PN, Bley T, Pavlov AI. Antioxidant activity and phenolic content of betalain extracts from intact plants and hairy root cultures of the red beetroot *Beta vulgaris* cv. Detroit dark red. Plant Foods for Human Nutrition. 2010;**65**(2):105-111. DOI: 10.1007/s11130-010-0156-6

[46] Zielińska-Przyjemska M, Olejnik A, Dobrowolska-Zachwieja A, Grajek W. In vitro effects of beetroot juice and chips on oxidative metabolism and apoptosis in neutrophils from obese individuals. Phytotherapy Research. 2009;**23**(1):49-55. DOI: 10.1002/ptr.2535

[47] Reddy MK, Alexander-Lindo RL, Nair MG. Relative inhibition of lipid peroxidation, cyclooxygenase enzymes, and human tumor cell proliferation by natural food colors. Journal of Agricultural and Food Chemistry. 2005;**53**(23):9268-9273. DOI: 10.1021/jf051399j

[48] Tesoriere L, Butera D, D'Arpa D, Di Gaudio F, Allegra M, Gentile C, Livrea MA. Increased resistance to oxidation of betalain-enriched human low-density lipoproteins. Free Radical Research. 2003;**37**(6):689-696. DOI: http://dx.doi.org/10.1080/1071576031000097490

[49] Chowdhury K, Banu LA, Khan S, Latif A. Studies on the fatty acid composition of edible oil. Bangladesh Journal of Science and Industrial Research. 2007;**42**(3):311-316. DOI: http://dx.doi.org/10.3329/bjsir.v42i3.669

[50] Dhavamani S, Rao YPC, Lokesh BR. Total antioxidant activity of selected vegetable oils and their influence on total antioxidant values in vivo: A photochemiluminescence based analysis. Food Chemistry. 2014;**164**:551-555. DOI: http://dx.doi.org/10.1016/j.foodchem.2014.05.064

[51] Nehir El S, Simsek S. Food technological applications for optimal nutrition: An overview of opportunities for the food industry. Comprehensive Reviews in Food Science and Food Safety. 2012;**11**(1):2-12. DOI: 10.1111/j.1541-4337.2011.00167.x

[52] Heaney RP. Factors influencing the measurement of bioavailability, taking calcium as a model. Journal of Nutrition. 2001;**131**(Suppl. 4):1344S-1348S

[53] Srinivasan VS. Bioavailability of nutrients: A practical approach to in vitro demonstration of the availability of nutrients in multivitamin-mineral combination products. Journal of Nutrition. 2001;**131**(Suppl. 4):1349S-1350S

[54] McClements DJ, Xiao H. Excipient foods: Designing food matrices that improve the oral bioavailability of pharmaceuticals and nutraceuticals. Food & Function. 2014;**5**(7):1320-1333. DOI: 10.1039/C4FO00100A

[55] Señorans JF, Ibáñez E, Cifuentes A. New trends in food processing. Critical Reviews in Food Science and Nutrition. 2003;**43**(5):507-526. DOI: http://dx.doi.org/10.1080/10408690390246341

[56] Harris WS, Miller M, Tighe AP, Davidson MH, Schaefer EJ. Omega-3 fatty acids and coronary heart disease risk: Clinical and mechanistic perspective. Atherosclerosis. 2008;**197**(1):12-24. DOI: http://dx.doi.org/10.1016/j.atherosclerosis.2007.11.008

[57] Berger A, Jones PJ, Abumweis SS. Plant sterols: Factors affecting their efficacy and safety as functional food ingredients. Lipids in Health and Disease. 2004;**3**(1):5-19. DOI: 10.1186/1476-511X-3-5

[58] Rudkowska I. Plant sterols and stanols for health ageing. Maturitas. 2010;**66**(2):158-162. DOI: http://dx.doi.org/10.1016/j.maturitas.2009.12.015

[59] Kim JY, Paik JK, Kim OY, Park HW, Lee JH, Jang Y. Effects of lycopene supplementation on oxidative stress and stress and markers of endothelial function in health men. Atherosclerosis. 2011;**215**(1):189-195. DOI: http://dx.doi.org/10.1016/j.atherosclerosis.2010.11.036

[60] Wang H, Leung LK. The carotenoid lycopene differentially regulates phase I and II enzymes in enzymes in dimethylbenz[α]anthracene-induced MCF-7 cells. Nutrition. 2010;**26**(11-12):1181-1187. DOI: 10.1016/j.nut.2009.11.013

[61] Dewell A, Hollenbeck PLW, Hollenbeck CB. Clinical review: A critical evaluation of the role of soy protein and isoflavone supplementation in the control of plasma cholesterol concentrations. Journal of Clinical Endocrinology & Metabolism. 2006;**91**(3):772-780. DOI: 10.1210/jc.2004-2350

[62] Brufau G, Boatella J, Rafecas M. Nuts: Source of energy and macronutrients. British Journal of Nutrition. 2006;**96**(Suppl. 2):S24-S28. DOI: 10.1017/BJN20061860

[63] Venkatachalam M, Sathe SS. Chemical composition of selected edible nut seeds. Journal of Agricultural and Food Chemistry. 2006;**54**(13):4705-4714. DOI: 10.1021/jf0606959

[64] Yang J, Liu RH, Halim L. Antioxidant and antiproliferative activities of common edible nut seeds. LWT – Food Science and Technology. 2009;**42**(1):1-8. DOI: http://dx.doi.org/10.1016/j.lwt.2008.07.007

[65] Simopoulos AP. The Mediterranean diets: What is so special about the diet of Greece? The scientific evidence. Journal of Nutrition. 2001;**131**(Suppl. 11):3065S-3073S

[66] Cabrera C, Lloris F, Giménez R, Ollala M, López MC. Mineral content in legumes and nuts: Contribution on the Spanish dietary intake. Science of Total Environment. 2003;**308**(1-3):1-14. DOI: http://dx.doi.org/10.1016/S0048-9697(02)00611-3

[67] Gómez-Ariza JL, Arias-Borrego A, Garcia-Barrera T. Multielemental fractionation in pine nuts (*Pinus pinea*) from different geographic origins by size-exclusion chromatography with UV and inductively coupled plasma mass spectrometry detection. Journal of Cromatography A. 2006;**1121**(2):191-199. DOI: http://dx.doi.org/10.1016/j.chroma.2006.04.025

[68] Scalbert A, Williamson G. Dietary intake and bioavailability of polyphenols. Journal of Nutrition. 2000;**130**(Suppl. 8):2073S-2085S

[69] Dreher ML, Maher CV, Kearney P. The traditional and emerging role of nuts in healthful diets. Nutrition Reviews. 1996;**54**(8):241-245. DOI: 10.1111/j.1753-4887.1996.tb03941.x

[70] Amaral JS, Casal S, Pereira J, Seabra R, Oliveira B. Determination of sterol and fatty acid compositions, oxidative stability, and nutritional value of six walnut (*Juglans regia* L.) cultivars grown in Portugal. Journal of Agricultural and Food Chemistry. 2003;**51**(26):7698-7702. DOI: 10.1021/jf030451d

[71] Prineas RJ, Kushi LH, Folsom AR, Bostick RM, Wu Y, Mann GV, Mirkin G, Mogadam M, Sabate J, Fraser GE. Walnuts and serum lipids [1]. New England Journal of Medicine. 1993;**329**(5):358-360. DOI: 10.1056/NEJM199307293290513

[72] Savage GP. Chemical composition of walnuts (*Juglans regia* L) grown in New Zealand. Plant Foods for Human Nutrition. 2001;**56**(1):75-82. DOI: 10.1023/A:1008175606698

[73] Amaral JS, Rui Alvez M, Seabra RM, Oliveira BPP. Vitamin E composition of walnut (*Juglans regia* L): A 3-year comparative study of different cultivars. Journal of Agricultural and Food Chemistry. 2005;**53**(13):5467-5472. DOI: 10.1021/jf050342u

[74] Reiter RJ, Manchester LC, Tan DX. Melatonin in walnut: Influence on levels of melatonin and total antioxidant capacity of blood. Nutrition. 2005;**21**(9):920-924. DOI: 10.1016/j.nut.2005.02.005

[75] Li L, Tsao R, Yang R, Liu C, Zhu H, Young JC. Polyphenolic profiles and antioxidant activities of heartnut (*Juglans ailanthifolia* var. *cordiformis*) and Persian walnut (*Juglans regia* L.). Journal of Agricultural and Food Chemistry. 2006;**54**(21):8033-8040. DOI: 10.1021/jf0612171

[76] Banel DK, Hu FB. Effects of walnut consumption on blood lipids and other cardiovascular risk factors: A meta-analysis and systematic review. American Journal of Clinical Nutrition. 2009;**90**(1):56-63. DOI: 10.3945/ajcn.2009.27457

[77] Feldman EB. The scientific evidence for a beneficial health relationship between walnuts and coronary heart disease. Journal of Nutrition. 2002;**132**(5):1062S-1101S

[78] Iwamoto M, Sato M, Kono M, Hirooka Y, Sakai K, Takeshita A. Walnuts lower serum cholesterol in Japanese men and women. Journal of Nutrition. 2000;**130**(9):171-176

[79] Salas-Salvadó J, Fernández-Ballart J, Ros E, Martínez-González MA, Fitó M, Estruch R, Corella D, Fiol M, Gómez-Gracia E, Arós F, Flores G, Lapetra J, Lamuela-Raventós R, Ruiz-Gutiérrez V, Bulló M, Basora J, Covas MI. Effect of a Mediterranean diet supplemented with nuts on metabolic syndrome status. Archives of Internal Medicine. 2008;**168**(22):2449-2458. DOI: 10.1001/archinte.168.22.2449

[80] Bertram JS, King T, Fukishima L, Khachik F. Enhanced activity of an oxidation product of lycopene found in tomato products and human serum relevant to cancer prevention. In: Sen CK, Sies H, Baeuerle PA, editors. Antioxidant and Redox Regulation of Genes. New York: Elsevier Inc.; 2000. pp. 409-424. DOI: http://dx.doi.org/10.1016/B978-0-12-636670-9.50028-3

[81] Campbell JK, Canene-Adams K, Lindshield BL, Boileau TW, Clinton SK, Erdman JW Jr. Tomato phytochemicals and prostate cancer risk. Journal of Nutrition. 2004;**134**(Suppl. 12):3486S-3492S

[82] Etminan M, Takkouche B, Caamaño-Isorna F. The role of tomato products and lycopene in the prevention of prostate cancer: A meta-analysis of observational studies. Cancer Epidemiology, Biomarkers & Prevention. 2004;**13**(3):340-345

[83] Giovannucci E. A review of epidemiologic studies of tomatoes, lycopene, and prostate cancer. Experimental Biology and Medicine. 2002;**227**(10):852-859. DOI: 10.1177/153537020222701003

[84] Kun Y, Lule US, Xiao-Lin D. Lycopene: Its properties and relationship to human health. Food Reviews International. 2006;**22**(4):309-333. DOI: http://dx.doi.org/10.1080/87559120600864753

[85] Erhardt JG, Meisner C, Bode JC, Bode C. Lycopene, β-carotene, and colorectal adenomas. American Journal of Clinical and Nutrition. 2003;**78**(6):219-224

[86] Jiménez-Colmenero F, Herrero A, Cofrades S, Ruiz-Capillas C. Meat and functional foods. In: Hui YH, editor. Handbook of Meat and Meat Processing. 2nd ed. New York: CRC Press Taylor & Francis Group; 2012. pp. 225-248. DOI: 10.1201/b11479

[87] Martín D, Ruiz J, Kivikari R, Puolanne E. Partial replacement of pork fat by conjugated linoleic acid and/or olive oil in liver pâtés: Effect on physicochemical characteristics and oxidative stability. Meat Science. 2008;**80**(2):496-504. DOI: http://dx.doi.org/10.1016/j.meatsci.2008.01.014

[88] Jiménez-Colmenero F, Sánchez-Muniz F, Olmedilla-Alonso B. Design and development of meat-based functional foods with walnut: Technological, nutritional and health impact. Food Chemistry. 2010;**123**(4):959-967. DOI: http://dx.doi.org/10.1016/j.foodchem.2010.05.104

[89] Arihara K. Functional foods. In: Jensen W, Devine C, Dikemann M, editors. Encyclopaedia of Meat Sciences. 1st ed. London, UK: Elsevier Science Ltd; 2004. pp. 492-499

[90] Hasler CM. Functional foods: Their role in disease prevention and health promotion. Food Technology. 1998;**52**(11):63-70

[91] Cassidy A, Brown JE, Hawdon A, Faughnan MS, King LJ, Millward J, Zimmer-Nechemias L, Wolfe B, Setchell KD. Factors affecting the bioavailability of soy isoflavones in humans after ingestion of physiologically relevant levels from different soy foods. Journal of Nutrition. 2006;**136**(1):45-51

[92] Chanteranne B, Branca F, Kaardinal A, Wähälä K, Braesco V, Ladroite P, Brouns F, Coxam V. Food matrix and isoflavones bioavailability in early post menopausal women: A European clinical study. Clinical Intervention in Aging. 2008;**3**(4):711-718

[93] Wootton-Beard PC, Ryan L. A beetroot juice shot is a significant and convenient source of bioaccessible antioxidants. Journal of Functional Foods. 2011;**3**(2):329-334. DOI: http://dx.doi.org/10.1016/j.jff.2011.05.007

[94] McDougall GJ, Stewart D. The inhibitory effects of berry polyphenols on digestive enzymes. Biofactors. 2005;**23**(4):189-195. DOI: 10.1002/biof.5520230403

[95] Santini A, Tenore GC, Novellino E. Nutraceuticals: A paradigm of proactive medicine. European Journal of Pharmaceutical Sciences. 2017;**96**:53-61. DOI: http://dx.doi.org/10.1016/j.ejps.2016.09.003

[96] Prescha A, Grajzer M, Dedyk M, Grajeta H. The antioxidant activity and oxidative stability of cold-pressed oils. Journal of the American Oil Chemists Society. 2014;**91**(8):1291-1301. DOI: 10.1007/s11746-014-2479-1

[97] Chen Q, Anders S, An H. Measuring consumer resistance to a new food technology: A choice experiment in meat packaging. Food Quality and Preference. 2013;**28**(2):419-428. DOI: http://dx.doi.org/10.1016/j.foodqual.2012.10.008

[98] Barnett J, Begen F, Howes S, Regan A, McConnon A, Marcu A, Verbeke W. Consumers' confidence, reflections and response strategies following the horsemeat incident. Food Control. 2016;**59**:721-730. DOI: http://dx.doi.org/10.1016/j.foodcont.2015.06.021

[99] Hung Y, Verbeke W, de Kok TM. Stakeholder and consumer opinions and interest in innovative processed meat products: Results from a qualitative study in four European countries. Food Control. 2016;**60**:690-698. DOI: 10.1016/j.foodcont.2015.09.002

Chapter 10

Liposomes as Matrices to Hold Bioactive Compounds for Drinkable Foods: Their Ability to Improve Health and Future Prospects

Marina Marsanasco, Nadia Silvia Chiaramoni and Silvia del Valle Alonso

Additional information is available at the end of the chapter

http://dx.doi.org/10.5772/intechopen.69117

Abstract

The aim of this chapter is to describe the use of bioactive compounds with beneficial effects on human health beyond their basic nutritional value. Bioactive compounds like vitamin E, vitamin C, and fatty acids (omega-3 and omega-6) have an important nutritional contribution and are related to the prevention of certain diseases with global impact such as cancer. However, the addition of vitamins in a food product is not easy: E is destroyed by UV-light, and C is dramatically reduced during heat processes. The use of liposomes as matrices to hold bioactive compounds appears to be a promising solution. Liposomes were made of natural soybean lecithin, which has a great nutritional importance, and more so combined with stearic acid or calcium stearate (CaS). Thus, this stabilize liposomes and contribute to the stability of bioactive compounds and to preserve their activity. The stability of bioactive compounds/liposomes incorporated into aqueous food must be demonstrated in properties such as oxidative tendency, morphology, size, and membrane packaging after heat treatment processes. But to make a product applicable at the commercial level, its texture and mouthfeel arising from the ingestion of drinkable foods are all-important to consumer's choice and sensory acceptability must not undergo any modification.

Keywords: bioactive compounds, liposomes, nutrition, healthy, food

1. Introduction: nutritional properties of soybean lecithin

Phosphatidylcholines can be divided into two types, which differ in their origin: soy phosphatidylcholine (SPC) and egg phosphatidylcholine, both naturally occurring and containing certain polyunsaturated fatty acids (PUFAs) such as linolenic acid (omega-3) and linoleic acid (omega-6), represented by 18:2 and 18:1, respectively [1, 2].

Both linolenic and linoleic acids are essential fatty acids, since they cannot be synthesized by the body and therefore must be obtained by the diet [1–4].

It has been reported that essential fatty acids are highly beneficial in the prevention of diseases such as cardiovascular diseases [2, 5–7], schizophrenia [8], and cancer [9]. In addition, these fatty acids have vasodilator, antihypertensive, anti-inflammatory, and anti-atherothrombotic properties [10].

PUFAs have great human nutritional importance. This is related to the existence of two families of PUFAs: the n-6 family and the n-3 family [4]. n-6 PUFAs are derived from linoleic acid, have two double bonds, and are characterized by having their first double bond at carbon number 6 [1], whereas n-3 PUFAs are derived from linolenic acid, have three double bonds, and are characterized by having their first double bond at carbon number 3. Linoleic acid is metabolized to arachidonic acid, whereas linolenic acid generates eicosapentaenoic acid and docosahexaenoic acid. All of them use the same metabolic pathways and compete for the same elongase and desaturase enzymes [1, 4].

In addition to being a source of energy, the n-6 and n-3 PUFA families are incorporated into cell membranes, where they are precursors of eicosanoids (prostaglandins, prostacyclins, thromboxanes, and leukotrienes), which are involved in numerous physiological processes such as blood clotting or inflammatory and immunological responses [1].

Among vegetable oils, flaxseed oil is considered to be the richest source of linolenic acid (57% of total fatty acids). Rapeseed, soybeans, wheat germ, and walnuts contain between 7 and 13% of the said fatty acid. Some authors consider vegetables (e.g., spinach, lettuce) as a good source of linolenic acid, although their fat content is quite low. Meat, particularly that of ruminants, and dairy products also provide this fatty acid. However, modern farming techniques have led to a decrease in the n-3 fatty acid content in meat (especially in lamb and beef) due to the almost generalized use of n-6-rich grain concentrates to feed cattle [1, 2, 11].

Soybean lecithin is considered a bioavailable source of choline, which was officially recognized as an essential nutrient by the Institute of Medicine in 1998 [12]. This nutrient is needed for the synthesis of neurotransmitters (acetylcholine), cell-membrane signaling (phospholipids), lipid transport (lipoproteins), and methyl-group metabolism (homocysteine reduction) [13, 14]. It plays important roles in brain and memory development in the fetus, and some researchers indicate that choline and methionine intake may be important in reducing the risk of neural tube defects. Studies have also shown that choline supplementation during critical periods of neonatal development can have long-term beneficial effects on memory. Besides, intake of choline has been associated with lower homocysteine levels. This effect is important because increased levels of homocysteine have been associated with greater risk for several chronic diseases and conditions, including cardiovascular disease, cancer, and cognitive decline and bone fractures [12].

In 2001, the FDA made a statement regarding the Dietary Reference Intakes for thiamine, riboflavin, niacin, vitamin B6, folate, vitamin B12, pantothenic acid, biotin, and choline (Food and Nutrition Board, Institute of Medicine (IOM), NAS, 1998, page 390), which stated that choline functions as a precursor for acetylcholine, phospholipids, and the methyl donor betaine [15]. Choline is found in a wide variety of foods like chicken, liver, soy flour, salmon, sockeye, egg, uncooked quinoa, wheat germ, milk, cauliflower, and peas. Among the most concentrated sources of dietary choline are liver, eggs, and wheat germ. In foods, choline is found in free and esterified form (such as phosphocholine, glycerophosphocholine, phosphatidylcholine, and sphingomyelin) [12, 14].

Phosphatidylcholines are obtained by separating the egg yolk, which is generally separated from the whole egg, and then if not used immediately, it is dried or frozen. Soybean lecithin is obtained during the degumming step of oil refining [16], which consists in treating the oil with water at a temperature of 70°C or vapor, so that the phospholipids are hydrated and become insoluble in the fatty phase. Subsequently, the oil is transferred from the mixing tank to a centrifuge, in which the phospholipids, which are hydrated in the excess water, are separated from the degummed oil. The lecithin obtained has commercial value and is especially used, due to its emulgence, in various food industries (chocolate, fine bakery, etc.) [17].

These "raw" lecithins are complex mixtures, which contain significant quantities of triacylglycerols [16]. Also, they may be a mixture of lipids composed largely of phosphatidylcholine, phosphatidylethanolamine, and phosphatidylinositol, combined with other substances such as triglycerides and fatty acids and carbohydrates [18]. The refined degrees of lecithin may contain these components in varying proportions and in combinations depending on the type of fractionation used.

Egg and soybean lecithins may be purified and/or modified to improve their properties [16]. For example, in the case of soybean lecithin, a purification process is required to obtain the highest percentage of phosphatidylcholine to be called soy phosphatidylcholine.

It should be considered that the cost of phospholipids isolated from natural sources is always lower than that of those obtained by synthetic or semisynthetic methods. For natural phospholipids, the more pure they are, the higher the price is [19, 20]. Egg lecithin may be further purified by extraction with ethanol. Solvents may be used to separate lecithin from these triacylglycerols. Soybean lecithin may be precipitated (de-oiled) by acetone and may be enriched in phosphatidylcholine by extraction with ethanol [16, 21].

Lecithin quality is defined by the essays suggested by the "American Oil Chemistry Society":

- Insoluble in acetone: estimates the content of phospholipids.
- Acidity index: measures the free fatty acid content.
- Peroxide index: measures the degree of oxidation.

- Viscosity.
- Gardner color scale.
- Insolubles in hexane: measures the content of solid impurities.

2. Improvement of soybean lecithin by addition of calcium stearate (CaS) or stearic acid

When designing supporting additive matrices for the food industry, it is very important to study membrane stability [2, 22]. The structure of the additives must remain without significant changes over time so activity of the encapsulated component can be assured.

Several authors have reported that cholesterol is a very useful membrane stabilizer, especially when oxidative stability is needed [23, 24]. The effects of cholesterol on the membrane have been very well documented. It is known, for example, that cholesterol modifies the lipid order in membranes: when the concentration of cholesterol in membranes is below 10%, the lipid order in the liquid phase increases, and the lipid order in the gel phase decreases [25]. Cholesterol can also establish hydrogen bonds, thus increasing mechanic resistance [23], and can decrease membrane permeability [26, 27].

All the above mentioned indicate that cholesterol is an interesting candidate to maintain membrane stability. However, cholesterol cannot be used in food industry, because it is very well documented that it is related to atherosclerosis and has a tendency to produce heart diseases [28–31].

Thus, to avoid the use of cholesterol, several other components have been studied as membrane stabilizers. One particular candidate is stearic acid [23]. Hsieh and coworkers studied stearic acid as a membrane stabilizer, by comparing its effect with the one induced by cholesterol [23].

SA is a fatty acid with an 18-carbon-long chain. Because of its hydrophobic character, it is located in between acyl chains of the phospholipids in the bilayer.

The authors reported that liposomes prepared with egg yolk phosphatidylcholine and stearic acid in a 1:0.25 molar ratio present the same encapsulation efficiency and oxidative stability as liposomes prepared with egg yolk phosphatidylcholine and cholesterol in the same molar ratio.

If stearic acid is replaced by calcium stearate, membrane packing requirements would be fulfilled, and also it will contribute with calcium to food-containing additive matrices, which is an additional benefit.

Liposome formulations based on soybean lecithin or soy phosphatidylcholine with stearic acid or calcium stearate have also been studied as food additives, to determine their efficiency to enrich aqueous food with antioxidant vitamins. The molar ratio reported by Hsieh and coworkers was so efficient for the mixture of SPC with stearic acid that this was considered as the ratio to be used. Results from our laboratory have shown that formulations of SPC with stearic acid or innovative addition of calcium stearate, in the same molar ratio, resulted in an improvement regarding oxidative stability and protection of thermolabile vitamins [32]. On the other hand, these formulations did not induce any unpleasant flavor when added to milk or orange juice, so they can be applied in food commercial products [32–35].

3. Processing issues in adding vitamins E and C protected with lecithin liposomes

This topic is of particular importance in relation to the nutritional value and quality aspects of processed foods as well as in relation to nutrition labeling. A number of general reviews have already been published because it is sought to determine the most suitable processing and time-temperature conditions, to achieve the desired objective, i.e., maximum retention of a specific vitamin or best retention of color or flavor consistent with microbiological stability and safety [36].

In the case of packaged liquid foods, different heat treatment processes can be applied. These include pasteurization, ultra-pasteurization, and ultrahigh temperature. Pasteurization is a heat treatment whose objective is to destroy non-sporulated pathogenic microorganisms and significantly reduce banal microbiota to offer the consumer a safe product with an acceptable shelf life to be consumed in a short term [37]. Ultra-pasteurization is a heat treatment in which the food is subjected for at least 2 s to a minimum temperature of 138°C by a thermal process of continuous flow and immediately cooled below 5°C and packed in a non-aseptic form in sterile and hermetically sealed packaging. Ultrahigh temperature is a process in which the food is subjected for 2–4 s to a temperature between 130 and 150°C, by a continuous flow thermal process, immediately cooled to less than 32°C, and packed under aseptic conditions in sterile packaging and hermetically sealed. This type of food has a shelf life of 5–6 months at room temperature and in closed packaging [18, 38]. The problem with these treatments is that they can generate losses of nutrients, especially of vitamins.

Processing with heat treatment like cooking conditions causes variable losses of vitamins. The losses of these nutrients are related with the cooking method and type of food and reckon on particular experimental arrangements during the culinary process, e.g., temperature, the presence of oxygen, light, moisture, pH, and, of course, duration of heat treatment. Vitamin C, retinol, folate, and thiamine are the most labile vitamins during culinary processes [39].

Concentrated juices are in high demand because they provide a significant amount of nutrients; however, during their elaboration and pasteurization process, they lose flavor, aroma, and nutritional contribution, mainly of vitamins. For this reason, the juices that are marketed in supermarkets present a considerable variation in their vitamin content, mainly of vitamin C (ascorbic acid) [40].

Vitamin C is one of the vitamins most sensitive to heat treatment. With regard to processed foods, there is a loss of vitamin activity that is related to the intensity of the heat treatment. Other factors that influence the loss of this vitamin activity by cooking are the pH, oxygen, surface of exposure to water, conditions of the heat treatment, and the presence of metals such as copper [37, 39, 41].

The losses of this vitamin vary between 40 and 60% in in-bottle sterilized milk, between 20 and 40% after ultrahigh temperature treatment [42, 43], and between 15 and 25% after a pasteurization treatment [44, 45]. Other researchers [46] showed that vitamin C is reduced by different types of pasteurization around 10% for long-term and low-temperature (LTLT) process and 50% for process which included 90°C during 30 min. The authors showed that vitamin C in baobab drink decreased with increasing temperature. These results coincide with that found in orange juice, where a loss of vitamin C of 10 ± 2.5% was observed after slow pasteurization or LTLT process (60 min at 65°C) and of 13.7 ± 1.9% after pasteurization for 45 min at 75°C. With a more extreme heat treatment for 30 min at 90°C, the vitamin loss increased, resulting in a final value of 23.3 ± 3.8% [47].

More current studies have demonstrated a high thermosensitivity of vitamin C against thermal processes that are related to the type of process and food in which this vitamin is found. For example, losses of this vitamin between 29 and 61.45% have been demonstrated in vegetable bleaching processes with temperatures between 94 and 98°C and cooking times between 90 s and 3 min. This shows that the presence of water further favors vitamin loss by leaching [39].

The loss of vitamin C is also related to the type of heat treatment [39, 41]. Other authors [48] demonstrated that the concentration of vitamin C was drastically reduced by various methods of steam cooking, conventional cooking, and high-pressure cooking.

With respect to vitamin E, which is a liposoluble vitamin, it is thermostable but readily oxidizes in the air [49], especially in the presence of ferric ion and other metals. Therefore, the use of some chemical substances such as hydrogen peroxide should be avoided, as it may lead to oxidation and, therefore, loss of vitamin activity [37]. In addition, vitamin E is destroyed by exposure to UV light and is lost, to a large extent, during the refining of oils [17].

During the processing and storage of food, meat and meat products, milk and derivatives, and cereals show few changes in the content of vitamin E. However, during the storage of vegetable foods, vitamin E has a weak antioxidant character, and in the presence of animal fats, it is much more active, especially if there are synergistic substances like vitamin C [37].

Considering the abovementioned, the application of liposomes is a promising solution to avoid losses of vitamins and promote their shelf life and protection [49, 50]. The use of liposomes to encapsulate and protect these vitamins and other bioactive compounds has a number of positive aspects [50]. For example, liposoluble vitamins such as vitamin E mix perfectly with the hydrophobic area of phosphatidylcholine. In addition, the absorption and bioavailability of this vitamin increase when it is encapsulated in liposomes. In particular, vitamin C encapsulated in liposomes retains 50% of its activity after 50 days in refrigerated storage, whereas non-encapsulated vitamin loses its activity after 19 days. Also, liposomes present an important protective effect over thermolabile vitamin C, shown by an antioxidant action after pasteurization [32, 34, 35, 49].

In the case of liposoluble vitamins, the importance of these food systems is that they can be added in aqueous foods [50, 51], such as orange juice, maintaining the stability and preserving the activity of vitamins [34].

4. Addition of improved liposomes containing bioactive compounds to food products: a case study

In order for a food to be considered functional, it must demonstrate (i) that it has a beneficial effect on one or more specific functions of the organism, beyond the usual nutritional effects; (ii) that it improves the state of health and well-being; and (iii) that it reduces the risk of an illness. This means that these foods must necessarily contain some of the so-called functional ingredients or bioactive compounds. It has been shown that, when implemented in aqueous foods, bioactive compounds generate a functional food which can promote health, physical ability, and mental state to benefit consumers of different ages [2, 52–54].

Bioactive compounds, including vitamins, antioxidants, minerals, dietary fibber, essential fatty acids, flavonoids, isothiocyanates, phenolic acids, plant stanols and sterols, polyols, prebiotics and probiotics, phytoestrogens, and soy protein, are the main components of functional foods [52, 53, 55]. Some nutrients have an important nutritional contribution and have been shown to be related to the prevention of certain diseases of great global impact such as cancer. This is the case of essential fatty acids as omega-3 and omega-6 and certain vitamins like vitamins E and C.

Vitamin E or α-tocopherol is the main liposoluble antioxidant in the body. It protects lipids against oxidative damage [56].

Also, it has a desirable effect when blood cholesterol decreases the incidence on atherosclerosis, and the cardiocirculatory system has a positive effect. An additional antioxidant vitamin is ascorbic acid or vitamin C. One of the biological roles of ascorbic acid is to participate in oxidation-reduction processes, blood coagulation, tissue regeneration, and building steroid hormones, inducing free radical inactivation. This vitamin also takes part in the inhibition of nitrosamine formation and participates in the collagen synthesis [17].

The importance of antioxidant vitamins is that several clinical studies have described beneficial effects in a variety of tumors, such as prostate, gastric, and lung tumors. This fact is based on experimental studies that highlight the role of free radicals as key factors associated with the development of cancer, and it is precisely the effectiveness of dietary antioxidants such as vitamin E or vitamin C that play an important role in the prevention of the development and progression of this disease [57–59].

However, most bioactive compounds such as fatty acids, carotenoids, tocopherols, flavonoids, polyphenols, phytosterols, and liposoluble vitamins have hydrophobic nature [52], which makes difficult their application in aqueous foods. Besides, it is not easy to maintain the stability of vitamins. In particular, in the case of functional foods with added vitamins, a number of factors must be taken into account to maintain their stability: their structure (whether they are hydrosoluble or liposoluble); their relation with diverse conditions as pH, the presence of oxygen and metals; the way in which vitamins are added to the food in question; and the heat treatment and storage conditions of the final product [34]. Vitamin E is liposoluble and destroyed by UV light [17], while Vitamin C is dramatically reduced by heat treatment processes [46, 49]. Thus, liposomes, which are microscopic spherical vesicles, composed of polar lipids that enclose liquid compartments within their structure and enable the encapsulation of both hydrophilic and lipophilic materials [20, 22, 27, 49, 58, 60], may be a promising solution for incorporating bioactive compounds [61] into foods regardless of their affinity for water and for generating a protection over them.

Liposomes are classified into small unilamellar vesicles (SUVs), large unilamellar vesicles (LUVs), and large multilamellar vesicles (MLVs), according to their size and lamellarity, the latter of which relates to the method of preparation [62, 63]. The process of forming MLVs consists in mixing the lipids in ethanol, which is then removed by evaporation. Subsequently, the dry lipid film is hydrated, maintaining the temperature above the phase transition temperature of the lipid mixture [2, 22, 27, 60]. So, these liposomes form spontaneously when the dry lipid film is hydrated with water or buffer [27, 62]. Typically, their size distribution ranges from 0.1 μm to a maximum value which may be up to 500 μm in diameter, and they contain hundreds of concentric lamellae [22, 27, 62]. **Figure 1** shows a MLV of soy phosphatidylcholine and calcium stearate with vitamins E and C.

Figure 2 shows the concentric lamellae from MLVs, where liposoluble compounds, such as vitamin E, are located within the lamellae and hydrosoluble ones, such as vitamin C, prefer the aqueous interface. The concentric lamellae are formed by phospholipids such as phosphatidylcholine, which has a phosphate with a choline head and a carbon chain or fatty acids as omega-3 and omega-6, formed by carbon chains with carboxylic acid heads.

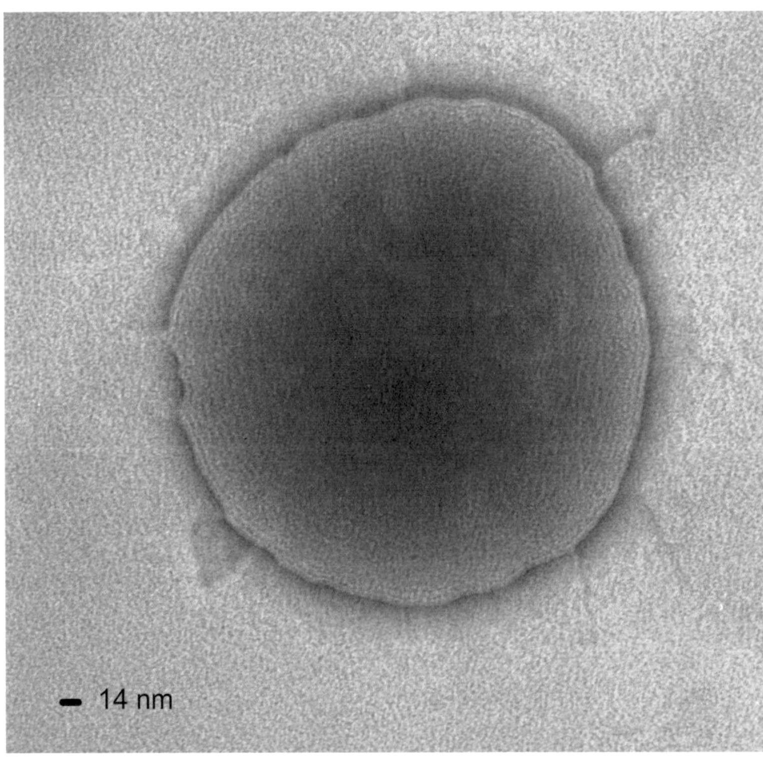

Figure 1. Transmission electron microscopy of soy phosphatidylcholine and calcium stearate liposomes (50 mM, molar ratio of 1:0.25) and vitamin C (90 mM) and vitamin E (5 mM).

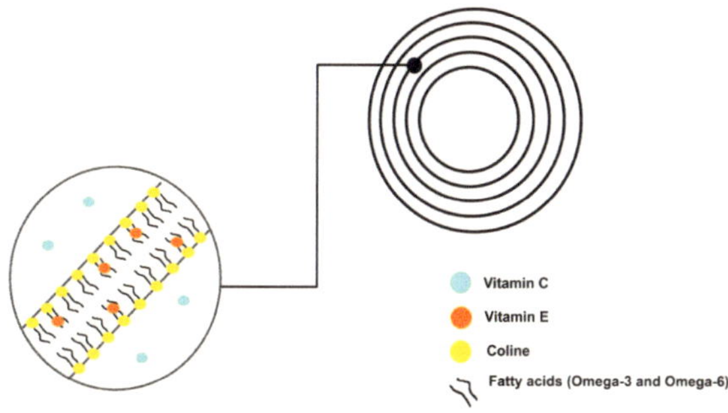

Figure 2. Concentric lamellae from MLVs and the location of vitamins E and C.

Liposomes have been employed as potential carriers to deliver food components and have many applications in food industry including protecting sensitive ingredients, increasing the bioavailability of nutrients and confining undesirable flavors. These types of matrices have been applied as food additives and have the ability to encapsulate vitamins, antioxidants, proteins, peptides, antimicrobials, essential oils, flavors, enzymes, minerals, and fatty acids [58, 60].

Liposomes have been used in the food industry for improving the flavor of ripened cheese using accelerated methods, for promoting antioxidant activity with the synergistic delivery of ascorbic acid and tocopherols in foods of functional food ingredients, and the stabilization of minerals (such as iron) in milk [58]. In respect to the industry of cheese, liposomal entrapment of enzymes offers advantages for cheese applications such as being prepared from ingredients naturally present in this product, because these vesicles can protect casein from early hydrolysis during the production of cheese [49].

Another example of the application is the encapsulation of calcium lactate encapsulation in lecithin liposomes to fortify soymilk with levels of calcium equivalent to those found in cow's milk [51]. Also, liposomes have been applied to encapsulation of lactase because they release lactase in the stomach and, therefore, remove the sweet taste of hydrolyzed milk [2].

Food grade phosphatidylcholine can be applied in food without the need for any clinical study. This aspect is particularly related to the regulation and regulation of food in each country. For example, in Argentina, soy lecithin is approved by local food regulations such as the Argentine Food Code and Resolutions of the Common Marked Group, being endorsed by control agencies such as ANMAT [18, 64].

The main objective of implementing a functional food is to generate a product with a high nutritional value that benefits the health of the population. People generally strive to consume a wide variety of foods and assure the ingestion of compounds such as antioxidants, vitamins, carotenoids, fiber, flavonoids, specific fatty acids, minerals, prebiotics and probiotics, phytoestrogens, soy protein, and vitamins, among others.

In the food industry, several matrices are being applied to encapsulate or associate bioactive compounds. These include liposomes, nanoemulsions, microemulsions, solid lipid nanoparticles, and polymeric nanoparticles [65, 66]. All the matrices that may be applied in the food industry should have a series of properties like stability, applicability, and sensory evaluation of bioactive compounds in the product [2], which must be considered when incorporating bioactive compounds, especially vitamins, to generate new functional foods. No one would be willing to invest in the development and production on a larger scale of a food that is not acceptable for potential consumers.

In the food industry, considering that phospholipids can be oxidized and that this can limit their shelf life, membrane stability and structure are important factors when designing liposomes [2, 22]. Also, it is very important that liposomes remain stable after pasteurization because the higher the stability, the higher the protection of vitamins [32] and bioactive compounds.

Our research group focuses on the structural study, oxidative stability, and application in food of different liposomal formulations with bioactive compounds (omega-3, omega-6, vitamins E and C) to develop a functional food in commercial pasteurized orange juice. In our studies, the design and strategy of the implementation of these liposomes are based on the use of soy phosphatidylcholine a natural lipid that contains linolenic acid (omega-3) and linoleic acid (omega-6). These essential fatty acids are being added as part of soybean lecithin in the proportion needed for 200 mL of orange juice (38.28 mg for linoleic acid and 3.46 mg for linolenic acid). Soybean lecithin is a commercial product available, described in the Argentine Food Code and approved by ANMAT and INAL (Argentinean Food Quality Organisms) [32–35].

Besides, the design of liposomes has been made for encapsulating bioactive compounds as vitamins E and C. For 200 mL of orange juice, 2 mL of liposome suspension (50 mM) with vitamins was added, which implies that the orange juice was fortified with 4.3 mg of vitamin E (5 mM), equivalent to 43% of the recommended daily intake and 31.70 mg of vitamin C, equivalent to 70.44% of the recommended daily intake according the Argentine Food Code [33, 35].

Liposomes to protect hydrophilic or lipophilic vitamins must not only possess a long circulation time but also maintain the encapsulated vitamins for longer times; this means that they should have low leakage rate. Also, part of our research involved stearic acid (SA) or calcium stearate (CaS) that have been added to stabilize membrane liposomes to contribute to maintaining the stability of bioactive compounds and preserving their activity [32–35].

Food products must also undergo thermal treatment so the structural and oxidative stability of liposomes must be taken into consideration in all of the food-process conditions. Stable liposomes should have conserved size, shape, and surface properties. Size is usually analyzed by light scattering, whereas shape and structure are usually studied by optical and transmission electron microscopy, respectively. Also, to assure that membrane surface is maintained during food manufacturing and processing, a fluorescent probe like merocyanine 540 can be used to monitor surface changes. To complement this, surface charge and oxidative stability can be analyzed by the zeta potential and ORAC method, respectively. Liposomes used in our work as vehicles showed significant stability in all of the parameters mentioned above and conserve an important protective effect over thermolabile vitamin C [32, 34, 35].

Results of our lab regarding the oxidative level in matrices holding liposomes-bioactive compounds obtained showed a high stability in this parameter. The liposomal formulations were resuspended in acetic acid 3% w/v, indicated as food model systems by Argentinean regulations as the Argentine Food Code. The three liposomal formulations without vitamins

http://dx.doi.org/10.5772/intechopen.69117

(**Table 1**) had the same oxidative stability by the ORAC method without significant differences regarding SPC (Dunnett Test, statistics not shown in **Table 1**) probably because of the low peroxidation of SPC [34].

When the vitamin C was incorporated in the three liposomal systems, it showed a significant higher value than the controls, related to the antioxidant activity after the pasteurization process [34]. In previous results, the percentage of encapsulation efficiency of vitamin C in these liposomes was determined and was c.a. 86% [32]. So, it is possible to infer that with the encapsulation efficiency data and the antioxidant activity these liposomes will protect efficiently most of the vitamin C and hence maintained its antioxidant activity after pasteurization against damage induced by the LTLT process [34, 35]. Noteworthy, liposomes will also exert the vitamin C protection.

Besides, to confirm their capability as commercial functional food, rheological behavior and sensory evaluation of liposomes/bioactive compounds should be performed. In our case, liposomes with bioactive compounds (omega-3, omega-6, vitamins E and C) were added to implement a functional orange juice with all of the above considerations. The sensory evaluation of liposomes in orange juice was performed by the overall acceptability and triangular tests with 40 and 78 potential consumers, respectively.

The three liposomal formulations, soy phosphatidylcholine (SPC), soy phosphatidylcholine and stearic acid (SPC:SA), and soy phosphatidylcholine and calcium stearate (SPC:CaS), studied remained stable even after pasteurization, as demonstrated by morphology, size, membrane packing, and high oxidative stability. Besides, all systems showed protection of the thermolabile vitamin C, which maintained its antioxidant activity after pasteurization. SPC and SPC:SA systems had a rheological behavior similar to a Newtonian fluid, whereas SPC:CaS had a pseudoplastic one; both stages considered excellent for larger-scale production. The incorporation of all liposomal formulation did not change the acceptability of orange juice. From all the aspects covered, it can be concluded that these liposomes with bioactive compounds, especially vitamin C, can be added to orange juice for commercial application with added commercial and nutritional value.

Liposomal formulation	Without vitamins	With vitamin C
SPC	100.30 ± 13.05	180.80 ± 22.95***
SPC:SA	95.75 ± 2.75	186.80 ± 26.55***
SPC:CaS	98.50 ± 4.04	206.80 ± 4.50***

Data correspond to ORAC assay in liposomal formulations (50 mM) in acetic acid 3% w/v after pasteurization. Data correspond to soy phosphatidylcholine (SPC), soy phosphatidylcholine and stearic acid (SPC:SA) 1:0.25 molar ratio, and soy phosphatidylcholine and calcium stearate (SPC:CaS) in 1:0.25 molar ratio without vitamins and with 90 mM of vitamin C. Each column represents the mean ± SD of four independent assays. Statistical comparison was made:

Between each system with vitamin/s with respect to the same system without vitamins (control) through the Dunnett test. Significant differences with respect to the control are shown as ***$p < 0.001$.

With respect to SPC in systems without vitamins with the Dunnett test, no significant differences were observed.

Table 1. Peroxidation assay of ORAC in matrix liposome-bioactive compounds.

5. Importance of bioactive compounds impact in consumers of different ages and economical levels

Nowadays, foods are not intended to only satisfy hunger and provide the necessary nutrients to humans but also to prevent nutrition-related diseases and improve the physical and mental well-being of consumers [67].

Diet quality issues in aging populations are of great concern. Functional foods should be considered as foods with health benefits beyond what is interpreted as nutrients and the challenge of bioavailability [54, 68]. It is very important of combining science with consumer desires when considering how to formulate foods that older consumers will actually purchase and eat [68].

Years of research have demonstrated that diet quality has a huge effect on physical and cognitive condition, bone and eye health, vascular function, and the immune system effectiveness. This can be challenging to achieve for reasons very well known like that aging is often accompanied by a loss of appetite and changes in taste and smell, all of which can lead to more limited food choices and lower intake of healthful foods. In other words, aging also often affects food choice and intake since it is accompanied by general oral health decline and a reduced ability to swallow. On top of this, many older adults experience mobility constraints, which make it difficult to shop for food, lift heavy jars, or even open containers. Also, low income is prevalent in aging populations, making it difficult for many older adults to access high-quality foods that in general tend to be more expensive [68].

Macronutrients, namely, omega-3 fatty acids and fiber, are a must in maintaining health during aging. Dietary fiber is known to be important for maintaining intestinal health and protecting against heart disease and other metabolic conditions. With lipids, epidemiological studies have found that higher intakes of omega-3 fatty acids provide greater protection against many conditions, including cardiovascular events (e.g., arrhythmias, cardiac death, and recurrent myocardial infarction), diabetes, and cognitive decline. The problem is that omega-3 fatty acids are very limited in regular diets of older adults, with the main sources like fatty fish, flax seeds, and walnuts. The health effects associated with this group of fatty acids are an important area of current investigation. With respect to the micronutrients, almost every dietary survey conducted over the past few decades has shown that older adults have inadequate intakes of some essential micronutrients. Moreover, subsets of older adults are often at greater risk of certain micronutrient deficiencies. For example, non-Hispanic black and low-income older adults typically experience micronutrient intake levels lower than the other groups of older adults. According to 2005–2006 data, 92% of adults over the age of 51 years are below the Estimated Average Requirement (EAR) for vitamin E; 67% are below the EAR for magnesium; 46% are below the EAR for vitamin C; 33% are below the EAR for zinc; and 32% are below the EAR for vitamin B6. Only 14.6% are above the al for calcium (1200 mg) [68].

Because of the difficulties in obtaining sufficient levels of vitamin E through the diet, many people are taking vitamin E supplements. The concentrations of certain tocopherols are actually lower in people taking supplements. Also, the larger problem is that negative consequences can occur when supplements are erroneously used as a substitute for food [68].

Bioactive compounds could be used to create functional foods for older adults that improve or maintain taste and smell, digestion, brain health, the immune system, bone and joint health, cardiovascular health, gut flora (i.e., probiotic foods), and eye health [68, 69]. The other important issue arising is if the health-food developers do not relate their products to what is important for consumers; then consumers will not use those products. In order to relate products to those factors that are important to consumers, companies must comprehensively understand aging consumers' needs and accept that understanding into food solutions that consumers want, need, and can afford is not even cost-profitable.

Product development—that is, translating aging consumers' needs into products on the shelf—is a very complex, time-consuming process. It involves everything from "culinary creation" (i.e., making a food that tastes good) to ensuring microbiological stability and regulatory compliance. For a product development, there are four essential "elements":

- Form is a key element of the decision-making. For aging boomer consumers, ease of use and legibility of preparation instructions are additional considerations, like developing new types of easy-to-open packages.
- Function is another key consideration, with the primary goal being to ensure that a product is safe regardless of consumer needs. For older adults, this means that the health benefit is validated with the targeted age group and that the products actually deliver those benefits specifically to older adults.
- Appeal (i.e., taste, texture, and appearance). If a product does not taste or look good, people will not eat it, regardless of its contents. Product development involves extensive sensory work to ensure that the intended benefits are delivered. For aging boomer consumers, additional considerations include vibrancy, potency, and consistency.
- Affordability (i.e., raw materials, manufacturability, distribution). This is a huge concern, especially in today's economic climate and especially for aging boomer consumers. Health food developers should optimize raw material usage, working with suppliers to ensure a cost-effective supply chain and minimizing manufacturing and distribution costs. Also, unit size is important. As people age, they tend to cook only for themselves [68].

The United States is one of the countries that have a clear goal of incorporating functional foods to prevent disease, so it is easy to find cereal bars intended for middle-aged women, supplemented with calcium to prevent osteoporosis or with soy protein to reduce the risk of breast cancer and folic acid to improve heart health. In Europe, "value added" signs are used, and in Germany confections are added with coenzyme Q10 and vitamin E. In Italy, supermarket gondolas offer omega-3 yogurts and vitamins, and in France, there is added sugar added with fructooligosaccharides to promote the development of beneficial intestinal flora [70].

Another author [71] reports that are in accordance with the opinion that consumers in general are hardly willing to compromise on the taste of functional foods for health. For that reason, it is important to evaluate the sensorial aspect in the functional foods. The overall conclusion indicated that consumer demand is undoubtedly in the functional foods market, but the industry must respond with good tasting in the products [72].

In relation to the importance of functional foods in the infant population, international studies from the World Health Organization have informed that 5.9 million children under the age of 5 years died in 2015. The problem is that children under 5 years of age who die annually in the world are from preventable diseases. Pneumonia, diarrhea, and malaria are the main causes of death if considered the period from the end of the neonatal stage through the first 5 years of life. Children are the most vulnerable because of malnutrition, which contributes for about 45% of all child deaths [73].

This problem affects then the socioeconomic opportunities that children will have in adulthood, thus increasing the healthcare maintenance. So, it is useless to convey that prevention is a must in this case. Besides, it is well documented that malnutrition causes a lot of problems in children like delayed growth in height, delayed development, weakening of defenses to infections, and, in the most severe cases, death. The problem in itself effect is far more serious in the first years of life due to the greater need for calories and nutrients and because it is a stage of rapid growth of the body [73, 74].

The publication Maternal and Child Health, developed in collaboration by UNICEF and the Argentine Society of Pediatrics, offers a general statistics about this problem in our country. In these studies, a percentage of people obtained is with unsatisfied basic needs which are from different urban regions which is 36.6% for groups of 0–2 years old and 34.1% for groups of 0–17 years old [75].

In Argentina, the National Nutrition and Food Program was created in compliance with the obligation of the state to ensure citizenship the right to a minimum of food intake and cover the requirements of nutritional benefits of children up to 14 years old, pregnant women, and disabled and elderly (70 years onward, in poverty).

In this way, in Argentina, enrichment of wheat flour, established by Law No. 25.630, enacted in July 2002, where this flour destined for consumption in the national market should be clearly highlighted with the rest of the nutrients and in which concentrations are each: ion (30 mg/kg), folic acid (2.2 mg/kg), thiamine (6.3 mg/kg), riboflavin (1.3 mg/kg), and niacin (13 mg/kg).

The problem is that there are other nutrients of importance for the normal development and functioning of children and that it is necessary that the intake of the nutrients be carried out in Argentina as well as in the worldwide level. Let us not forget that adequate food intake during the first 2 years of life is fundamental. Given the rapid growth of children, which conditions high nutritional requirements, coupled with a limited intake capacity in volume, this stage presents in itself a high nutritional vulnerability [76].

6. Evaluation of the functional and sensory properties of improved liposomes with vitamins in food products

The sensorial analysis allows knowing the organoleptic properties of the food because it is realized through the senses. Sensory evaluation is innate in man since from the moment that a product is tested, a judgment is made about it, whether it likes or dislikes it, and describes and recognizes its characteristics of taste, smell, texture, etc. [77, 78].

When a food market requires so, a certain product must meet requirements for nutrition, hygiene, safety, quality, and sensory aspects, to be accepted by the consumer. It is from all such properties that sensory analysis of foods is an effective tool for quality control. In such a way, sensory evaluation always gives the same global sensory characteristics and acceptability of a food [77, 78].

There are different sensory methods of evaluations. In general, they can be descriptive, discriminatory, and acceptable and preferable. Discriminatory tests should be used when it is necessary to determine whether two samples are significantly different. It is possible that two samples have chemically different formulations, but the sensory perception of the people is unable to perceive the difference. The development of products is based on this possibility, since in reformulating the ingredients of the food are sought that the consumer does not detect any difference [78, 79].

These tests are widely used in the industry, in quality control procedures, in the study of impact by changes in formulation or process, as well as in the ability of consumers to discriminate between two similar products [79].

The affective tests are those in which the evaluator expresses his/her subjective reaction to the product, indicating if he/she likes or dislikes it, accepts or rejects it, or prefers it or not to another. The main purpose of affective methods is to evaluate the response (reaction, preference, or acceptance) of actual or potential consumers of a product. It is necessary, first, to determine whether one wishes to evaluate simply preference or degree of satisfaction (taste or disgust) or whether one also wants to know what is the acceptance of the product among consumers [77]. The choice of the test to be performed will depend on the objectives of the test. The measure of sensory acceptability is a logical and necessary step before launching a product to the market. No one would be willing to invest in a new product that will be sensory unpleasant [80].

These self-swelling mixtures, obtained in large quantities, can be added to the final commercial product online. Based on this property of liposomes, those containing bioactive compounds (PUFAs, vitamins E and C) were added to commercial orange juice (1:100 ratio) of an Argentinean trademark (Citric® of El Carmen S.A.) was selected for a sensory evaluation. These liposomes with bioactive compounds were prepared the day before the sensory evaluation was programmed, then pasteurized, and finally added to the commercial orange juice (1:100 ratio). Samples obtained were kept refrigerated at 4°C until the sensory evaluation was performed. Specific care was taken so that commercial orange juice always kept the physicochemical, microbiological, and sensory characteristics. If any variation in the flavor exists, it would be from the addition of liposomes. For both sensory tests, samples were given to each evaluator in disposable regular cups of 200 mL; and each cup contained 30 mL of product. Each evaluator was provided with mineralized water and unsalted crackers as flavor neutralizers [34, 81].

Two different tests were used to study the addition of liposomes with bioactive compounds in the commercial orange juice. The first test was the triangle test, which was performed to compare the differences between commercial orange juice with and without liposomes. To analyze similarities between samples, 78 evaluators were selected [34, 81].

Consumers of commercial orange juice (men and women over 18 years old) were selected and instructed in the test [58]. In each test, two samples were the same, and the third one was different (product with and without liposomes). The orange juice was previously brought up to room temperature, and the randomness of the samples was ensured during the whole procedure. Each evaluator tested nine samples to accomplish the requirements of the triangular test for the three formulations. Each evaluator was requested to drink water to neutralize flavors between samples from the same triangle, as requested in a sensory evaluation procedure. Before changing from one sensory triangle to the next, they were also asked to eat a cracker and drink water to avoid sensory fatigue. To end, each evaluator completed a card with their personal inputs [34, 81].

The other test applied was the affective test, and 40 consumers of the commercial orange juice were selected, men and women over 18 years old [77]. The sensory acceptability of the evaluator is faced with unknown samples to judge. The orange juice sample was kept at room temperature, and the randomness was ensured while the test lasted. Hedonic rating scales associated with score used were as follows: (1) I really dislike it; (3) I dislike it; (5) I neither dislike nor like it; (7) I like it; and (9) I really like it. The evaluators are faced with these possibilities or intermediate ones [81]. As stated before in between samples, evaluators were induced to drink water to abstain sensory fatigue [34].

In the triangular test, the outcome results for the orange juice, containing the liposomes with bioactive compounds, considering favorable/total were 43/78 for SPC, 35/78 for SPC:SA, and 38/78 for SPC:CaS. By favorable answers of the evaluator, they found the difference between the commercial product with and without liposomes/bioactive compounds. Applying the statistical table for the triangular test, and considering a significant level of 0.10 for 78 evaluators, the minimum number of correct answers for samples that showed significant differences is 32 [59]. From the above, it is concluded that there are significant differences between commercial juice with or without liposomes and bioactive compounds [34].

With respect to the affective test, although the significant differences were obtained in the triangular test, the addition of liposomes with bioactive compounds did not change the acceptability of the product. These results are reflected in **Table 2**, where the three added formulations showed no significant differences with respect to commercial juice. The results obtained showed that all three liposomal formulations are potentially applicable in the product [34].

Test knowing the samples					
COJ with SPC	**COJ**	**COJ with SPC:SA**	**COJ**	**COJ with SPC:CaS**	**COJ**
5.93 ± 1.61	5.88 ± 1.81	7.03 ± 1.51	7.08 ± 1.44	7.00 ± 1.76	6.90 ± 1.49

Data correspond to soy phosphatidylcholine (SPC), soy phosphatidylcholine and stearic acid (SPC:SA) in 1:0.25 molar ratio, and soy phosphatidylcholine and calcium stearate (SPC:CaS) in 1:0.25 molar ratio. Qualifications of 40 panelists for commercial orange juice (COJ) with or without liposomes with 5 mM of vitamin E and 90 mM of vitamin C, for each type of formulation. Statistics were performed using the test for paired samples between each commercial orange juice sample with and without liposomes. No significant differences were obtained.

Table 2. Total assay acceptability of liposomal formulations knowing the samples in commercial orange juice (COJ).

7. Future trends

It should be a community efforts trend. By community it is meant an equal commitment effort from two different levels, governmental and science food developers.

For the development of a new food product, it is important to involve the food industry in devising solutions to certain problems, such as the nutrition and health at different stages of life. In this aspect, the application of functional foods has a promising future considering that it promotes and benefits the health beyond the nutritional contribution.

There is a general consensus with Singh [53] concept that several challenges, including discovering of beneficial compounds, establishing optimal intake levels, and developing adequate food delivering matrix and product formulations, need to be addressed.

The implementation of matrices such as liposomes for the transport of bioactive compounds can facilitate the ingestion of these in diverse foods, especially the aqueous ones, being able to be offered to the sectors of the population of risk such as children and elderly people. These types of matrices must have some stability either by adding bioactive compounds or by being present in the food. There is another aspect of great importance that should not be left aside, and that is the taste of functional foods for health. This issue is intrinsically highly speculative, risky, and deemed to yield a niche market strategy. This conclusion entails a challenging future for food product designers, food technologists, and sensory scientists dealing still with one of the fastest growing segments of the food market.

The juices being so well accepted by young children and adults can result in useful tools to be fortified with iron, calcium, vitamin C, vitamin E, and other antioxidants. Its natural content or the result of its fortification in ascorbic acid facilitates the absorption of iron from vegetables and legumes and on health improvement as a final future goal.

Acknowledgements

This work was funded by the following sources: Universidad Nacional de Quilmes-Project PUNQ974/11; Consejo Nacional de Investigaciones Científicas y Técnicas (CONICET), Project PIP-CONICET # 11220110100214; Ministerio Nacional de Ciencia y Tecnología (MINCyT), Project MINCyT-CAPES N°; and BR/11/01 (2012–2013). Besides, we are grateful to Dario Alti for the design in **Figure 2**.

Author details

Marina Marsanasco, Nadia Silvia Chiaramoni and Silvia del Valle Alonso*

*Address all correspondence to: silviadelvalle@gmail.com

Laboratorio de Biomembranas, Grupo de Biología Estructural y Biotecnología (GBEyB), IMBICE-CONICET, Universidad Nacional de Quilmes, Buenos Aires, Argentina

References

[1] Carrero JJ, Martín-Bautista E, Baró L, Fonollá J, Jiménez J, Boza JJ, López-Huertas E. Alimentos funcionales. Efectos cardiovasculares de los ácidos grasos omega-3 y alternativas para incrementar su ingesta. Nutrición Hospitalaria. 2005;**20**(1):63-69

[2] Hadian Z. A review of nanoliposomal delivery system for stabilization of bioactive omega-3 fatty acids. Electronic Physician. 2016;**8**(1):1776-1785. DOI: 10.19082/1776

[3] López LB, Suárez MM. Fundamentos de Nutrición normal. 1st ed. El Ateneo; Buenos Aires, Argentina. 2013. 446 p

[4] Lands B. Review. Historical perspectives on the impact of n-3 and n-6 nutrients on health. Progress in Lipid Research. 2014;**55**:17-29. DOI: 10.1016/j.plipres.2014.04.002

[5] Earnest CP, Hammar MK, Munsey M, Mikus CR, David RM, Bralley JA, Church TS. Microencapsulated foods as a functional delivery vehicle for omega-3 fatty acids: A pilot study. Journal of the International Society of Sports Nutrition. 2009;**6**(12):1-6. DOI: 10.1186/1550-2783-6-12

[6] Leaf A, Weber PC. Cardiovascular effects of n-3 fatty acids. The New England Journal of Medicine. 1998;**318**(9):549-557. DOI: 10.1056/NEJM198803033180905

[7] Lee KW, Lip GYH. The role of omega-3 fatty acids in the secondary prevention of cardiovascular disease. QJM: An International Journal of Medicine. 2003;**96**(7):465-480. DOI: 10.1093/qjmed/hcg092

[8] Sivrioglu EY, Kirli S, Sipahioglu D, Gursoy B, Sarandôl E. The impact of w-3 fatty acids, vitamins E and C supplementation on treatment outcome and side effects in schizophrenia patients treated with haloperidol: An open-label pilot study. Progress in Neuro-Psychopharmacology & Biological Psychiatry. 2007;**31**(7):1493-1499. DOI: 10.1016/j.pnpbp.2007.07.004

[9] Jenski LJ, Zerouga M, Stillwell W. Omega-3 fatty acid-containing liposomes in cancer therapy. Proceedings of the Society for Experimental Biology and Medicine. 1995;**210**(3):227-233

[10] Baguma-Nibasheka M, Brenna JT, Nathanielsz PW. Delay of preterm delivery in sheep by omega-3 long-chain polyunsaturates. Biology of Reprodution. 1999;**60**(3):698-701

[11] Rogers LK, Valentine CJ, Keim SA. DHA supplementation: Current implications in pregnancy and childhood. Pharmacological Research. 2013;**70**(1):13-19. DOI: 10.1016/j.phrs.2012.12.003

[12] Zeisel SH, da Costa KA. Choline: An essential nutrient for public health. Nutrition Reviews. 2009;**67**(11):615-623. DOI: 10.1111/j.1753-4887.2009.00246.x

[13] Penry JT, Manore MM. Choline: An important micronutrient for maximal endurance-exercise performance? International Journal of Sport Nutrition and Exercise Metabolism. 2008;**18**(2):191-203

[14] Corbin KD, Zeisel SH. Choline metabolism provides novel insights into non-alcoholic fatty liver disease and its progression. Current Opinion in Gastroenterology. 2012;**28**(2):159-165. DOI: 10.1097/MOG.0b013e32834e7b4b

[15] U.S. Department of Health and Human Services, U.S. Food & Drug administration. [Internet].2015.Availablefrom:http://www.fda.gov/Food/IngredientsPackagingLabeling/LabelingNutrition/ucm073599.htm [Accessed: February 25, 2017]

[16] Hasenhuettl GL. Chapter 2: Synthesis and commercial preparation of food emulsifiers. In: Food Emulsifiers and Their Applications. 2nd ed. Springer; New York, USA. 2008. pp. 11-37

[17] Primo Yúfera E. Química de los Alimentos. 1st ed. Síntesis, S.A.; Madrid, Spain. 1997. 462 p

[18] Administración Nacional de Medicamentos, Alimentos y Tecnología médica. Argentine Food Code. [Internet]. 2017. Available from: http://www.anmat.gov.ar/alimentos/normativas_alimentos_caa.asp [Accessed: February 25, 2017]

[19] Li J, Wang X, Zhang T, Wang C, Huang Z, Luo X, Deng Y. A review on phospholipids and their main applications in drug delivery systems. Asian Journal of Pharmaceutical Sciences. 2015;**10**(2):81-98. DOI: 10.1016/j.ajps.2014.09.004

[20] Yokota D, Moraes M, Pinho SC. Characterization of lyophilized liposomes produced with non-purified soy lecithin: A case study of casein hydrolysate microencapsulation. Brazilian Journal of Chemical Engineering. 2012;**29**(2):325-335. DOI: 10.1590/S0104-66322012000200013

[21] Palacios LE, Wang T. Egg-yolk lipid fractionation and lecithin characterization. Journal of the American Oil Chemists' Society. 2005;**82**(8):571-578. DOI: 10.1007/s11746-005-1111-4

[22] Keller B. C. Liposomes in nutrition. Trends in Food Science & Technology. 2001;**12**(1):25-31. DOI: 10.1016/S0924-2244(01)00044-9

[23] Hsieh YF, Chen TL, Wang YT, Chang JH, Chang HM. Properties of liposomes prepared with various lipids. Journal of Food Science. 2002;**67**(8):2808-2813. DOI: 10.1111/j.1365-2621.2002.tb08820.x

[24] Mosca M, Ceglie A, Ambrosone L. Effect of membrane composition on lipid oxidation in liposomes. Chemistry and Physics of Lipids. 2011;**164**(2):158-165. DOI: 10.1016/j.chemphyslip.2010.12.006

[25] Engberg O, Hautala V, Yasuda T, Dehio H, Murata M, Slotte JP, Nyholm TKM. The affinity of cholesterol for different phospholipids affects lateral segregation in bilayers. Biophysical Journal. 2016;**111**(3):546-556. DOI: 10.1016/j.bpj.2016.06.036

[26] Lasic DD. Novel applications of liposomes. Trends in Biotechnology. 1998;**16**(7):307-321. DOI: 10.1016/S0167-7799(98)01220-7

[27] Pradhan B, Kumar N, Saha S, Roy A. Liposome: Method of preparation, advantages, evaluation and its application. Journal of Applied Pharmaceutical Research. 2015;**3**(3):1-8

[28] Furberg D, Adams HP Jr, Applegate WB, Byington RP, Espeland MA, Hartwell T, Hunninghake DB, Lefkowitz DS, Probstfield J, Riley WA. Effects of lovastatin on early carotid atherosclerosis and cardiovascular events. Circulation. 1994;**90**(4):1679-1687

[29] La Rosa JC, Hunninghake D, Bush D, Criqui MH, Getz GS, Gotto AM Jr, Grundy SM, Rakita L, Robertson RM, Weisfeldt ML. The cholesterol facts. A summary of the evidence relating dietary fats, serum cholesterol, and coronary heart disease: A joint statement by the American heart Association and the National Heart, Lung, and Blood Institute. The Task Force on Cholesterol Issues, American Heart Association. Circulation. 1990;**81**(5):1721-1733

[30] Sacks FM, Pfeffer MA, Moye LA, Rouleau JL, Rutherford JD, Cole TG, Brown L, Warnica JW, Arnold JM, Wun CC, Davis BR, Braunwald E. The effect of pravastatin on coronary events after myocardial infarction in patients with average cholesterol levels. The New England Journal of Medicine. 1996;**335**(14):1001-1009

[31] Shepherd J, Cobbe SM, Ford I, Isles CG, Lorimer AR, Macfarlane PW, McKillop JH, Packard CJ. Prevention of coronary heart disease with pravastatin in men with hypercholesterolemia. West of Scotland Coronary Prevention Study Group. The New England Journal of Medicine. 1995;**333**(20):1301-1307

[32] Marsanasco M, Márquez AL, Wagner JR, Alonso S del V, Chiaramoni NS. Liposomes as vehicles for vitamins E and C: An alternative to fortify orange juice and offer vitamin C protection after heat treatment. Food Research International. 2011;**44**:3039-3046. DOI: 10.1016/j.foodres.2011.07.025

[33] Marsanasco M, Márquez AL, Wagner JR, Chiaramoni NS, Alonso S del V. Bioactive compounds as functional food ingredients: Characterization in model system and sensory evaluation in chocolate milk. Journal of Food Engineering. 2015;**66**:55-63. DOI: 10.1016/j.jfoodeng.2015.05.007

[34] Marsanasco M, Piotrkowski B, Calabró V, Alonso S del V, Chiaramoni NS. Bioactive constituents in liposomes incorporated in orange juice as new functional food: Thermal stability, rheological and organoleptic properties. Journal of Food Science and Techonolgy. 2015;**52**(12):7828-7838. DOI: 10.1007/s13197-015-1924-y

[35] Marsanasco M, Calabró V, Piotrkowski B, Chiaramoni N, Alonso S del V. Fortification of chocolate milk with omega-3, omega-6, and vitamins E and C by using liposomes. European Journal of Lipid Science and Technology. 2016;**118**(9):1271-1281. DOI: 10.1002/ejlt.201400663

[36] Holdsworth SD. Optimisation of thermal processing. A review. Journal of Food Engineering. 1985;**4**:89-116. DOI: 10.1016/0260-8774(85)90014-7

[37] Ordóñez JA, editor. Tecnología de los Alimentos. Volumen I, componentes de los alimentos y procesos. 1st ed. Síntesis, S.A.; Madrid, Spain. 1998. 365 p

[38] Mastellone P. Ayudando a conocer el mundo de la leche. 1st ed. Estrada; Buenos Aires, Argentina. 2000. 101 p

[39] Lešková E, Kubíková J, Kováčiková E, Košická M, Porubská J, Holčíková K. Vitamin losses: Retention during heat treatment and continual changes expressed by mathematical models. Journal of Food Composition and Analysis. 2006;**19**(4):252-276. DOI: 10.1016/j.jfca.2005.04.014

[40] Ayala Soto JG, Hernández Ochoa LR, Gutiérrez Méndez N, Chávez FD. Comparación del perfil vitamínico en jugos ultrapasteurizados de manzana y su impacto térmico de degradación. Revista Iberoamericana de las Ciencias Biológicas y Agropecuarias. 2015;**3**(6):1-22

[41] Gudden S, Yadava U. Effect of Heat Processing on the Vitamin C of Some Fruits. 1st ed. LAP LAMBERT Academic Publishing; Germany. 2012. 92 p

[42] Ford JE, Porter JWG, Thompson SW, Toothill J, Edward-Webb J. Effects of ultra-hightemperature (UHT) processing and of subsequent storage on the vitamin content of milk. Journal of Dairy Research. 1969;**36**(3):447-454. DOI: 10.1017/S0022029900012966

[43] Haddad GS, Loewenstein M. Effect of several heat treatments and frozen storage on thiamine, riboflavin, and ascorbic acid content of milk. Journal of Dairy Science. 1983;**66**(8):1601-1606. DOI: 10.3168/jds.S0022-0302(83)81980-8

[44] Mottar J, Naudts M. Quality of UHT milk compared with pasteurised and in-bottle sterilised milks. Lait. 1979;**59**:476-488

[45] Rechcigl M. Handbook of Nutritive Value of Processed Food (Volume I). Food for Human Use. 1st ed. CRC Press; Boca Raton, Fl, USA. 1982. 679 p

[46] Abioye AO, Abioye VF, Ade-Omowaye BIO, Adedeji AA. Kinetic modelling of ascorbic acid loss in baobab drink at pasteurization and storage temperatures. Journal of Environmental Science, Toxicology and Food Technology. 2013;**7**(2):17-23

[47] Lima M, Heskitt BF, Burianek LL, Nokes SE, Sastry SK. Ascorbic acid degradation kinetics during conventional and ohmic heating. Journal of Food Processing Preservation. 1999;**23**(5):421-434. DOI: 10.1111/j.1745-4549.1999.tb00395.x

[48] Francisco M, Velasco P, Moreno DA, García-Viguera C, Cartea ME. Cooking methods of *Brassica rapa* affect the preservation of glucosinolates, phenolics and vitamin C. Food Research International. 2010;**43**(5):1455-1463. DOI: 10.1016/j.foodres.2010.04.024

[49] Emami S, Azadmard-Damirchi S, Peighambardoust SH, Valizadeh H, Hesari J. Liposomes as carrier vehicles for functional compounds in food sector. Journal of Experimental Nanoscience. 2016;**11**(9):737-759. DOI: 10.1080/17458080.2016.1148273

[50] Zabodalova L, Ishchenko T, Skvortcova N, Baranenko D, Chernjavskij V. Liposomal beta-carotene as a functional additive in dairy products. Agronomy Research. 2014; **12**(3):825-834

[51] Quirós-Sauceda AE, Ayala-Zavala JF, Olivas GI, González-Aguilar GA. Edible coatings as encapsulating matrices for bioactive compounds: A review. Journal of Food Science and Technology. 2014;**51**(9):1674-1685. DOI: 10.1007/s13197-013-1246-x

[52] Kris-Etherton PM, Hecker KD, Bonanome A, Coval SM, Binkoski AE, Hilpert KF, Griel AE, Etherton TD. Bioactive compounds in foods: Their role in the prevention of cardiovascular disease and cancer. The American Journal of Medicine. 2002; **113**(9):71-88.

[53] Singh H. Nanotechnology applications in functional foods; opportunities and challenges. Preventive Nutrition and Food Science. 2016;**21**(1):1-8. DOI: 10.3746/pnf.2016.21.1.1

[54] Bigliardi B, Galati F. Innovation trends in the food industry: The case of functional foods. Trends in Food Science & Technology. 2013;**31**:118-129. DOI: 10.1016/j.tifs.2013.03.006

[55] Guiné RPF, Lima MJ. Overview and developments regarding functional foods and beverages. Current Nutrition & Food Science. 2008;**4**(4):298-304. DOI: 10.2174/157340108786263720

[56] Atkinson J, Epand RF, Epand RM. Tocopherols and tocotrienols in membranes: A critical review. Free Radical Biology and Medicine. 2008;**44**(5):739-764. DOI: 10.1016/j.freeradbiomed.2007.11.010

[57] Machlin LJ, Bendich A. Free radical tissue damage: Protective role of antioxidant nutrients. The FASEB Journal. 1987;**1**(6):441-445

[58] Liu RH. Potential synergy of phytochemicals in cancer prevention: Mechanism of action. Journal of Nutrition. 2004;**134**(12):3479-3485

[59] Paul R, Ghosh U. Effect of thermal treatment on ascorbic acid content of pomegranate juice. Indian Journal of Biotechnology. 2012;**11**:309-313

[60] Khanniri E, Bagheripoor-Fallah N, Sohrabvandi S, Mortazavian AM, Khosravi-Darani K, Mohammadia R. Application of liposomes in some dairy products. Critical Reviews in Food Science and Nutrition. 2016; **56**(3): 484-493. DOI: 10.1080/10408398.2013.779571

[61] Farhang B, Kakuda Y, Corredig M. Encapsulation of ascorbic acid in liposomes prepared with milk fat globule membrane-derived phospholipids. Dairy Science & Technology. 2012;**92**:353-366. DOI: 10.1007/s13594-012-0072-7

[62] Lasic DD. The mechanism of vesicle formation. Biochemical Journal. 1988;**256**(1):1-11. DOI: 10.1042/bj2560001

[63] Akbarzadeh A, Rezaei-Sadabady R, Davaran S, Woo Joo S, Zarghami N, Hanifehpour Y, Samiei M, Kouhi M, Nejati-Koshki K. Liposome: Classification, preparation, and applications. Nanoscale Research Letters. 2013;**8**(1):102. DOI: 10.1186/1556-276X-8-102

[64] Resolutions of the Common Marked Group N° 34/10 [Internet]. 2017. Available from: BPF http://www.anmat.gov.ar/webanmat/Legislacion/Alimentos/Resolucion_Conjunta_199-2011_y_641-2011.pdf

[65] Wang S, Su R, Nie S, Sun M, Zhang J, Wu D, Moustaid-Moussa N. Application of nanotechnology in improving bioavailability and bioactivity of diet-derived phytochemicals. The Journal of Nutritional Biochemistry. 2014;**25**:363-376. DOI: 10.1016/j.jnutbio.2013.10.002

[66] Mohammadi M, Ghanbarzadeh B, Hamishehkar H. Formulation of nanoliposomal vitamin D3 for potential application in beverage fortification. Advanced Pharmaceutical Bulletin. 2014;**4**(2):569-575

[67] Siró I, Kápolna E, Kápolna B, Lugasi A. Functional food. Product development, marketing and consumer acceptance—A review. Appetite. 2008;**51**:456-467. DOI: 10.1016/j.appet.2008.05.060

[68] Pray L, Boon C, Miller EA, Pillsbury L. IOM (Institute of Medicine). Providing Healthy and Safe Foods as We Age: Workshop Summary. 1st ed. The National Academies Press; Washington D.C., USA. 2010. 193 p. DOI: 10.17226/12967

[69] Rangel-Huerta OD, Pastor-Villaescusa B, Aguilera CM, Gil A. A systematic review of the efficacy of bioactive compounds in cardiovascular disease: Phenolic compounds. Nutrients. 2015;**7**:5177-5216. DOI: 10.3390/nu7075177

[70] Alvídrez-Morales A, González-Martínez BE, Jiménez-Salas Z. Tendencias en la producción de alimentos: Alimentos funcionales. Revista de Salud Pública y Nutrición. 2002;**3**(3):1-6

[71] Verbeke W. Functional foods: Consumer willingness to compromise on taste for health? Food Qualityand Preference. 2006;**17**:126-131. DOI: 10.1016/j.foodqual.2005.03.003

[72] Hilliam M. Future for dairy products and ingredients in the functional foods market. Australian Journal of Dairy Technology. 2003;**58**:98-103

[73] World Health Organization [Internet]. 2016. Available from: http://www.who.int/mediacentre/factsheets/fs178/en/ [Accessed: February 25, 2017]

[74] Boerma T, Mathers C, AbouZahr C, Chatterji S, Hogan D, Stevens G, Mahanani WR, Ho J, Rusciano F, Humphreys G. Health in 2015: From MDGs, Millennium Development Goals to SDGs, Sustainable Development Goals. 1st ed. World Health Organization; France. 2015. 216 p

[75] Rodrigo A. La nutrición en las primeras etapas de la vida, su relación con el aprendizaje. Nutrición, desarrollo y alfabetización. In: Morasso M del C, Duro E, editors. Nutrición, desarrollo y alfabetización. 2nd ed. UNICEF; Buenos Aires, Argentina. 2010. p. 25-75

[76] Mangialavori G, Abeyá Gilardon E, Biglieri Guidet A, Durán P, Kogan L. Feeding Children under 2 Years. Results of the National Survey of Nutrition and Health. Buenos Aires, Argentine: Ministry of Health; 2010. pp. 1-61

[77] Anzaldúa-Morales A. La evaluación sensorial de los alimentos en la teoría y la práctica. 1st ed. Acribia; Zaragoza, Spain. 1994. 220 p

[78] Singh-Ackbarali D, Maharaj R. Sensory evaluation as a tool in determining acceptability of innovative products developed by undergraduate students in food science and technology at the University of Trinidad and Tobago. Journal of Curriculum and Teaching. 2014;**3**(1):10-18. DOI: 10.5430/jct.v3n1p10

[79] Olivas-Gastélum R, Nevárez-Moorillón GV, Gastélum-Franco MG. Las pruebas de diferencia en el análisis sensorial de los alimentos. Tecnociencia Chihuahua. 2009;**3**(1):1-7

[80] Sancho J, Bota E, de Castro JJ. Introducción al análisis sensorial de los alimentos. 1st ed. Edicions Universitat Barcelona; Barcelona, Spain. 1999. 336 p

[81] Meilgaard M, Civille GV, Carr TB. Sensory Evaluation Techniques. 3rd ed. CRC Press; Florida, USA. 1999. 464 p

Section 3

Tools to Evaluate the Effects of Bioactive Ingredients, Functional Foods and Diet

Chapter 11

Food Metabolomics: A New Frontier in Food Analysis and its Application to Understanding Fermented Foods

Oluwafemi Ayodeji Adebo, Patrick Berka Njobeh,
Janet Adeyinka Adebiyi, Sefater Gbashi and
Eugenie Kayitesi

Additional information is available at the end of the chapter

http://dx.doi.org/10.5772/intechopen.69171

Abstract

The emergence of food metabolomics, otherwise known as foodomics, has opened new frontiers and possibilities for scientists to characterize and simultaneously determine and obtain the comprehensive profile of the food metabolome. Qualitative and quantitative determinations of this metabolome offer insights into the underlying processes involved and details about the content of the food analytes. This had seemed technically challenging and impossible over time, but can now be done due to the advent of sophisticated analytical equipment and chemometric tools. The application of this technique offers enormous opportunities to obtain detailed information that can be correlated to various properties, functionalities and potentials in fermented foods. This chapter thus evaluated and documented studies presented in the literature on the food metabolomics study of fermented foods, with a view of appraising its prospects, applications and subsequent utilization in the study of fermented foods.

Keywords: foodomics, food metabolomics, fermentation, fermented foods, chemometrics

1. Introduction

Fermentation continues to be a viable food processing technique all over the world. This might be attributed to the ease and simplicity of the process and its numerous other benefits, including providing variety in foods, improving palatability and aesthetic value, detoxification and imparting desirable sensorial properties [1–3]. Furthermore, it plays significant role in conferring health promotion and functional benefits to fermented foods. In line with this are different studies on fermented foods reporting their ability to reduce diarrhea, malnutrition, encourage child growth and development, exhibit nutraceutical and functional effects including being antidiabetic, antihypertensive, chemoprotective, reducing oxidative stress, cardiovascular diseases and possessing probiotic properties [1–8]. Sequel to these benefits and the ever growing market for functional foods, fermented food products are positioned as food sources that can improve consumer well-being and reduce the risk of diseases.

Although fermentation like other food processing techniques is needed for the transformation of food prior to consumption, it results in structural changes, formation, modification and/or degradation of compounds and an increase or decrease in these constituents could occur. Characterization and comprehensive monitoring of the metabolic, physicochemical, biochemical and structural changes occurring during the fermentation process have thus been relatively difficult. The advent of food metabolomics, also known as "foodomics" enables scientists to obtain detailed and comprehensive molecular profile of thousands of metabolites in foods, all in a single run [9, 10]. Food metabolomics thus presents a holistic approach of providing insight, resolving and identifying the complexities and multifunctionality of fermentation and its subsequent food products.

According to Cifuentes [9] and Garcia-Canas [11], food metabolomics is a valuable and promising tool for food processors and scientists to understand the metabolome of food, including its biochemistry and composition. Being one of the "omics" technology, it offers enormous opportunities to obtain detailed information that can be correlated to the functional and nutraceutical composition of foods. This chapter thus provides an overview of food metabolomics studies that have applied this to fermented foods in the literature and its prospects for further use.

2. Fundamentals of food metabolomics

Metabolomics itself is designated to mean a comprehensive analysis, study, identification and quantification of "as many small metabolites" as possible in a system at a specific time and condition through the use of omics technologies [12–18]. Related to this and taking a cue from earlier authors [9, 11, 19, 20], food metabolomics or foodomics can thus be defined as the study of "as many small metabolites" in food under a specific condition and time through the application of omics technologies. It is a discipline involving the combination of food, nutrition, advanced analytical and data processing techniques and bioinformatics. According to Wishart [21], metabolomics permits the simultaneous characterization of a variety of compounds and metabolites and thus offer food and nutrition scientists the privilege to acquire comprehensive and detailed molecular composition of food. This feature makes metabolomics applicable to different aspects of food science including food safety, food quality, functional foods, food microbiology, food processing and nutrition (**Figure 1**). Sequel to the potential embedded in food metabolomics, scientists are gradually utilizing advanced analytical strategies as opposed to the traditional and classical existing methodologies, which does not provide the much-needed information to understand the complexities in food. Such complexities are however compounded in fermented foods, containing a variety of nutrients, compounds and

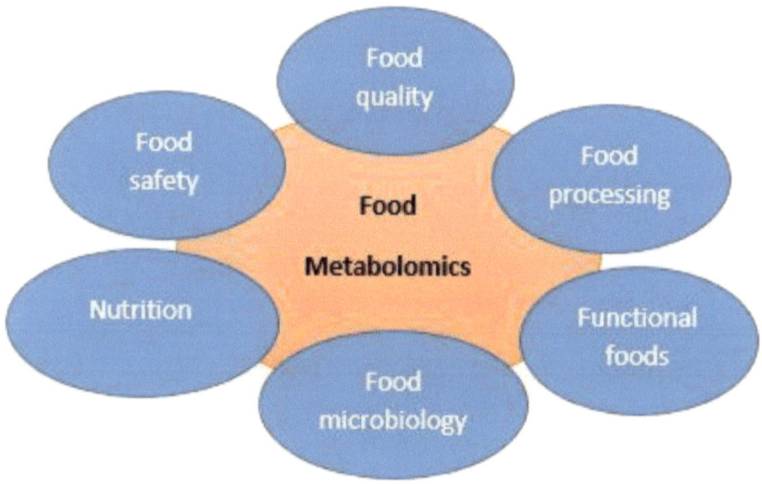

Figure 1. Different aspects of food metabolomics.

volatiles with diverse concentrations, chemical structures, affinity and polarities. Food metabolomics thus provides the opportunity for understanding this multifaceted analyte.

As with another metabolomics study, food metabolomics analyses can generally be classified into either targeted or untargeted. The targeted analysis focuses on a specific group of intended metabolites with such requiring subsequent quantification and identification [18, 22]. They are thus more detailed and require greater levels of extraction and purification prior to analysis. In contrast to targeted analysis, untargeted metabolomics analysis is broader and focused on the detection of a variety of metabolites to obtain fingerprints or patterns without essentially quantifying or identifying specific metabolites [16, 23, 24].

3. The process of food metabolomics analysis

Every metabolomics analysis consists of a sequence of steps prior to obtaining the data [16, 19, 24]. Not all the steps, depicted in **Figure 2** are not, however, necessary for food metabolomics or any other metabolomics studies. Major factors that determine the selection of steps include the type of study (targeted or untargeted), sample form (solid, liquid) and the available instrumentation and detection technique [gas chromatography-mass spectrometry (GC-MS), liquid chromatography-mass spectrometry (LC-MS), nuclear magnetic resonance (NMR), etc.] [16]. A description of these steps is nevertheless summarized in the ensuing sections of this chapter.

3.1. Sample preparation

Sample preparation is essential and vital in any analysis. This is needed to prepare the sample into a "ready state" form, release the analyte (metabolites) available, reduce experimental error

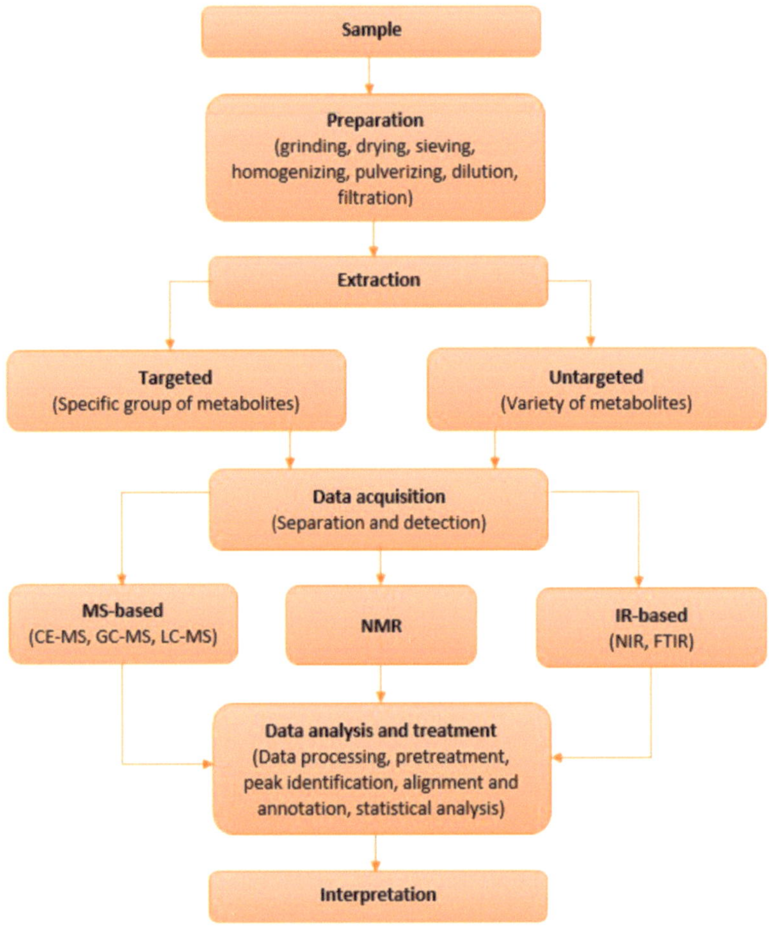

Figure 2. Schematic presentation of the food metabolomics process.

and ensure the analytical procedure is reproducible. Grinding, size reduction and homogenization are some of the needed steps prior to analysis to ensure proper mixing and present a sample that is a true representative. The concentration of samples is also important with freeze-drying and use of liquid nitrogen commonly used in food metabolomics studies of fermented foods. This not only concentrates the metabolites but also reduces the possibility of losing heat labile components during conventional oven drying techniques. Both freeze drying and liquid nitrogen have been applied in the preparation of fermented foods in food metabolomics studies for *cheonggukjang* [25, 26], *meju* [27], *doenjang* [28] and cocoa beans [29]. Nevertheless, care must be taken to avoid the introduction of any form of unwanted variability throughout this step, which might result in significant experimental discrepancy, that would surpass biological variance. Sampling conditions and time should also be controlled to limit inconsistency in results.

3.2. Extraction

Among the many steps for food metabolomics studies, extraction is a vital and important one. Considering the varying and diverse constituents and composition of fermented foods, including but not limited to amino acids, organic acids, phytochemicals, sugars, minerals, nucleic acids, vitamins and other volatile compounds, extraction may be somewhat tricky. Hence, extraction techniques to be utilized would be largely dependent on the form of study (targeted or untargeted), characteristics, number and quantity of metabolites of interest [15, 30, 31]. Extraction protocol would not thus be an express decision but rather influenced by the focus of analysis (study).

For targeted analysis, a suitable purification scheme and the use of appropriate internal standards is important [15]. This might not be the case for untargeted analysis due to the need to target as many metabolites as possible. Extraction techniques commonly used for fermented foods are solvents (methanol, chloroform, ethanol, acetonitrile) [27, 28, 32], similar to those used in other metabolomics studies [15, 17, 24, 31]. When the focus of the study is on specific non-volatile metabolites, derivatization may be required prior to analysis on GC-MS. This is necessary to make sure the samples are thermostable, increase volatility and improve the detectability of the analyte [16, 18].

3.3. Data acquisition

Data acquisition in form of separation and detection of metabolites is a key step in metabolomics studies. It essentially requires advanced analytical techniques, considering the complexity, diversity and number of metabolites to characterize in food [16, 18]. Separation techniques commonly used for food and fermented foods include high-performance liquid chromatography (HPLC), ultra-performance liquid chromatography (UPLC), gas chromatography (GC), capillary electrophoresis (CE) and ion mobility spectrometry (IMS) [16, 18, 21, 33]. Detection techniques include mass spectrometry (MS), NMR, high-resolution magic angle spinning (HRMAS) NMR, Fourier transform (FT) NMR, near infrared spectroscopy (NIR) and Fourier transform infrared spectroscopy (FTIR) [15, 16, 18, 21]. A detailed review and working principle of these separation and detection techniques have been presented in the literature and can be consulted for further reading [15, 18, 21, 24, 30, 31, 34, 35].

In the studies of fermented foods using metabolomics, most separation are either done by GC or LC (for polar compounds), while detection is done majorly by MS with few other studies reporting the use of NMR and FTIR. A major consideration and factor in the use of GC and LC is their higher sensitivity and separation. While GC–MS is usually utilized for the determination of primary metabolites i.e., carbohydrates, amino acids, organic acids, fatty acids and phytochemicals, LC–MS are frequently employed for secondary metabolites including alkaloids, flavonoids, phenolic acids, peptides, polyamines and saponins [31, 36]. As indicated by Tugizimana et al. [24], developments towards the enhancement of chromatographic include the use of multi-dimensional separation systems such as two-dimensional liquid chromatography (LC $\times$ LC) and two-dimensional gas chromatography (GC $\times$ GC). Furthermore, the use of better MS platforms including time of flight (TOF), Orbitrap, MS $\times$ MS/MS^n provides better resolution, higher scan speeds, detailed fragmentation information, higher resolution, selectivity and better molecular specificity as seen with the Pegasus HRT GC. For separation, MS is most preferred and coupled with either GC or LC, it allows for the comprehensive evaluation and discrimination of compounds [18]. It should, however, be noted that due to the varying behaviors, polarity, volatility, structure, configuration, solubility and molecular weight of different metabolites in fermented foods, a single data acquisition technique for the detecting and separating all these components is quite impossible. A combination of different techniques would rather provide a better analytical potential for a full metabolomics study.

3.4. Data analysis and treatment

Metabolomic studies are quite synonymous with the generation of a large amount of data, that may be somewhat confusing at first. Subsequent analysis of such high-throughput data can be roughly divided into two: pretreatment and analysis [37]. Handling these huge data would require an automated software for quantification and identification [24]. Pretreatment basically involves alignment, normalization, compound identification, centering, transformation, scaling, removing baseline artefacts and peak picking [16, 24, 38, 39], in order to convert the raw data set into a form that can be utilized for subsequent analysis. Succeeding analysis of the cleaned data in food metabolomics studies are majorly done using different chemometric tools, to provide a description and understanding of the variations and/or similarities in the metabolites. Wold [40] has defined chemometrics as a branch of science concerned with the data analysis (extracting information from data), ensuring that the data set contains maximum information using several mathematical multivariate data analysis (MVDA) tools.

Depending on the purpose of the food metabolomics study, there are three major categories of MVDA. These are exploratory/informative, classification/discrimination and regression/prediction [16, 38, 41]. While informative analyses are focused on identification and quantification to obtain sample intrinsic information (such as the development of metabolite databases and the discovery of biomarkers), discriminative analyses are majorly aimed at finding differences between samples/treatments [16, 42]. In contrast, predictive models are focused on quantification and prediction of a variable that may be difficult to quantify [16, 38]. MVDA tools commonly used in food metabolomics studies include artificial neural networks (ANN), principal component analysis (PCA), orthogonal projection to latent structures-discriminant analysis (OPLS-DA), partial least square discriminant analysis (PLS-DA), principal component regression (PCR), hierarchical cluster analysis (HCA), canonical correlation analysis (CCA) and others [16, 38, 43]. Detailed strategies, algorithms and explanation on these MVDA techniques have been described in detail elsewhere [24, 39, 43–46].

4. Food metabolomics of fermented foods

Food metabolomics has been applied and adopted in the study of different foods in the literature [11, 16–18, 21, 33, 47, 48]. Specifically, for fermented foods, which is the focus of this chapter, it is conventionally used to observe, monitor metabolite changes occurring during fermentation and to investigate the composition of such fermented food. Such knowledge has assisted in providing a comprehensive understanding of the fermentation process and probably predict sensory, nutritional, functionality and nutraceutical quality of the final fermented product. Few studies presented in the literature on food metabolomics studies of fermented foods are summarized in **Table 1**. This section of this chapter would thus focus on the documented changes in metabolite groups and the use of metabolomics in understanding the modifications occurring during the fermentation process of these foods.

Metabolites produced during the fermentation of a Korean cuisine called *cheonggukang*, have been investigated by several authors [25, 26, 49–51] (**Table 1**). Using ^{1}H NMR, Choi et al. [25] observed a decrease in sugars and citric acid with fermentation time. Acetic acid, phenylalanine and tyrosine however increased with time. Baek et al. [49], reported a total of 5 sugar alcohols, 10 sugars, 7 organic acids and 20 amino acids in the same product after obtaining it using different *Bacillus* sp. with subsequent analysis on gas chromatography-time of flight mass spectrometry (GC-TOF-MS). Most of the amino acids showed increasing amounts with time, sugars and sugar alcohols (arabitol, ribitol, sorbitol, myoinositol and lactitol) showed decreases, whereas there were variations in organic acids. Similar occurrences and variations in amino acids, organic acids and also fatty acids, carbohydrates, soyasaponins, isoflavonoids and nucleosides were observed using different metabolomics techniques [26, 50, 51].

Chen et al. [52] reported the occurrence of 28 metabolites including 13 amino acids, six organic acids, three organic bases and sucrose in fermented crab paste as analyzed on NMR. Using PCA and OPLS-DA the authors were able to observe a decline in taurine, betaine, trigonelline, trimethylamine-N-oxide and inosine with an accumulation of sugars and hypoxanthine. 53 compounds including organic acids, alcohols, sugars, amino acids were identified from the metabolomic profiling of *daju* fermented with *Bacillus licheniformis* [53]. Using NMR and PCA, the authors observed a decomposition of polymers such as protein, starch and cellulose to smaller monomers and accumulation of saccharides. *Doejang*, a Korean delicacy has been studied using food metabolomics techniques (**Table 1**). Characterization and profiling on ^{1}H NMR, GC-MS, GC-TOF-MS and data analysis on PCA, PLS-DA revealed the presence of amino acids, sugars and sugar derivatives and organic acids in *doenjang* [28, 54]. Using PCA, Yang et al. [54] was able to discriminate *doenjang* samples fermented for different days and reported increasing levels of amino acids, with no significant change in sugars and variation in the levels of fatty acids. Likewise, Lee et al. [28] observed an increase in monosaccharides, sugar alcohols and most amino acids during fermentation of *doenjang*. Variations were also observed in the levels of organic acids, fatty acids isoflavones and soyasaponins [28]. Kang et al. [27], reported a decrease in the concentration of citric acid during fermentation, with variations in the quantities of peptides, amino acids, nucleosides, organic acids and urea cycle intermediates were reportedly altered throughout the fermentation process.

Using both GC-MS, high-performance liquid chromatography coupled to a diode array detector (HPLC-DAD) in combination with hierarchical cluster analysis (HCA), a strong correlation was observed between volatiles, flavonoids and polyphenolic compounds of two types of wheat dough [55]. The authors observed a general increase in polyphenol content of the wheat doughs, but a diverse metabolite profile in the two wheat substrates used. Likewise, Mayorga-Gross et al. [29] investigated the metabolites changes occurring during cocoa fermentation on an ultra-high performance liquid chromatography with electrospray ionization quadrupole time of flight mass spectrometry (UPLC-ESI$^+$-Q-TOF-MS) system and adopted a PLS-DA model for data processing. The clustering of ions according to retention times and mass spectrum on the PLS-DA model yielded a total of 37 discriminating metabolites. Sugars,

Produce	Raw material	Metabolite group	Metabolite forms	Data acquisition method	Data processing technique	Reference
Cheonggukjang	Soybean	Al[a]↑↓	1, 2, 3, 4-tetrakis[(trimethylsilyl)oxy]-butane, 2, 3-bis (trimethylsilyl)-butane, δ-tocopherol, γ-tocopherol, D-ribitol, tyramine, glycerol, hydroxylamine, phytol	^{1}HNMR*,CE-TOF-MS*, GC-FID[δ],GC-TOF-MS*,LC-MS/MS*	PCA, PLS-DA	[25, 26, 49–51]
		Am[b]	1,3-diamino-propane, phenethylamine, putrescine, tryptamine, serotonin, spermidine			
		AA[c]	α-aminobutyric, β-alanine, γ-aminobutyric (GABA), *g*-aminobutyric, 2, 6-diaminopimelate, alanine, aminoadipate, arginine, asparagine, aspartic, betanine, choline+, citrulline, DL-2-aminobutyric, DL-asparagine, DL-cysteine, DL-glutamine, DL-homoserine, DL-leucine, DL-methionine, DL-*N*-acetyl-serine, DL-ornithine, DL-phenylalanine, DL-threonine, DL-tryptophan, DL-valine, glutamic, glutamate, glycine, histidine, homotyrosine, homovaline, hydroxyproline, isoleucine, leucine, lysine, L-arginine, L-aspartic, L-cysteine, L-histidine, L-isoleucine, L-lysine, L-serine, L-tyrosine, methionine, *N-a*-acetylornithine, *N*-acetyl-glutamic acid, ornithine, phenylalanine, proline, pyroglutamate, pyroglutamic, serine, threonine, tryptophan, tyrosine, valine			
		SUG, SUGDs[d]	δ-trehalose, arabinose, arabitol, D-fructose, D-galactosamine, D-glucosamine, D-lactose, D-maltose, D-pintol, D-ribose, D-xylobiose, D-xylose, fructose, fructose-6-phosphate, galactose, galactinol, glucose, glucose-6-phosphate, inositol, isomaltose, lactate, lactitol maltose, mannose, mannotriose, melibiose, myo- ribitol, *N*-acetyl-raffinose, ribose, sorbitol, sucrose, xylose			
		FA[e]	Arachidic, behenic, linoleic, linolenic, myristic, oleic, palmitic, palmitoleic, stearic			
		IFVN[f]	6″-*O*-acetyldaidzin, 6″-*O*-acetylgenistin, 6″-*O*-malonylglycitin, daidzin, glycitin, genistin, quercetin-tri-*O*-β-glucopyranoside			
		NTs[g]	Adenine, adenosine, cytidine, cytosine, dihydrouracil, guanine, guanosine, hypoxanthine, thymine, uracil, xanthine			

Produce	Raw material	Metabolite group	Metabolite forms	Data acquisition method	Data processing technique	Reference
		OA[h]	2-hrdoxyisobutyric, 2-hydroxy-glutaric, 3-methyl-2-[(trimethylsilyl)oxy]-pentanoic acid, acetic, benzenepropanoic, calcium pantothenate, *cis*-aconitate, citric, citrilamic, D-galacturonic, DL-isocitric, DL-lactic, Dl-malic, formic, fumaric, galactaric, gluconic, glutamic, glutaric, glycerate, glycolic, itaconic, lactic, malic, malinic, malonic, *n*-octadecanoic, oxalic, palmitic, phenylpyruvate, quinate, saccharic, shikimic, succinic, succinate, tartaric, *trans*-aconitic acid, *trans*-caffeic, *trans*-sinapic, trimethylsilyl, 3, 5-bis(trimethylsilyl)-3-methylvalerate			
		SSAPN[i]	A3, Bg, I, II, IV, V			
		V[j]	1, 3-diamino-propane, phenethylamine, putrescine, tryptamine, serotonin, spermidine			
			Choline, nicotinic acid			
		O[k]	3-amino-2-one-piperidin, allantonate, glycero-3-phosphate, mevalonolactone, phosphoric acid, R-(−)-1-amino-2-propanol, trigonelline, urea			
Crab paste	Crab	AA[c]	Alanine, arginine, glutamate, glycine, histidine, isoleucine, leucine, methionine, phenylalanine, tryptophan, tyrosine, valine	^{1}HNMR[*]	PCA, OPLS-DA	[52]
		OA[h]	Acetate, formate, fuarate, lactate, succinate, taurine			
		OB[l]	Betaine, trimethylamine (TMA), trimethylamine-*N*-oxide			
		PUR, PYR[m]	2-pyridinemethanol, adenosine diphosphate (ADP), hypoxanthine, inosine, trigonelline			
		SUG[d]	Sucrose			
Daqu	Barley and peas	Al[a]	Ethanol, glycerol, isopropanol	^{1}HNMR[δ]	PCA	[53]
		AA[c]	2-Aminobutyrate, cysteine, glutamate, glycine, glycylproline, homoserine, isoleucine, proline, serine, threonine			

Produce	Raw material	Metabolite group	Metabolite forms	Data acquisition method	Data processing technique	Reference
		SUG, SUGDs[d]	Arabinitol, fructose, galactitol, galactose, glucose, gluticol, lactose, maltose, mannitol, myo-inositol, ribose, sucrose			
		OA[h]	2-hydroxyisobutyrate, 2-phosphoglycerate, acetate, glycerate, glycolate, isobutyrate, lactate, pyruvate, succinate, taurine			
		OB[l]	Betaine, *cis*-aconitate			
		O[k]	Acetone, allantoin, ascorbate, choline, ethylene glycol, galactonate, maltate, malonate, *N*-nitrosodimethylamine, *O*-phosphocholine, *O*-phosphoserine, oxypurinol, propionate, propylene glycol, *S*-sulfocysteine, urea			
Doenjang	Soybean	AA[c]	γ-aminobutyric, alanine, aminoaldiphic, aminobutyric, asparagine, aspartic, glutamine, glutamic, glycine, histidine, isoleucine, leucine, lysine, methionine, ornithine, phenylalanine, proline, pyroglutamic, sarcosine, serine, thioproline, threonine, tryptophan, tyrosine, valine	^{1}HNMR[δ]	PCA, PLS-DA	[28, 54]
		SUG, SUGDs[d]	α-glucose, β-glucose, arabinose, arabitol, erythrose, fructose, galactonic, galactose, glucitol, glucose, glycerol, glucosamine, inositol, mannitol, mannose, maltose, melibiose, *myo*-inositol, raffinose, ribitol, ribonic acid, sucrose, xylitol	GC-TOF-MS[*], UPLC–Q-TOF-MS[*]		
		FA[e]	Arachidic, behenic, caproc, eicosanic, eicosadienoic, lauric, linoleic, linolenic, magaric, myristic, oleic, palmitic, palmitoleic, pentadecyclic, stearic, tricosanoic			
		IFVN[f]	Acetyldaidzin, acetylgenistin, acetylglycitin, daidzin, daidzein, genistin, glycitin, glycitein, malonyldaidzin, malonyglycitin, malonygenistin			
		OA[h]	2-ketoglutaric, acetic, carbonic, citric, formic, fumaric, glucaric, glycolic, lactate, lactic, maleic, malic, malomic, malonic, manelic, oxalic, pipecolic, propionic, pyroglutamic, succinic, vanilic			
		SSAPN[i]	γg, γa, Bd, Be, I, II, III, IV, V			

Produce	Raw material	Metabolite group	Metabolite forms	Data acquisition method	Data processing technique	Reference
		O[k]	Choline, phosphocholine			
Fermented cereal	Wheat	Al[a]	1-decanol, 1-dodecanol, 1-octanol, 1, 2-dodecanediol, 7-methyl-4-octanol, dimethyl-1-octanol, ethylalcohol, hexanol, isoamylalcohol, methyl-2-buten-1-ol, methyl-3-heptanol, octadien-2-ol, octen-3-ol, pentanol, phenethylalcohol	SPME-GC-MS[*]	HCA	[55]
		C[n]	1, 1, 3-trimethyl-3-cyclohexene-5-one, 6-methyl-5-hepten-2-one, acetoin, decadienal, dodecanal heptanal, hexanal, methylpentanal, nonadienal, nonanone, octanone, octenal, pentanal	HPLC-DAD[δ]		
		H[o]	1, 2-dimethyl-benzene, 1, 3-hexadiene, 2-ethyl-furan, 2-penthyl-furan, 2-methyldecane, 3-methyl-dodecane, 4-methyl-dodecane, 5-methyldodecane, 10-methylnonadecane, 10-methyl-eicosane, furanone			
		OA[h]	2-methylbutanoic, 3-methylbutanoic, acetic, dodecanoic, pentanoic, hexanoic, heptanoic			
		O[k]	Ester			
Fermented cocoa beans	Cocoa beans	CTH, CTHd[p]	Epicatechin, *O*-hexoside-proanthocyanidin A5′, *O*-pentoside-proanthocyanidin A5, procyanidin	UPLC-ESI-QTOF-MS[*]	PCA, PLS-DA,	[29]
		O[k]	Tripeptide, sucrose			
Fermented tea	Green tea, black tea	AA[c]	Glutamine, glutamic acid, glucoside, histamine, leucine, phenylalanine, proline, theanine, theanine-glucoside, tyrosine, tryptophan, valine	^{1}HNMR[*], UHPLC-QTOF-MS[*]	PCA	[56, 57]
		Ak[q]	Caffeine, choline, glycerophosphocholine, theobromine			
		CTH, CTHd[p]	3-galloylprocyanidin B1, Cathechin, epiafzelechin, epicatechin-3-gallate, epicatechin, epicatechin gallate, epigallocatechin, epigallocatechin gallate, epigallocatechin methylgallate, theaflavin-3-gallate, theaflavin 3, 3′-digallate, theaflavin-3′-gallate, theasinensin A, theasinensin F, pigallocatechin-3-gallate, procyanidin B1, procyanin B2			

Produce	Raw material	Metabolite group	Metabolite forms	Data acquisition method	Data processing technique	Reference
		FVNG, VOG[r]	Apigenin-6, 8-C-diglucoside, apigenin 6-C-glucoside 8-C-arabinoside, apigenin-6-C-arabinoside-8-C-glucoside, isoquercitrin, isovitexin, kaempferol 3-O-galactosylrutinoside, kaempferol 3-O-glucosylrutinoside, kaempferol-3-O-galactoside, kaempferol-3-O-glucoside, kaempferol-3-O-rutinoside myricetin 3-galactoside, quercetin-3-O-galactoside, quercetin 3-O-glucosylrutinoside, rutin			
		L[s]	LysoPC, MG			
		NTs[g]	(S)-5′-deoxy-5′(methylthio)adenosine, 5′-deoxy-5′(methylthio)adenosine, adenine, guanosine, inosine			
		OA[h]	3-O-*p*-coumaroylquinic, 4-O-*p*-coumaroylquinic, *p*-coumaric, caffeoylshikimic, theogallin			
		SUG[d]	α-glucose, β-glucose, sucrose			
		O[k]	Caffeine, gallic acid, *N*-(1-deoxy-1-fructosyl)leucine, *N*-(1-deoxy-1-fructosyl)tyrosine, *N*-vinyl-2-pyrrolidone, *O*-demethylfonsecin, theanine, unknown compounds			
Fermented milk	Milk	AA[c]↑↓	3-aminobutyric, alanine, arginine, asparagine, aspartic, GABA, glutamine, glycine, isoleucine, methionine, threonine	CE-TOF-MS[*]		[58]
		Am[b]↓	Cyclohexylamine			
		OA[h]↓↑	2-oxoglutaric, citric, isocitric			
		PUR↑↓	Adenine, guanine, hypoxanthine			
		P[t]↑	Ala-Pro, Leu-Pro, Pro-Pro, Val-Leu, Val-Pro, Val-Pro-Pro			
		SUG[d]↓	Fructose 1, 6-diphosphate			
		V[j]↓	Pyridoxamine			
Fermented soymilk	Soymilk	AA[c]↓	Phenylalanine, tyrosine	^{1}HNMR[*]	PCA	[59]

Produce	Raw material	Metabolite group	Metabolite forms	Data acquisition method	Data processing technique	Reference
		OA^{h}↑↓	Acetic, citric, fumaric, lactate, lactic, malic, oxalacetic, succinic,			
		SUG^{d}↓	Raffinose, stachyose, sucrose			
		O^{k}	Choline			
Fermented soybean	Soybean	AA^{c}↑↓	Aspartic, GABA, glutamic, glycine, pyroglutamic, serine, threonine	GC-TOF-MS$^{\delta}$, LC-ESI-MS$^{\delta}$	PCA, PLS-DA	[60]
		SUG, SUGDs^{d}↑↓	Arabitol, fructose, galactose, maltose, mannitol, myo-inositol, ribose, sorbitol, tagatose			
		FA^{e}↑	Palmitic, pentadecanoic, stearic			
		IFVN^{f}↑↓	8-hydroxydaidzein, acetyldaidzin, acetylglycitin, acetylgenistin, daidzein, aidzin, genistin, glycitein, glycitin, hydroxygenistein, hydroxyglycitein			
		NT^{g}↑	Uracil			
		OA^{h}↓	Cinnamic, citric, malomic			
		SSAPN^{i}	I			
Gochujang	Wheat/rice	AA^{c}	Alanine, GABA, glycine, glutamic, isoleucine, leucine, phenylalanine, proline, pyroglutamic, serine, threonine, tyrosine, valine	UPLC-Q-TOF-MS*, GC-TOF-MS*	PCA, PLS-DA	[61]
		Ak^{q}, DPH^{u}	Alnustone, dihydrocapsaicin, capsaicin			
		IFVN^{f}, FLVD^{v}	Apigenin-diglucoside, daidzein, glycitein, genistein, hydroxydaidzen, kaempferol, luteolin-diglucoside			
		L^{w}	Lyso (PC16:0, PC18:1, PC18:2)			
		SUG^{d}	Adonitol, arabinose, erythritol, fructose, fumaric, gentibiose, glucitol, gluconic, glucose, glycerol, glyceryl-glucoside, lactose, inositol, myo-inositol, xylose, xylitol			
		OA^{h}	Citric, malic, malonic, phosphoric, propanoic, succinic			
		SSAPN^{i}	I, III, V			

Produce	Raw material	Metabolite group	Metabolite forms	Data acquisition method	Data processing technique	Reference
		O^{k}	glyceryl-glucoside, unknown compounds			
Kimchi	Vegetables	AA^{c}↑↓	δ-aminobutyric, alanine, asparagine, aspartic, glycine, glutamic, glutamine, leucine, ornithine, proline, threonine, valine	GC-MS*	PCA, PLS-DA	[62]
		SUG, $SUGDs^{d}$↑↓	D-fructose, galactose, glucose, glycerol, mannitol, myo-inositol, sucrose, xylose			
		OA^{h}↑↓	1-Propene-1, 2, 3-tricarboxylic acid, 2-keto-L-gluconic acid, 2, 3, 4-trihydroxybutyric acid, citric, fumaric, gluconic, isocitric, lactic, malic, octadecanoic, palmitic, pentanedioic, propanoic, pyrotartaric, ribonic, succinic			
		O^{k}↑↓	Adenine, Urea			
Koji	Rice	AA^{c}↑↓	Alanine, aspartic, GABA, glutamic, glycine, isoleucine, leucine, lysine, methionine, ornithine, phenylalanine, proline, pyroglutamic, serine, threonine, tryptophan, tyrosine, valine	GC-TOF-MS*, UHPLC-LTQ-IT-MS/MS*	PCA, PLS-DA	[63]
		FA^{e}↑↓	Hydroxy-oxo-octadecenoic, linoleic, linolenic, oleic, palmitic, pinellic, stearic			
		$FLVN^{v}$↑↓	Apigenin-C-glucosyl-C-arabinoside, chrysoeriol-hexoside, chrysoeriol-rutinoside, isovitexin-O-glucoside, tricin, tricin-7-O-rutinoside, tricin-O-glucoside			
		LPL^{x}↑↓	Lyso (PE14:0, PC14:0, PC18:3, PC16:1, PE18:2, PC18:2, PE16:0, PC16:0, PC18:1)			
		OA^{h}↑↓	Citric, fumaric, gluconic, glyceric, kojic, lactic, malic, malonic, shikimic, succinic, oxalic,			
		PA^{y}↑↓	4-hydroxybenzoic acid, ferulic acid			
		SUG, $SUGDs^{d}$↑↓	Erythritol, fructose, glucose, glycerol, maltose, *myo*-inositol, pentitol, sorbitol, xylose, xylitol,			

Produce	Raw material	Metabolite group	Metabolite forms	Data acquisition method	Data processing technique	Reference
V^{j}↑↓	Nicotinic acid					
O^{k}↑↓	Bacillibactin, unknowns					
Makgeolli	Rice	AA^{c}	Alanine, asparagine, glutamic, glutamine, glycine, leucine, lysine, ornithine, proline, pyroglutamic, tryptophan, tyrosine	$GC\text{-}MS^{*}$	OPLS-DA	[64]
		Al^{a}↑	4-hydroxyphenylethanol			
		OA^{h}↑	2-hydroxyglutaric, citric, lactic, malic, succinic			
		SUG,↑↓ $SUGDs^{d}$	Erythritol, fructose, glucose, glycerol, *myo*-inositol, ribose			
		O^{k}↑↓	1, 2-propanediol, phosphoric			
Meju	Soybean	AA^{c}↑↓	γ-aminobutyric, acetylornithine, alanine, arginine, citrulline, glutamic, glutamine, histidine, isoleucine, leucine, lysine, methionine, ornithine, phenylalanine, proline, pyroglutamic, threonine, tryptophane, tyrosine valine	$UPLC\text{-}Q\text{-}TOF\ MS^{*}$	PLS-DA	[27]
		NTs↑↓	Adenine, hypoxanthine, uracil, xanthine	OPLS-DA		
		OA^{h}↑↓	Citric, pipecolic			
		P^{t}↑↓	Glu-Gln, Glu-Tyr, Leu-Gln, Leu-Glu, Glu-Phe, Leu-Pro, Ser-Pro, Val-Glu, Val-Thr, Val-Leu, Leu-Val-Pro-Pro			
Miso	Soybean	AA^{c}↑↓	Arginine, aspartate, glutamate, glutamine, lysine, phenylalanine, pyroglutamic,	$LC\text{-}MS^{\delta}$	PCA	[65]
		OA^{h}↑	Citric			
		O^{k}	Fructosyl-leucine, fructosyl-phenylalanine			
Myeolchi-aekjeot	Fish	AA^{c}↑↓	Alanine, arginine, aspartate, glutamate, glutamic, glutamine, glycine, isoleucine, leucine, serine, threonine	$^{1}HNMR^{*}$		[66]
			Betanine, choline, creatine, inosine, methyl amines			

Produce	Raw material	Metabolite group	Metabolite forms	Data acquisition method	Data processing technique	Reference
		Am[b], NTs[g]↑↓				
		OA[h]↑↓	Acetate, lactate,			
		SUG,↑↓ SUGDs[d]	Glucose, glycerol			
Saeu-jeot	Shrimp	AA[c]↑	Alanine, arginine, asparagine, aspartate, glutamate, glutamine, glycine, isoleucine, leucine, lysine, methionine, phenylalanine, proline, pyroglutamate, serine, threonine, tryptophan, tyrosine, valine	^{1}HNMR*	CCA	[67]
		Am[b]↑↓	Dimethylamine, trimethylamine			
		OA[h]↑	Acetate, butyrate, lactate			
		SUG,↑↓ SUGDs[d]	Glucose, glycerol			

a—alcohols, b—amines, c–amino acids, d–carbohydrates, sugars and sugar derivatives, e–fatty acids, f–isoflavonoids, g–nucleotides, h–organic acids, i–soyasaponins, j–vitamins, k–others (not classified), l–organic bases, m–purines and pyrimidines, n–carbonils, o–hydrocarbons, p–catechin and catechin derivatives, q–alkaloids, r–flavonol glycosides and flavone glycosides, s–lipids, t–peptides, u–diphenylheptanoid, v–flavonoids, w–lipids, x–lysophospholipids, y–phenolic acids, ↑–increase in metabolites, ↓–decrease in metabolites, ↑↓–both increase and decrease in metabolites, *–non-targeted/profiling metabolomics, $^{\delta}$–targeted metabolomics, CE-TOF-MS–capillary electrophoresis time of flight mass spectrometry, CCA–canonical correspondence analysis, FTIR–Fourier transform infrared spectroscopy, GC-FID–gas chromatography-flame ionization detector, GC-MS–gas chromatography-mass spectrometry, GC-TOF-MS–gas chromatography-time of flight mass spectrometry, ^{1}HNMR–proton nuclear magnetic resonance, HPLC-DAD–high performance liquid chromatography-diode array detector, LC-MS/MS–liquid chromatography tandem-mass spectrometry, OPLS-DA–orthogonal partial least square discriminant analysis, PCA–principal component analysis, PLS-DA–partial least square discriminant analysis, SPME-GC-MS–solid phase microextraction-gas chromatography-mass spectrometry, UPLC-ESI-QTOF-MS–ultra high performance liquid chromatography with electrospray ionization quadrupole time of flight mass spectrometry, UHPLC-LTQ-IT-MS/MS –ultra high pressure liquid chromatography linear ion trap-high resolution Orbitrap mass spectrometry, UPLC-Q-TOF MS–ultra performance liquid chromatography quadrupole time of flight mass spectrometry.

Table 1. Summary of food metabolomics studies of fermented foods reported in literature.

flavanols, anthocyanins were observed to decreased with fermentation time, while most oligopeptides initially increased, with a later decrease during fermentation [29].

Lee et al. [56] and Tan et al. [57] studied metabolic changes during tea fermentation. Using ^{1}H NMR, UPLC-Q-TOF-MS and PCA, these authors were able to differentiate partially and fully fermented tea according to their fermentation patterns. The authors observed a decrease in caffeine epicatechin, epigallocatechin, caffeine, quinate, theanine and sucrose, whereas gallic acid and glucose levels increased [56]. Alanine levels remained constant with caffeine being a major discriminator. A similar decrease in catechin, epigallocatechin in fermented tea was observed in another study, though levels of flavanols rapidly increased but later decreased [57]. Varying increases and decreases in the levels of flavonol and flavone glycosides, phenolic acids, alkaloids and amino acids were also recorded by these authors [57].

Other similar studies on the food metabolomics studies of fermented foods that have been reported in the literature include foods from milk [58, 59] soybean [27, 28, 65] and cereals [61, 63, 64] (**Table 1**). Others include *kimchi* [62], *myeolchi-aekjeot* [66] and *saeu-jeot* [67]. These fermented food products, their corresponding metabolites and trend in terms of increases or decreases in reported metabolites are summarized in **Table 1**.

5. Role of food metabolomics in the development of functional foods

Sequel to the relevance and importance of consuming functional foods for improved health, concerted efforts by relevant stakeholders in academia and food industry have been geared towards the development and delivery of functional foods to the populace. In this regard, food metabolomics as a technique is vital in the efficient and proper evaluation of such products and subsequent elucidation of the metabolite profile. Through the selection of appropriate techniques in combination with adequate MVDAs, a thorough understanding of the effects of processing parameters and different optimization steps during the development of such functional foods is possible. Successive data generated, could thus be interpreted in terms of the functionality and other health benefits such product would confer to intending consumers.

6. Future prospects

Fermented foods have distinct ecological niches that present an opportunity to use new approaches that take advantage of advances in 'omics' to understand and characterized them. Considering the wide range of these fermented foods in the world and the number of yet to be characterized and identified components, subsequent analysis of these components needs to be explored to further advance and contribute to existing knowledge. While the future of food metabolomics will involve the development of better analytical techniques, efforts should also be made at developing standardized databases of data from fermented foods.

Currently, metabolomics studies on fermented foods are still limited compared to plants. If harnessed well, the application of food metabolomics would play an invaluable role in the development of strategies for improving the safety, quality, shelf life and overall composition of fermented foods. In the nearest future with concerted efforts, food metabolomics could be used as an effective alternative and/or complement traditional sensory evaluation for fermented foods. Since metabolites impact sensory qualities, food metabolomics can clarify the influence of fermentation on biomarkers responsible for sensorial qualities.

7. Conclusion

There is a steady growing interest in food metabolomics, due to its application and capability in providing high throughput data and a platform for detailed understanding on the fermentation process. The feasibility of food metabolomics approach also suggests its viability for future progress in food science, nutrition and other related fields. This also coincides with the recent sensitization and encouragement of the consumption of functional and nutraceutical foods that can reduce the risks of degenerative diseases and ensure healthy nutrition. Although considerable progress has been made in the field of food metabolomics and its application in understanding fermented foods as demonstrated in this chapter, challenges of fully interpreting the complex data generated from the sophisticated equipment used still needs to be addressed and simplified. Nonetheless, food metabolomics has provided a medium that will greatly improve our understanding of the diversity of fermented foods and even more potential to explore their functionality. Since the delivery of most functional foods to the populace is through the industry, subsequent adoption of this technology would translate to a better understanding of processes and its influence on product quality. This could thus save costs, time and labor that might have been expended in conventional analytical techniques, that would provide less information.

Acknowledgements

This work was supported via the Global Excellence and Stature (GES) Fellowship of the University of Johannesburg (UJ) granted to the main author (Adebo, O.A). This work was also partly supported by the National Research Foundation (NRF) Center of Excellence (CoE) in Food Security co-hosted by the University of Pretoria (UP) and the University of Western Cape (UWC), South Africa.

Author details

Oluwafemi Ayodeji Adebo*, Patrick Berka Njobeh, Janet Adeyinka Adebiyi, Sefater Gbashi and Eugenie Kayitesi*

*Address all correspondence to: oluwafemiadebo@gmail.com and eugeniek@uj.ac.za

Department of Biotechnology and Food Technology, University of Johannesburg, Johannesburg, Gauteng, South Africa

References

[1] Galati A, Oguntotinbo FA, Moschetti G, Crescimanno M, Settanni L. The cereal market and the role of fermentation in cereal-based food production in Africa. Food Reviews International. 2014;**30**:317–337. DOI: 10.1080/87559129.2014.929143

[2] Adebiyi JA, Obadina AO, Adebo OA, Kayitesi E. Fermented and malted millet products in Africa: Expedition from traditional/ethnic foods to industrial value added products. Critical Reviews in Food Science and Nutrition. DOI: 10.1080/10408398.2016.1188056

[3] Adebiyi JA, Obadina AO, Mulaba-Bafubiandi AF, Adebo OA, Kayitesi E. Effect of fermentation and malting on the physicochemical properties of millet flour and biscuit. Journal of Cereal Science. 2016;**70**:132–139. DOI: 10.1016/j.jcs.2016.05.026 DOI:10.1016/j.jcs.2016.05.026#doilink

[4] Jay JM, Loessner MJ, Golden DA. Modern Food Microbiology. 7th ed. India: Springer; 2005. DOI: 10.1007/b100840

[5] Saavedra JM. Use of probiotics in pediatrics: Rationale, mechanisms of action, and practical aspects. Nutrition in Clinical Practice. 2007;**22**:351–365. DOI: 10.177/0115426507022003351

[6] Anukam KC, Reid G. African traditional fermented foods and probiotics. Journal of Medicinal Food. 2009;**12**:1177–1184. DOI: 10.1089/jmf.2008.1063

[7] Wang CY, Wu SJ, Fang JY, Wang YP, Shyu YT. Cardiovascular and intestinal protection of cereal pastes fermented with lactic acid bacteria in hyperlipidemic hamsters. Food Research International. 2012;**48**:428–434. DOI: 10.1016/j.foodres.2012.05.006

[8] Taylor JRN, Duodu KG. Effects of processing sorghum and millets on their phenolic phytochemicals and the implications of this to the health-enhancing properties of sorghum and millet food and beverage products. Journal of the Science of Food and Agriculture. 2015;**95**:225–237. DOI: 10.1002/jsfa.6713

[9] Cifuentes A. Food analysis and foodomics. Journal of Chromatography A. 2009;**1216**:7109. DOI: 10.1016.chroma.2009.09.018

[10] Hu C, Xu G. Mass-spectrometry-based metabolomics analysis for foodomics. Trends in Analytical Chemistry. 2013;**52**:36–46. DOI: 10.1016/j.trac.2013.09.005

[11] Garcia-Canas V, Simo C, Herrero M, Ibanez E, Cifuentes A. Present and future challenges in food analysis: Foodomics. Analytical Chemistry. 2012;**84**:10150–10159. DOI: 10.1021/ac301680q

[12] Niwa T. Metabolic profiling with gas chromatography–mass spectrometry and its application to clinical medicine. Journal of Chromatography. 1986;**379**:313–345

[13] Nicholson JK, Lindon JC, Holmes E. Metabolomics: Understanding the metabolic responses of living systems to pathophysiological stimuli via multivariate statistical analysis of biological NMR spectroscopic data. Xenobiotica. 1999;**29**:1181–1189. DOI: 10.1080/004982599238047

[14] Fiehn O. Metabolomics—The link between genotypes and phenotypes. Plant Molecular Biology. 2002;**48**:155–171. DOI: 10.1023/A:1013713905833

[15] Fukusaki E, Kobayashi A. Plant metabolomics: Potential for practical operation. Journal of Bioscience and Bioengineering. 2005;**100**:347–354. DOI: 10.1263/jbb.100.347

[16] Cevallos-Cevallos JM, Reyes-De-Corcuera JI, Etxeberria E, Danyluk MD, Rodrick GE. Metabolomic analysis in food science: A review. Trends in Food Science and Technology. 2009;**20**:557–566. DOI: 10.1016/j.tifs.2009.07.002

[17] McGhie TK, Rowan DD. Metabolomics for measuring phytochemicals, and assessing human and animal responses to phytochemicals, in food science. Molecular and Nutrition Food research. 2012;**56**:147–158. DOI: 10.1002/mnfr.201100545

[18] Mozzi F, Ortiz ME, Bleckwedel J, De Vuyst L, Pescuma M. Metabolomics as a tool for the comprehensive understanding of fermented and functional foods with lactic acid bacteria. Food Research International. 2013;**54**:1152–1161. DOI: 10.1016/j.foodres.2012.11.010

[19] Herrero M, Garcıa-Canas V, Simo C, Cifuentes A. Recent advances in the application of CE methods for food analysis and foodomics. Electrophoresis. 2010;**31**:205–228. DOI: 10.1002/elps.200900365

[20] Capozzi F, Bordoni A. Foodomics: A new comprehensive approach to food and nutrition. Genes and Nutrition. 2013;**8**:1–4. DOI: 10.1007/s12263-012-0310-x

[21] Wishart DS. Metabolomics: applications to food science and nutrition research. Trends in Food Science and Technology, 2008;**19**:482–493. DOI: 10.1016/j.tifs.2008.03.003

[22] Ramautar R, Demirci A, Jong, GJD. Capillary electrophoresis in metabolomics. Trends in Analytical Chemistry. 2006;**25**:455. DOI: 10.1016/j.trac.2006.02.004

[23] Monton MRN, Soga T. Metabolome analysis by capillary electrophoresis-mass spectrometry. Journal of Chromatography A. 2007;**1168**:237–246. DOI: 10.1016/j.chroma.2007.02.065

[24] Tugizimana F, Paiter L, Dubery I. Plant metabolomics: a new frontier in phytochemical analysis. South African Journal of Science. 2013;**109**:1–11. DOI: 10.1590/sajs.2013/20120005

[25] Choi HK, Yoon JH, Kim YS, Kwon DY. Metabolomic profiling of *Cheonggukjang* during fermentation by ^{1}H NMR spectrometry and principal components analysis. Process Biochemistry. 2007;**42**:263–266. DOI: 10.1016/j.procbio.2006.07.014

[26] Park MK, Cho IH, Lee S, Choi HK, Kwon DY, Kim YS. Metabolite profiling of *Cheonggukjang*, a fermented soybean paste, during fermentation by gas chromatography-mass spectrometry and principal component analysis. Food Chemistry. 2010:**122**:1313–1319. DOI: 10.1016/j.foodchem.2010.03.095

[27] Kang HJ, Yang HJ, Kim MJ, Han ES, Kim HJ, Kwon DY. Metabolomic analysis of meju during fermentation by ultra performance liquid chromatography-quadrupole-time of flight mass spectrometry (UPLC-Q-TOF MS). Food Chemistry. 2011;**127**:1056–1064. DOI: 10.1016/j.foodchem.2011.01.080

[28] Lee SY, Lee S, Lee S, Oh JY, Jeon EJ, Ryu HS, Lee CH. Primary and secondary metabolite profiling of *doenjang*, a fermented soybean paste during industrial processing. Food Chemistry. 2014;**165**:157–166. DOI: 10.106/j.foodchem.2014.05.089

[29] Mayorga-Gross AL, Quiros-Guerrero LM, Fourny G, Vaillant F. An untargeted metabolomic assessment of cocoa beans during fermentation. Food Research International. 2016;**89**:901–909. DOI: 10.1016/j.foodres.2016.04.017

[30] Villas-Boas SG, Mas S, Akesson M, Smedsgaard J, Nielsen J. Mass Spectrometry in metabolome analysis. Mass Spectrometry Reviews. 2005;**24**:613–646. DOI: 10.1002/mas.20032

[31] t'Kindt R, Morreel K, Deforce D, Boerjan W, Van Bocxlaer J. Joint GC-MS and LC-MS platforms for comprehensive plant metabolomics: Repeatability and sample pre-treatment. Journal of Chromatography B. 2009;**1152**:3572–3580. DOI: 10.1016/j.jchromb.2009.08.041

[32] Namgung HJ, Park HJ, Cho IN, Choi HK, Kwon DY, Shim SM, Kim YS. Metabolite profiling of doenjang, fermented soybean paste, during fermentation. Journal of the Science of Food and Agriculture. 2010;**90**:1926–1935. DOI: 10.1002/jsfa.4036

[33] Cevallos-Cevallos JM, Reyes-De-Corcuera JI. Metabolomics in food science. Advances in Food and Nutrition Research. 2012;**67**:1–24. DOI: 10.1016/B978-0-12-394598-3.00001–0

[34] Chin E, Slupsky CM. Applications of metabolomics in food science: Food composition and quality, sensory and nutritional attributes. In: Weimer BC, Slupsky CM. (Eds), Metabolomics in Food and Nutrition. 2013. England: Woodhead Publishing. pp. 217–230. DOI: 10.1533/9780857098818.2.217

[35] Corsaro C, Cicero N, Mallamace D, Vasi S, Naccari C, Salvo A, Giofre SV, Dugo G. HR-MAS and NMR towards Foodomics. Food Research International. 2016;**89**:1085–1094. DOI: 10.1016/j.foodres.2016.09.033

[36] Singh D, Lee S, Lee Ch. Metabolomics for empirical delineation of the traditional Korean fermented foods and beverages. Trends in Food Science and Technology. 2017;**61**:103–115. DOI: 10.1016/j.tifs.2017.01.001

[37] Katajamaa M, Orešiè M. Data processing for mass spectrometry-based metabolomics. Journal of Chromatography A. 2007;**1158**: 318–328. DOI: 10.1016/j.chroma.2007.04.021

[38] Cubero-Leon E, Penalver R, Maquet A. Review on metabolomics for food authentication. Food Research International. 2014;**60**:95–107. DOI: 10.1016/j.foodres.2013.11.041

[39] Tugizimana F, Steenkamp PA, Piater LA, Dubery IA. A conversion on data mining strategies in LC-MS untargeted metabolomics: Pre-processing and pre-treatment steps. Metabolites. 2016;**6**:1–18. DOI: 10.3390/metabo6040040

[40] Wold S. Chemometrics; what do we mean with it, and what do we want from it? Chemometrics and Intelligent Laboratory Systems. 1995;**30**:109–115. DOI: 10.1016/0169-7439(95)00042-9 DOI:10.1016/0169-7439(95)00042-9#doilink

[41] Trygg J, Gullberg J, Johansson AI, Jonsson P, Moritz T. Chemometrics in metabolomics—An introduction. In: Saito K, Dixon R, Willmitzer L. (Eds), Plant metabolomics. 2006. Berlin Heidelberg: Springer. Vol. 57. pp. 117–128. DOI: 10.1007/3-540-29782-0_9

[42] Wishart DS, Tzur D, Knox C, Eisner R, Guo AC, Young N, Cheng D, Jewell K, Arndt D, Sawhney S, Fung C, Nikolai L, Lewis M, Coutouly MA, Forsythe I, Tang P, Shrivastava S, Jeroncic K, Stothard P, Amegbey G, Block D, Hau DD, Wagner J, Miniaci J, Clements M, Gebremedhin M, Guo N, Zhang Y, Duggan GE, Macinnis GD, Weljie AM, Dowlatabadi R, Bamforth F, Clive D, Greiner R, Li L, Marie T, Sykes BD, Vogel HJ, Querengesser L. HMDB: The human metabolome database. Nucleic Acids Research. 2007;**35**:521–526. DOI: 10.1093/nar/gkl923

[43] Berrueta LA, Alonso-Salces RM, Hébergerb K. Supervised pattern recognition in food analysis. Journal of Chromatography A. 2007;**1158**:196–214. DOI: 10.1016/j.chroma.2007.05.024 DOI:10.1016/j.chroma.2007.05.024#doilink

[44] van der Werf MJ, Jellema RH, Hankemeier T. Microbial metabolomics: replacing trial-and-error by the unbiased selection and ranking of targets. Journal of Industrial Microbiology and Biotechnology. 2005;**32**:234–252. DOI: 10.1007/s10295-005-0231-4

[45] Broadhurst D, Kell D. Statistical strategies for avoiding false discoveries in metabolomics and related experiments. Metabolomics. 2006:**2**:171–196. DOI: 10.1007/s11306-006-0037-z

[46] Kemsley EK, Le Gall G, Dainty JR, Watson AD, Harvey LJ, Tapp HS, Colquhoun IJ. Multivariate techniques and their application in nutrition: A metabolomics case study. British Journal of Nutrition. 2007;**98**:1–14. DOI: 10.1017/S0007114507685365

[47] Kwon SW. Profiling of soluble proteins in wine by nano-high performance liquid chromatography/tandem mass spectrometry. Journal of Agriculture and Food Chemistry. 2004;**52**:7258–7263. DOI: 10.1021/jf048940g

[48] Moco S, Bino RJ, Vorst O, Verhoeven HA, de Groot J, van Beek TA., Vervoort J, de Vos CH. A liquid chromatography-mass spectrometry-based metabolome database for tomato. Plant Physiology. 2006;**141**:1205–1218. DOI: 10.1104/pp.106.078428

[49] Baek JG, Shim SM, Kwon DY, Choi HK, Lee CH, Kim YS. Metabolite profiling of *cheonggukjang,* a fermented soybean paste, inoculated with various *Bacillus* strains during fermentation. Bioscience Biotechnology and Biochemistry. 2010;**74**:1860–1868. DOI: 10.1271/bbb.100269

[50] Kim J, Choi JN, John KMM, Kang D, Son GH, Kim YS, Choi HK, Kwon DY, Lee CH. Correlation between antioxidative activities and metabolite changes during *cheonggukjang* fermentation. Bioscience, Biotechnology, and Biochemistry. 2011;**75**:732–739. DOI: 10.1271/bbb.100858

[51] Kim J, Choi JN, John KMM, Kusano M, Oikawa A, Saito K, Lee CH. GC−TOF-MS- and CE−TOF-MS-based metabolic profiling of *cheonggukjang* (fast-fermented bean paste) during fermentation and its correlation with metabolic pathways. Journal of Agricultural and Food Chemistry. 2012;**60**:9746–9753. DOI: 10.1021/jf302833y

[52] Chen D, Ye Y, Chen J, Yan X. Evolution of metabolomics profile of crab paste during fermentation Food Chemistry. 2016;**192**:886–892. DOI: 10.1016/j.foodchem.2015.07.098

[53] Yan Z, Zheng XW, Han BZ, Yan YZ, Zhang X, Chen JY. ^{1}H NMR-based metabolomics approach for understanding the fermentation behaviour of *Bacillus licheniformis*. Journal of the Institute of Brewing. 2015;**121**:425–431. DOI: 10.1002/jib.238

[54] Yang SO, Kim MS, Liu KH, Auh JH, Kim YS, Kwon DY, Choi HK. Classification of fermented soybean paste during fermentation by ^{1}H nuclear magnetic resonance spectroscopy and principal component analysis. Bioscience Biotechnology and Biochemistry. 2009;**73**:502–507. DOI: 10.1271/bbb.80467

[55] Ferri M, Serrazanetti DI, Tassoni A, Baldissarri M, Gianotti A. Improving the functional and sensorial profile of cereal-based fermented foods by selecting *Lactobacillus plantarum* strains via a metabolomics approach. Food Research International. 2016;**89**:1095–1105. DOI: 10.1016/j.foodres.2016.08.044

[56] Lee JE, Lee BJ, Chung JO, Shin HJ, Lee SJ, Lee CH, Hong YS. ^{1}H NMR-based metabolomic characterization during green tea (*Camellia sinensis*) fermentation. Food Research International. 2011;**44**:597–604. DOI: 10.1016/j.foodres.2010.12.004

[57] Tan J, Dai W, Lu M, Lv H, Guo L, Zhang Y, Zhu Y, Peng Q, Lin Z. Study of the dynamic changes in the non-volatile chemical constituents of black tea during fermentation processing by a non-targeted metabolomics approach. Food Research International. 2016;**79**:106–113. DOI: 10.1016/j.foodres.2015.11.018

[58] Hagi T, Kobayashi M, Nomura M. Metabolome analysis of milk fermented by -aminobutyric acid-producing *Lactococcus lactis*. Journal of Dairy Science. 2016;**99**:994–1001. DOI: 10.3168/jds.2015–9945

[59] Yang SO, Kim SH, Cho S, Lee JH, Kim YS, Yun SS, Choi HK. Classification of fermented soymilk during fermentation by ^{1}H NMR coupled with principal component analysis and elucidation of free-radical scavenging activities. Bioscience Biotechnology and Biochemistry. 2009;**73**:1184–1188. DOI: 10.1271/bbb.80743

[60] Lee S, Seo MH, Oh DK, Lee CH. Targeted metabolomics for *Aspergillus oryzae* mediated biotransformation of soybean isoflavones, showing variations in primary metabolites. Bioscience Biotechnology and Biochemistry. 2014;**78**:167–174. DOI: 10.1080/09168451.2014.877827

[61] Lee DE, Shin GR, Lee S, Jang ES, Shin HW, Moon BS, Lee CH. Metabolomics reveal that amino acids are the main contributors to antioxidant activity in wheat and rice *gochujangs* (Korean fermented red pepper paste). Food Research International. 2016;**87**:10–17. DOI: 10.1016/j.foodres.2016.06.015

[62] Park SE, Yoo SA, Seo SH, Lee KI, Na CS, Son HS. GC-MS based metabolomics approach of Kimchi for the understanding of Lactobacillus plantarum fermentation characteristics. LWT-Food Science and Technology. 2016;**68**:313–321. DOI: 10.1016/j.lwt.2015.12.046

[63] Lee DE, Lee S, Jang ES, Shin HW, Moon BS, Lee CH. Metabolomic profiles of *Aspergillus oryzae* and *Bacillus amyloliquefaciens* during rice koji fermentation. Molecules. 2016;**21**:1–15. DOI: 10.3390/molecules21060773

[64] Seo SH, Park SE, Yoo SA, Lee KI, Na CS, Son HS. Metabolite profiling of *Makgeolli* for the understanding of yeast fermentation characteristics during fermentation and aging. Process Biochemistry. 2016;**51**:1363–1373. DOI: 10.1016/j.procbio.2016.08.005

[65] Yoshida H, Yamazaki J, Ozawa S, Mizukoshi T, Miyano H. Advantage of LC-MS metabolomics methodology targeting hydrophilic compounds in the studies of fermented food samples. Journal of Agricultural and Food Chemistry. 2009;**57**:1119–1126. DOI: 10.1021/jf803235m

[66] Jung JY, Lee HJ, Chun BH, Jeon CO. Effects of temperature on bacterial communities and metabolites during fermentation of *myeolchi-aekjeot,* a traditional Korean fermented anchovy sauce. PLoS One. 2016;**11**:1–20. DOI: 10.1371/journal.pone.0151351

[67] Lee SH, Jung JY, Jeon CO. Effects of temperature on microbial succession and metabolite change during *saeu-jeot* fermentation. Food Microbiology. 2014;**38**:16–25. DOI: 10.1016/j.fm.2013.08.004

Chapter 12

Models to Evaluate the Prebiotic Potential of Foods

Jailane de Souza Aquino, Kamila Sabino Batista,
Francisca Nayara Dantas Duarte Menezes,
Priscilla Paulo Lins,
Jessyca Alencar de Sousa Gomes and
Laiane Alves da Silva

Additional information is available at the end of the chapter

http://dx.doi.org/10.5772/intechopen.69174

Abstract

The interest in studying the prebiotic effect of foods is increasing due to the way in which the consumption of these foods influences the gut microbiota and how the metabolic activity of the microbiota affects the health and well-being of the host. Several in vitro and in vivo studies have been developed to elucidate the prebiotic effect of foods, and particularly in in vivo studies, the physiological dynamics of this effect has been studied in healthy or diseased individuals. In this chapter, the main in vitro and in vivo models developed for the study of the prebiotic potential of foods will be approached, which can be used by those planning to advance in this field of research.

Keywords: functional foods, prebiotics, chronic diseases, animal models, intestinal microbiota

1. Introduction

Modern society has changed its standard of living every decade and today, health is becoming an increasingly important personal and social value. Prevention of health problems is prioritized due to the costs associated with curative medicine, especially chronic diseases, which can be prevented by a healthier lifestyle [1]. In addition to the practice of physical activity, adequate nutrition is an essential aspect influencing a person's health status. Consumers are more aware that their food choices can have consequences for their health and maintenance of a healthy lifestyle [2, 3].

Food matrixes are composed of several nutrient or non-nutrient substances that interact in a complex way. In this perspective, foods have the basic function of feeding, some of which present health benefits that go beyond nutrition, such as functional foods. Functional foods may exert physiological benefits and/or reduce the risk of chronic diseases, in addition to basic nutritional functions, and may be similar in appearance to conventional foods and consumed as part of a regular diet [4].

Prebiotics are among functional foods, which are defined as a component of the edible product, in which its health benefit must be measurable and not due to its absorption in the blood stream or due to the sole action of the component, but it should be evidenced that the simple presence of the prebiotic component and the formulation in which it is inserted alter the composition or activity of the microbial flora in the target host by modulating it [4], for stimulating the proliferation of a select group of beneficial colon bacteria and suppressing the proliferation of micro-organisms harmful to health [5].

To be considered prebiotic, food or its components must: (i) resist the processes of host digestion, absorption, and adsorption; (ii) be fermented by the microbiota that colonize the gastrointestinal tract (GI); and (iii) selectively stimulate the growth and/or activity of one or a limited number of bacteria within the gastrointestinal tract, altering the colonic microbiota in favor of a healthier composition [3, 4].

Prebiotics found in natural sources such as vegetables, roots, fruits, milk, and honey are non-digestible carbohydrates such as resistant starch (RS), galacto-oligosaccharides (GOS), fructooligosaccharides (FOS), xylooligosaccharides (XOS), pectic oligosaccharides (POS), and various oligosaccharides that provide carbohydrates fermentable by the beneficial colon micro-organisms [6, 7]. Among these, probiotic micro-organisms such as bacteria belonging to the genus *Lactobacillus* and *Bifidobacterium*, as well as *Streptococcus, Saccharomyces cerevisiae, Escherichia coli,* and *Bacillus* spp. stand out, which have been studied on a smaller scale. These bacteria are fermentative, obligatory, or facultative anaerobes, and their inherent biological characteristics allow them to prevail over potential pathogenic micro-organisms in the digestive tract [8].

Probiotic micro-organisms are currently defined as live micro-organisms, which when consumed in adequate amounts provide a positive health effect on the host [9]. Butel [10] suggests three modes of action of probiotics, which influence the host's health. One of the first suggested modes of action is called "barrier" effect or resistance to colonization against pathogenic bacteria due to the production of broad-spectrum inhibition bacteriocins, metabolites such as acid lactic and short-chain fatty acids—SCFA (e.g., acetate, butyrate, propionate)—which induce a decrease in pH, being favorable for bacterial growth, or biosurfactants with antimicrobial activity. The improvement of the barrier function in the gut mucosa may be due to the increase of the mucus layer or to the production of defensins and proteins of tight junctions.

In addition to prebiotic and probiotic foods, symbiotic foods, in which probiotic and prebiotic are combined, have been increasingly developed due to the favorable adaptation of the probiotic to the prebiotic substrate before consumption, which may increase the beneficial effects of each of them [11, 12].

In this context, the modulation of the gut microbiota by diet has been studied [13, 14]. The composition and metabolism of the colonic microbiota can be influenced by the type of diet, nutrient balance (mainly carbohydrates, proteins, and fats), and the amount of diet ingested [15]. The impact of diet on microbiota composition is determined by tolerance of gut conditions and by the competition for substrates among microbial species, which demonstrate different capabilities to utilize dietary substrates, promoting the competition for substrates available in the large intestine, playing an important role in defining microbiota composition [16]. The healthy microbiota can be defined as the normal microbiota that maintains and promotes well-being and absence of diseases, especially of the gastrointestinal tract. The colon is the most densely populated part of the gastrointestinal tract and houses about 500 different bacterial species. These bacteria, each with its own spectrum of metabolic activities, make the colon the most metabolically active organ in the human body [17].

The gut microbiota influences the metabolic processes, preventing and modulating chronic diseases such as obesity, diabetes, insulin resistance, and cardiovascular diseases [18] because it interferes in several systems such as cardiovascular [19], nervous [20, 21], immune [22], endocrine [23], and the gastrointestinal system itself.

From this perspective, the prebiotic effect of foods can be studied from in vitro systems or from in vivo models using healthy and diseased animals or humans. Each model has advantages and disadvantages, which will be discussed in the next sections of this chapter.

2. Types of prebiotics

Dietary fibers (DF) are bioactive components, which may have prebiotic activity, present in plants, defined as the edible part of plants or analogous carbohydrates resistant to digestion and absorption in the small intestine of humans, with complete or partial fermentation in the large intestine [24, 25]. Regarding water solubility, DFs are classified as soluble (SDF) and insoluble (IDF). IDFs include cellulose, lignin, and some hemicelluloses and pectins [26, 27]. SDFs, however, comprise the majority of pectins, gums, mucilages, and hemicelluloses [28, 29].

The concept of DF has been expanded to include functionally similar substances such as RS, inulin, FOS, and GOS. GOS or FOS may have beneficial effects such as anti-adhesion or direct immunomodulation that do not require fermentation and are therefore called additional biological activities not related to their effects on the gut microbiota [30]. There are several prebiotics with various origin and chemical properties. Inulin, FOS, GOS, lactulose, and polydextose are recognized as established prebiotics, whereas isomaltooligosaccharides (IMO), XOS, and lactitol are categorized as emerging prebiotics. In addition, resistant starch-rich whole grains are considered prebiotic in nature, and it is assumed that their consumption leads to many health benefits [31]. The fermentability of dietary fibers such as oat b-glucan, flaxseed gum, and fenugreek gum suggests their potential prebiotic application in promoting human health [31]. The main technological applications of prebiotics and the potential beneficial health effects on consumers of these foods are described in **Figure 1**.

Plant-derived polysaccharides arrive unchanged in the colon, being degraded by micro-organisms living in the human GI tract to SCFA (**Figure 2**). The degradation of complex oligosaccharides (pectin, cellulose, hemicellulose, and resistant starches) involves a strong metabolic alignment among diverse micro-organisms that makes up the intestinal microbiota, but these mechanisms are still not fully understood [24, 32].

In addition to DF, phenolic compounds (PC) or polyphenols may also benefit the gut microbiota, as up to 90% of plant PCs reach the colon and are used as substrates for the microbial production of small phenolic acids [33]. In turn, these biotransformed compounds modulate the microbial population in the gastrointestinal tract and are used as substrates for the production of SCFA [33, 34]. Results have reported that there is a possible interference of PC in

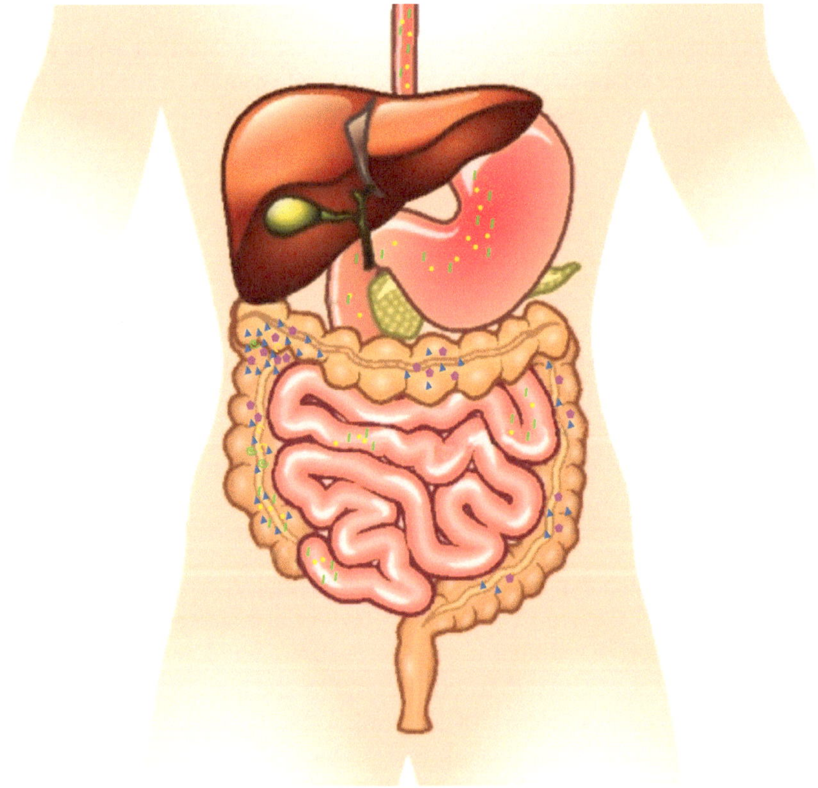

Figure 1. Degradation of dietary fibers and phenolic compounds by the gut microbiota. Dietary fibers (/) and phenolic compounds (•) reach the colon (mainly in the proximal part) and suffer a primary degradation by bacteria (▲) to oligosaccharides and monosaccharides (@) and small phenolic acids (c), respectively. Then, these compounds are used by the gut microbiota for the production of SCFA (⬟), which increase the number of beneficial intestinal bacteria.

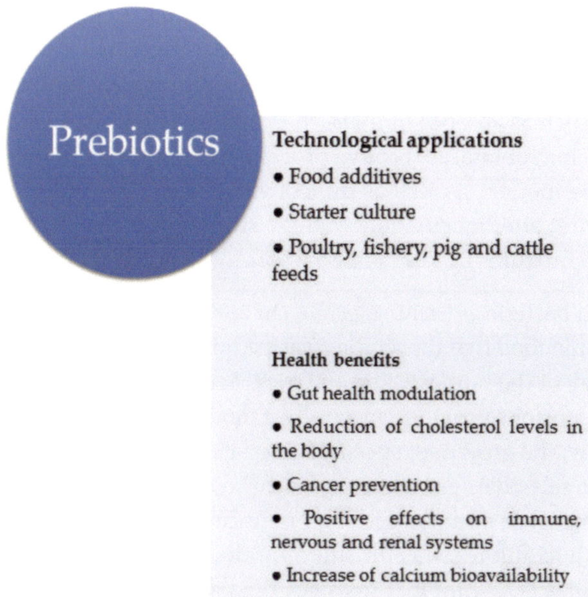

Figure 2. Some technological applications of prebiotics and health benefits from consumption.

the increase of viable *Bifidobacterium* and *Lactobacillus* cells in the intestine (in vivo model) and feces of animals or humans (in vitro model) [35, 36].

PC are secondary metabolites derived from pentoses-phosphate, shikimic acid, and phenylpropanoid pathways in plants. They are divided into four main classes according to their chemical structure: flavonoids (including flavonols, flavanols, flavanones, flavones, anthocyanidins, chalcones, dihydrochalcones, dihydroflavonols, and isoflavones), lignans, stilbenes, and tannins. They have numerous reported physiological properties, such as vasodilators, anti-thrombotic, anti-inflammatory, anti-apoptotic, hypolipemic, or anti-atherogenic properties [37].

Prebiotics should be ingested daily as a way of ensuring continuous effect on the intestinal microbiota. However, recommendations for daily doses will depend on the type of food containing the prebiotic compound (naturally or added) or the isolated prebiotic compound consumed as a nutraceutical or prebiotic administrated by gavage (orogastric) or added to diet. The consumption of 5–8 g per day of inulin, FOS, or RS has been shown to significantly increase fecal bifidobacteria [38, 39]. In another study, rats received daily oral administration (gavage) of FOS (3 g/kg) or GOS (4 g/kg) for 5 weeks [40].

Other studies have added prebiotics to diets for rodents such as Sprague-Dawley rats that consumed a high-fat diet and diet added of 10% oligofructose [41] or rats that consumed AIN-G diet added with 10% inulin or oligofructose [42]. Healthy or diabetic Wistar rats consumed basal diet supplemented with XOS (10%) or FOS (10%) or a combination of XOS (5%) and FOS (5%) [43].

3. Use of in vitro models in the study of the prebiotic potential of foods

In vitro modeling is useful for investigating the prebiotic potential of foods as it is less expensive, does not require sophisticated handling techniques, and allows simulating fermentation processes that occur along the large intestine and have few ethical limitations. However, they present limitations such as absence of interaction between neuroendocrine and immunological systems with the microbiota; absorptive processes, secretions, and defense systems are not incorporated into the models, as well as difficulty in controlling changes in the structures of microbial communities after inoculation. In these studies, it is possible to use pure microbial populations, known mixtures or fecal material [44].

The groups of colon bacteria present selective characteristics regarding the substrates available, and it is recommended that the studies use the mixed microbial culture, which simulates the microbial ecology of the human intestinal tract. Fermentation in anaerobic batches inoculated with fecal suspensions provides an excellent mode for small-scale screening of new substrates. Until recently, the growth of specific bacteria in such fermentations was measured by counting colonies on selective agar. This approach, however, has several disadvantages (time-consuming, labor intensive, and non-recovery of uncultivable organisms). As a result, molecular techniques such as fluorescence in situ hybridization (FISH) were developed to study microbial communities [13, 45]. FISH involves the use of genus-specific and in some cases species-specific fluorescently labeled oligonucleotide probes. Hybridization of the probe that has its own specificity to recognize a particular group of bacteria to the complementary target sequence within bacterial cells results in fluorescently labeled cells that can be visualized and enumerated using fluorescence microscopy [45].

Generally, food or a substrate prebiotic extracted from the test food itself is lyophilized and supplemented in different concentrations to Man, Rogosa and Sharpe (MRS) medium; the negative control is represented by the MRS medium without the addition of the test food or substrate, and the positive control is represented by inulin [46, 47] or fructooligosaccharide [17, 48], which are recognized prebiotics. Frequently, experiments include the MRS medium with addition of glucose as the carbon source, which also serves as a control. After media are defined, probiotic micro-organism strains such as *Lactobacillus* or *Bifidobacterium* are incubated and the samples are incubated under ideal conditions for the selected micro-organisms. Thereafter, viable cell counts and metabolism monitoring of these micro-organisms (quantification of short-chain fatty acids and pH, among other parameters) are performed to confirm the prebiotic property of the food [47, 49]. SCFAs are saturated aliphatic organic acids that have from one to six carbon atoms, such as acetate (C2), propionate (C3), and butyrate (C4), and are the final products of bacterial fermentation processes.

Recently, many byproducts of the food industry have been studied as cheap and alternative sources of prebiotics [6, 49, 50]. The prebiotic effect of cashew apple (*Anacardium occidentale* L.) agro-industrial byproduct powder on different potentially probiotic Lactobacillus strains (*L. acidophilus* LA-05 and *L. casei* L-26 and *L. paracasei* L-10) was cultivated in broth containing cashew apple powder (20 or 30 $g.L^{-1}$), glucose (20 $g.L^{-1}$), or FOS (20 $g.L^{-1}$). The cell viability of Lactobacillus strains (counts of viable cells) and changes in pH values, production of organic acids, and consumption of sugars in growth media were monitored for 48 h. The cultivation of *Lactobacillus* strains in broth containing glucose, FOS, or cashew apple powder resulted in high counts of viable cells, decreased pH, production of organic acids, and consumption of sugars over time, revealing intense bacterial metabolic activity and prebiotic activity [50]. Thuaytong and Anprung [51] used 1% (v/v) of prepared *L. acidophilus* LA-5, and *Bifidobacterium lactis* BB-12 was transferred into MRS broth, which was composed of 1% (w/v) glucose or 1% (w/v) inulin or 1% (w/v) prebiotic (guava samples), and demonstrated that both red guava and white pulp induced similar growth of prebiotic bacteria in glucose-containing medium.

The study by Gómez et al. [49] confirmed the prebiotic effects caused by a refined product containing POS that promoted the growth of beneficial bacteria and the increase of SCFA concentrations. In a study carried out by Sousa et al. [52], yacon flour revealed a potential prebiotic activity in the growth of probiotic strains *Enterococcus faecium* 32, *Bifidobacterium animalis* Bo, *L. acidophilus* Ki, and *L. casei* L26, probably due to its content in FOS. Teixeira et al. [47] evaluated the influence of Amazonian tubers *Dioscorea trifida, Calathea allouia,* and *Dioscorea altissima* on the growth of *Lactobacillus acidophillus* bacteria and observed that the best in vitro result was for *D. trifida* fiber, which stimulated the bacterial growth without significant difference from commercial inulin.

Another in vitro model that is being used to evaluate the prebiotic activity of foods is the fermentation of animal or human feces added to the test food or extract [13, 53] and it is also used for the purpose of evaluating the metabolism of fecal micro-organisms.

The beneficial health effects of prebiotics are related to their influence on the gut microbiota composition, stimulation of growth, metabolism, and activities of lactic acid bacteria, bifidobacteria, and other emergent strains such as *Roseburia intestinales* and *Faecalibacterium prausnitzii*) [7].

Quinoa (*Chenopodium quinoa* W.) and amaranth (*Amaranthus caudatus* L.) submitted to in vitro digestion and together with a control (without external carbon source) were used as carbon sources in batch cultures with fecal human inocula. After 48 h of incubation, both substrates stimulated in a similar proportion the growth of certain numerically predominant bacterial groups in the human gut microbiota, including *Bifidobacterium* spp., *Lactobacillus-Enterococcus, Atopobium, Bacteroides-Prevotella, Clostridium coccoides-Eubacterium rectale, F. prausnitzii,* and *Roseburia intestinalis* assessed by FISH, in addition to total SCFAs (acetate, propionate, and butyrate) with a decrease in pH, suggesting that these pseudocereals can have prebiotic potential [13].

Broad beans (*Vicia faba*) and lupin seeds (*Lupinus albus*) were submitted to in vitro digestion used as carbon sources in anaerobic batch cultures to evaluate their impact on the gut microbiota composition (by FISH) and on their metabolic products (lactate and SCFAs). The fermentation of the lupine seeds resulted in a higher total amount of SCFA than the bean fermentation, and in both, there was a decrease in the pH of the fermentation medium. In addition, legume fermentation increased microbial fecal batch cultures, such as *Bifidobacterium spp., Lactobacillus-Enterococcus, Atopobium, Bacteroides-Pretovella, C. coccoides-E. rectale, F. prausnitzii,* and *R. intestinalis* [54].

The prebiotic potential of POS obtained by orange peel wastes was assessed by in vitro fermentation using human fecal inocula. For comparative purposes, similar experiments were performed using orange pectin and commercial FOS as substrates for fermentation. POS particularly increased the amount of bifidobacteria and lactobacilli (assessed by FISH) so that the ratio between the counts of both genera and the total cell number increased from 17 in the inocula to 27% after fermentation. SCFA generation from POS fermentation was similar to that observed with FOS [49].

Sugar beet pulp (*Beta vulgaris* L.) and lemon peel wastes (*Citrus limon* L.) were used to obtain two mixtures of POS and in comparison, FOS and commercial pectins were assessed by in vitro fermentation and FISH using human fecal inocula. The joint populations of bifidobacteria and lactobacilli increased from 19 up to 29, 34, and 32% in cultures with pectic oligosaccharides from lemon peel wastes, beet pulp, and FOS, respectively. *Faecalibacterium* and *Roseburia* also increased their counts with all substrates (especially with pectic oligosaccharides from lemon peel wastes). The highest concentrations of organic acids were observed in media containing oligosaccharides, and these results confirm that pectic oligosaccharides present better prebiotic properties than pectins and are similar or better than FOS [6].

The prebiotic effect of oligosaccharides recovered and purified from caprine whey was evaluated by in vitro fermentation under anaerobic conditions using batch cultures at 37°C with human feces (by FISH). In this research, growth of *Bifidobacterium* spp. was significantly higher with purified oligosaccharides compared to the negative control. Lactic and propionic acids were the main SCFAs produced. These findings indicate that oligosaccharides naturally extracted from caprine whey or cheese whey (byproduct) could be used as new and valuable sources of prebiotics naturally produced in the lactating mammary gland of domestic species

Food	Main results	References
Oligosaccharides from Pitaya (*Hylocereus undatus* (Haw.))	↑ Resistance to gastric acidity ↑ Growth of *Bifidobacterium* and *Lactobacillus*	[56]
Byproducts of date pits (*Phoenix dactylifera* L. var.Medjoul) and apple bagasse (*Malus domestica* var. rayada)	Fermentation by colonic bacteria produced AGCC (formate, succinate, acetate, propionate, and butyrate)	[57]
Pomegranate peel (*Punica granatum*)	Fermentation of pomegranate peel flour by colonic bacteria generated acetic, propionic, and butyric acids	[36]
water-soluble xylan from wheat bran (XOS extraction)	↑ Growth of *L. brevis, B. adolescentis,* and the *Weissella* spp. on XOS ↑ Lactic acid and acetic acid production after 48-h incubation.	[58]
raw and roasted almonds (*Prunus amygdalus*)	Predigested raw and roasted almonds promoted the growth of *Lactobacillus acidophilus* (La-14) and *Bifidobacterium breve* (JCM 1192), and no significant differences were found between these two nuts	[59]
Apple pectin (*Malus domestica*)	↑ *Bifidobacterium, Lactobacillus,* and *Streptococcus* (including *Enterococcus*) in feces; ↓ *C. perfringens*, enterobacteria and Pseudomonas; ↑ Fecal concentrations of SFCA	[60]

Table 1. In vitro studies on the prebiotic potential of foods.

and not obtained by enzymatic reaction (trans-galactosylation) from lactose, although numerous papers and patents mostly refer to specific GOS [55].

Other studies evaluating the prebiotic potential of food using in vitro models are described in **Table 1**.

4. Use of in vivo models to study the prebiotic potential of foods

It has been well established that the colon microbiota has a deep influence on health. The study of the prebiotic potential in humans would be considered as a gold standard in case of absence of ethical and practical limitations, which may make the research unreliable or limited, in addition to the high dropout rates of study participants. Thus, animal models become an alternative to study the prebiotic potential of foods, since they allow direct access to intestinal contents as well as to organs and tissues [61].

Usually, the animal models used for the study of gut microbiota are swine [62], zebrafish [63], and more widely in rodents such as rats [47], hamsters [64], and mice [53], especially when the potential prebiotic of foods is evaluated.

Teixeira et al. [47] confirmed the prebiotic potential of Amazonian tubers by adding them to the diet of Wistar rats for 28 days, evaluating the pH and microbiota present in feces collected from the animals' caecum. Samal et al. [65] evaluated the prebiotic potential of Jerusalem artichoke (*Helianthus tuberosus* L.) added at different concentrations to the diet of rats for 12 weeks and observed that the consumption promoted beneficial effects on immunity, intestinal morphometry, and hindgut fermentation of rats. Supplementation with 2.5% of insoluble fibers from pineapple peel decreased the daily production of fecal ammonia, shortened gastrointestinal transit time, and increased the total amounts of SCFA in the caecal content as well as the growth of gut microflora such as *Lactobacillus* spp. and *Bifidobacterium* spp. in hamsters [64].

Not only should the gut microbiota be evaluated in in vivo studies but also other variables such as pH, feces humidity, and SCFAs production, which is directly related to the selective bacterial fermentation of prebiotics [66, 67]. In the large intestine, 95% of SCFA produced are rapidly absorbed by colonocytes, whereas the remaining 5% are expelled in the feces [68]. These microbial metabolites can be used as sources of energy by the host and can also act as regulators of energy consumption and metabolism [69]. pH acidification can also be an indicator of fermentation of prebiotic components of foods in the colon by endogenous bacteria and production of organic acids directly responsible for this process [70, 71]. In addition, the preservation of the intestinal epithelium in healthy rats or its recovery in diseased rats may provide evidence of the prebiotic potential, as observed by Hu et al. [72] and Moura et al. [73].

Bränning et al. [74] evaluated the potential prebiotic of blueberry husks added in diet as a substitute for digestible starch. The consumption of diet containing blueberry husk by rats for 5 days resulted in higher amounts of propionic acid and butyric acid in the distal colon and feces, respectively, when compared to rats that were fed a control diet without fibers. Both acids are essential substrates for colonic epithelial cells, improving gut health, and a surplus of substrates which also have metabolic effects. However, blueberry husk has antimicrobial effects, as observed by the decreased counts of lactobacilli, bifidobacteria, and enterobacteriaceae, and the larger pool of succinic acid may be a consequence of these antimicrobial effects. In this model, blueberry husks do not demonstrate prebiotic properties.

Rodríguez-Cabezas et al. [39] evaluated the synergistic effect of two dietary fibers with different fermentation patterns, FOS (Beneo Ò-95) and RS (FibersolÒ-2), administrated to healthy rats or in trinitrobenzenesulphonic acid (TNBS) colitic rats. Treatment groups (n = 20) received FOS (2 g/rat/day), RS (2 g/rat/day), or the mixture of both (37.5 FOS and 62.5% RS) (2 g/rat/day) incorporated in drinking water during 2 weeks. In healthy rats, the administration of the combination of FOS and RS induced changes in the intestinal microbiota and increased lactobacilli and bifidobacteria in caecum and colonic contents. In addition, treatment increased the moisture content and decreased the pH of caecum and colon. Furthermore, its administration upregulated the expression of trefoil factor-3 and mucin 2 (MUC-2) in comparison with untreated rats, thus improving the intestinal barrier function and increasing the propionate, butyrate, and total SCFA colonic contents. The beneficial effects observed with this combination were confirmed in the healthy or colitis rats.

Study model	Foods	Main results	References
Female rats Wistar	Cocoa fibers (*Theobroma cacao* L.)	↑Bifidobacterium and Lactobacillus; ↑ SCFA production and ↓ cecal and fecal pH	[76]
Male golden Syrian hamsters	Pineapple peel (*Ananas comosus* L. Merr.)	Modulation of the activities of fecal bacterial enzymes; ↓ ammonia contents in caecum and feces; ↑ concentration in the caecum of SCFA	[64]
Male rats Wistar	Passion fruit peel (*Passiflora edulis*)	Positive effect on SCFA production, but no change in gut microbiota was observed	[77]
Male rats Wistar	FOS and PC of strawberry (*Fragaria ananassa*)	↓ Cecal pH and ↓ production of putrefactive SCFA (sum of isobutyric, isovaleric, and valeric acids)	[78]
Male guinea pigs	FOS of Yacon (*Smallanthus sonchifolius* Poepp. & Endl)	↑ Cecal SFCA concentration	[79]
Male BALB/c mice	GOS of Chinese roots (Deshipu stachyose granules)	Growth of beneficial intestinal bacteria (Lactobacilli and Bifidobacteria) and inhibition of pathogenic bacteria (*Clostridium perfringens*) Effects on intestinal peristalsis promotion and bowel function improvement (constipation treatment)	[80]

Table 2. In vivo models for prebiotic food assessment.

Young adult male rats were fed ad libitum with purified control diet (CONT) containing 5% w/w cellulose (insoluble fiber) or diet containing 10% w/w cellulose (CELL), FOS, oat beta-glucan (GLUC), or apple pectin (PECT) for 4 weeks. Comparing CONT and CELL, caecal concentrations of fermentation products increased from 1.4 to 2.2 times in GLUC, FOS, and PECT, and colonic concentrations increased from 1.9 to 2.5 times in GLUC and FOS; however, no consistent changes in SCFA receptor gene expression were detected. The main fermentation products detected were acetate, propionate, butyrate, and succinate, and the differences in amounts of fermentation products among soluble fibers may reflect different fermentation patterns and/or different fermentation rates and turnover. This research concluded that the presence of soluble fermentable fiber appears to be more important than its source [75].

Other studies evaluating the prebiotic potential of foods using in vivo models are described in **Table 2**.

5. Prebiotics and other beneficial effects on health

The modulations of the intestinal microbiota and SCFA production are associated with many beneficial effects about the ingestion of prebiotics and isolated or added to foods, such as regulation of various physiological processes (e.g., inflammation) and metabolic processes (e.g., lipid and glucose metabolism), thus contributing to the treatment or prevention of chronic non-degenerative diseases [38].

Rats treated with prebiotics had a reduction of plasma pro-inflammatory cytokines, reduction of hepatic inflammatory expression, and oxidative stress markers [81]. Everard et al. [82] showed that diet enriched with prebiotics led to an improvement in glucose tolerance, increase in amount of L-cells, and associated parameters (expression of intestinal pro-glucagon mRNA and plasma glucagon-like peptide-1 levels or GLP-1) in addition to reduction in body fat accumulation, oxidative stress, and level of inflammation in obese rats.

Salazar et al. [69] supplemented 15 obese women with a mixture of inulin and oligofructose for 3 months and observed that prebiotics had a bifidogenic effect, but the elimination of SCFA in feces did not show a significant correlation with the serum concentration of lipids.

A prospective longitudinal cohort study with 1592 workers with metabolic syndrome found that there was an inverse association between consumption of insoluble fibers and increase in systolic and diastolic blood pressure, total cholesterol (TC), triglycerides (TG), apolipoprotein B100, and TG/high-density lipoprotein (HDL) ratio; however, the ingestion of soluble fibers was inversely associated only with triglycerides and apolipoprotein B100. Thus, the prevalence of metabolic syndrome was lower in participants who ingested larger amounts of insoluble fibers [83]. In contrast, a meta-analysis by Wu et al. [84] that included 18 cohort studies with 672,408 participants confirmed that dietary intake of soluble or insoluble fibers (especially from cereals and fruits) has a similar inverse effect associated with the risk of coronary heart disease.

Barbalho et al. [85] reported that the supplementation of passion fruit peels to healthy Wistar rats contributed to the elevation of HDL levels and the decrease in glycemia, TG, and TC levels of these animals compared to the control group. Such results would be associated with the soluble dietary fiber present in passion fruit peels, such as mucilage and pectins, which form a viscous gel that retains water and reduce the sensation of hunger, body weight, plasma levels of TC, TG, and low-density lipoprotein (LDL) and increase the excretion of cholesterol and bile salts in feces and HDL levels.

Obese rats fed with hyperlipid diet and diet added of lyophilized jabuticaba peel (rich in anthocyanins) exhibited increased HDL and improved insulin resistance, suggesting that the diet added of this byproduct may have a protective effect against cardiovascular diseases by increasing HDL levels [86].

Amaya-Cruz et al. [87] evaluated the effect of dietary fibers and polyphenols from guava (*Pisidium guajava*), peach (*Prunus persica*), and mango (*Mangifera indica*) byproducts on obesity-related hyperglycemia and hepatic steatosis in Wistar rats. Mango and peach byproducts presented better soluble/insoluble fiber ratio and high amount of polyphenols, which may have attenuated the development of hepatic steatosis and hyperglycemia in rats. In guava byproducts, they found great amount of soluble dietary fibers and condensed tannins, which may be related to the greater anti-obesogenic effect on animals, when compared to control rats and to those treated with other byproducts.

Changes in the intestinal microbiota may also influence the homeostasis of the immune [35], renal [88], and nervous systems [89], as well as the development and progression of pathophysiological processes such as hypertension [90] and colorectal cancer [91]. A mixture of non-digestible GOS ingested by mice for 3 weeks prior to induction of inflammatory neuropathology and anxiety improved anxiety and inflammation through decreased expression of IL-1b cytokine and 5-HT2AR serotonin receptor in the frontal cortex compared to the control group [92]. Healthy men and women daily supplied with FOS or GOS for 3 weeks showed decreased response to cortisol awakening, protecting against the risk of depression [93]. Rats with chronic kidney disease (CKD) fed for 3 weeks with RS diets had a delay in CKD progression and increased creatinine clearance when compared to CKD mice that received amylopectin [94].

6. Innovations in food processing with added prebiotics

The inclusion of prebiotics in industrialized foods has become a viable and healthy alternative, since there is a great demand of consumers for functional foods that can help in maintaining health. Moreover, the food industry can obtain numerous advantages from the addition of prebiotics in food products, such as improvement of sensory characteristics, better balance of the nutritional composition, and longer shelf-life [67]. In general, prebiotics are added to bakery products, breakfast cereals, beverages (e.g., fruit juices, coffee, cocoa, and tea), dairy products, table spreads, butter-based products, and desserts (ice cream, puddings, jellies, and chocolates) [67, 95]. Prebiotics also have gelling properties (e.g., inulin), which maintain the emulsion stability, provide spreadable texture, and water retention (e.g., inulin and FOS), thus allowing the development of processed foods with low fat content, with pleasant taste and texture [67, 96].

However, some important characteristics of the manufacturing process, such as low pH, high temperatures, and conditions favoring the Maillard reaction must be taken into account when choosing the prebiotic to be added to foods in order to avoid the formation of anti-nutritional compounds detrimental to the sensory quality of the final product and consumer health as well as the partial or total reduction of their action. Among prebiotics commonly used in the food industry, GOS are more stable at high temperatures and low pH mainly due to the beta bonds of their structure, which provide greater hydrolysis stability compared to FOS and inulin [96]. A type of RS known as RS3 can be added to fried battered products to increase the content of dietary fibers and avoid reducing moisture and the absorption of fats, since RS3 is very resistant to frying temperatures [97].

7. Concluding remarks

The importance of the consumption of prebiotics is unquestionable and they should be part of healthy diet. Prebiotics exert various technological functions in food and many health benefits not only related to the modulation of the intestinal microbiota but also to other beneficial physiological actions in various organs and systems of healthy or diseased men/animals. In this sense, the development of foods added due to prebiotics by the industry can be advantageous due to the demand and profitability of this market, as well as for consumers who will have healthy foods available that can be readily consumed for the prevention or treatment of diseases, thus reducing public health costs. However, there is no consensus on the recommended quantity of specific prebiotics for consumption in the diet, and this limitation is a major challenge regarding the different in vitro and in vivo models used to test the prebiotic potential of foods.

Both in vivo and in vitro models have helped advances of researches aimed at evaluating the prebiotic potential of foods through the composition and metabolism of the intestinal microbiota and their interactions. However, it is noteworthy that there are no ideal models, and the most adequate are those based on the study objectives and using association of complementary techniques.

Author details

Jailane de Souza Aquino*, Kamila Sabino Batista, Francisca Nayara Dantas Duarte Menezes, Priscilla Paulo Lins, Jessyca Alencar de Sousa Gomes and Laiane Alves da Silva

*Address all correspondence to: lalaaquino@hotmail.com

Experimental Nutrition Laboratory, Department of Nutrition, Health Sciences Center, Federal University of Paraíba (UFPB), Brazil

References

[1] Goetzke B, Nitzko S, Spiller A. Consumption of organic and functional food. A matter of well-being and health? Appetite. 2014;77:96-105

[2] Bachl T. Wellness trend benefits markets. In: Consumers' Choice '07. Berlin: BVE; 2007. pp. 9-12

[3] Chrysochou P. Food health branding: The role of marketing mix elements and public discourse in conveying a healthy brand image. Journal of Marketing Communications. 2010;**16**(1-2):69-85

[4] Food Quality and Standards Service, Food and Agriculture Organization. FAO Technical Meeting on Prebiotics. Food Quality and Standards Service (AGNIS), Food and Agriculture Organization of the United Nations; 2007

[5] Sreenivas KM, Lele SS. Prebiotic activity of gourd family vegetable fibres using in vitro fermentation. Food Bioscience. 2013;**1**:26-30

[6] Gómez B, Gullón B, Yáñez R, Schols H, Alonso JL. Prebiotic potential of pectins and pectic oligosaccharides derived from lemon peel wastes and sugar beet pulp: A comparative evaluation. Journal of Functional Foods [Internet]. 2016;**20**:108-121. DOI: 10.1016/j.jff.2015.10.029

[7] Scott KP, Martin JC, Duncan SH, Flint HJ. Prebiotic stimulation of human colonic butyrate-producing bacteria and bifidobacteria, in vitro. FEMS Microbiology Ecology. 2014;**87**(1):30-40

[8] Donelli G, Vuotto C, Mastromarino P. Phenotyping and genotyping are both essential to identify and classify a probiotic microorganism. Microbial Ecology in Health and Disease [Internet]. 2013;**24**:1-8. Available from: http://www.pubmedcentral.nih.gov/articlerender.fcgi?artid=3758930&tool=pmcentrez&rendertype=abstract

[9] Joint FAO/WHO Working Group Report. Guidelines for the Evaluation of Probiotics in Food. London, Ontario, Canada: World Health Organization, Food and Agriculture Organization; 2002

[10] Butel MJ. Probiotics, gut microbiota and health. Médecine et Maladies Infectieuses [Internet]. 2014;**44**(1):1-8. DOI: 10.1016/j.medmal.2013.10.002

[11] He Z, Wang X, Li G, Zhao Y, Zhang J, Niu C, et al. Antioxidant activity of prebiotic ginseng polysaccharides combined with potential probiotic Lactobacillus plantarum C88. International Journal of Food Science & Technology. 2015;**50**:1673-1682

[12] Rodrigues FC, Castro ASB, Rodrigues VC, Fernandes SA, Fontes EAF, de Oliveira TT, et al. Yacon flour and Bifidobacterium longum modulate bone health in rats. Journal of Medicinal Food [Internet]. 2012;**15**(7):664-670. Available from: http://www.ncbi.nlm.nih.gov/pubmed/22510044

[13] Estévez BG, Gullón P, Tavaria F, Yáñez R. Assessment of the prebiotic effect of quinoa and amaranth in the human intestinal ecosystem. Food & Function [Internet]. 2016;**7**:3782-3788. Available from: http://pubs.rsc.org/en/Content/ArticleLanding/2016/FO/C6FO00924G

[14] Maurer AD, Eller LK, Hallam MC, Taylor K, Reimer RA. Consumption of diets high in prebiotic fiber or protein during growth influences the response to a high fat and sucrose diet in adulthood in rats. Nutrition and Metabolism (London). 2010;**7**:77

[15] Scott KP, Gratz SW, Sheridan PO, Flint HJ, Duncan SH. The influence of diet on the gut microbiota. Pharmacological Research [Internet]. 2013;**69**(1):52-60. DOI: 10.1016/j.phrs.2012.10.020

[16] Flint HJ, Duncan SH, Scott KP, Louis P. Links between diet, gut microbiota composition and gut metabolism. Proceedings of the Nutrition Society [Internet]. 2015;**74**(1):13-22. Available from: http://www.ncbi.nlm.nih.gov/pubmed/25268552

[17] Gibson GR, Probert HM, Van Loo J, Rastall RA, Roberfroid MB. Dietary modulation of the human colonic microbiota: Updating the concept of prebiotics. Nutrition Research Reviews. 2004;**17**(2):259-275

[18] Marques TM, Cryan JF, Shanahan F, Fitzgerald GF, Ross RP, Dinan TG, et al. Gut microbiota modulation and implications for host health: Dietary strategies to influence the gut-brain axis. Innovative Food Science and Emerging Technologies. 2014;**22**:239-247

[19] Ufnal M, Pham K. The gut-blood barrier permeability—a new marker in cardiovascular and metabolic diseases? Medical Hypotheses [Internet]. 2017;**98**:35-37. DOI: 10.1016/j.mehy.2016.11.012

[20] Principi N, Esposito S. Gut microbiota and central nervous system development. Journal of Infection [Internet]. 2016;**73**(6):536-546. DOI: 10.1016/j.jinf.2016.09.010

[21] Luna RA, Foster JA. Gut brain axis: Diet microbiota interactions and implications for modulation of anxiety and depression. Current Opinion in Biotechnology [Internet]. 2015;**32**:35-41. DOI: 10.1016/j.copbio.2014.10.007

[22] Kabouridis PS, Pachnis V. Emerging roles of gut microbiota and the immune system in the development of the enteric nervous system. Journal of Clinical Investigation. 2015;**125**(3):956-964

[23] Sandrini S, Aldriwesh M, Alruways M, Freestone P. Microbial endocrinology: Host-bacteria communication within the gut microbiome. Journal of Endocrinology [Internet]. 2015;**225**(2):R21-R34. Available from: http://joe.endocrinology-journals.org/cgi/doi/10.1530/JOE-14-0615

[24] Candela M, Maccaferri S, Turroni S, Carnevali P, Brigidi P. Functional intestinal microbiome, new frontiers in prebiotic design. International Journal of Food Microbiology [Internet]. 2010;**140**(2-3):93-101. Available from: http://linkinghub.elsevier.com/retrieve/pii/S0168160510002485

[25] American Dietetic Association. Position of the American Dietetic Association: Health implications of dietary fiber. Journal of the American Dietetic Association [Internet]. 2008;**108**(10):1716-1731. Available from: http://www.sciencedirect.com/science/article/pii/S0002822308015666

[26] Cupersmid L, Fraga APR, de Abreu ES, Pereira IRO. Linhaça: Composição química e efeitos biológicos. e-Scientia. 2012;**5**(2):33-40

[27] Porfírio E, Henrique VSM, de Abreu Reis MJ. Elaboração de farofa de grãos, sementes oleaginosas e castanha de caju: composição de fi bras, ácidos graxos e aceitação [Development of a "farofa"(*) containing grains, oilseeds and cashew nut: Fatty acid & fibre composition and acceptance]. Brazilian Journal of Food Technology. 2014;**17**(3):185

[28] Ayala-Zavala JF, Vega-Vega V, Rosas-Domínguez C, Palafox-Carlos H, Villa-Rodriguez JA, Siddiqui MW, et al. Agro-industrial potential of exotic fruit byproducts as a source of food additives. Food Research International [Internet]. 2011;**44**(7):1866-1874. Available from: http://linkinghub.elsevier.com/retrieve/pii/S0963996911001086

[29] Kosmala M, Zduńczyk Z, Jus¨kiewicz J, Jurgoński A, Karlińska E, Macierzyński J, et al. Chemical composition of defatted strawberry and raspberry seeds and the effect of these dietary ingredients on polyphenol metabolites, intestinal function, and selected serum parameters in rats. Journal of Agricultural and Food Chemistry. 2015;**63**(11):2989-2996

[30] Roberfroid MB, Delzenne NM. Dietary fructans. Annual Review of Nutrition. 1998;**18**(1): 117-143

[31] Patel S, Goyal A. The current trends and future perspectives of prebiotics research: A review. 3 Biotech. 2012;**2**(2):115-125

[32] Ganapathy V, Thangaraju M, Prasad PD, Martin PM, Singh N. Transporters and receptors for short-chain fatty acids as the molecular link between colonic bacteria and the host. Current Opinion in Pharmacology. 2013;**13**(6):869-874

[33] Parkar SG, Trower TM, Stevenson DE. Fecal microbial metabolism of polyphenols and its effects on human gut microbiota. Anaerobe [Internet]. 2013;**23**:12-19. DOI: 10.1016/j.anaerobe.2013.07.009

[34] da Silva JK, Cazarin CBB, Colomeu TC, Batista ÂG, Meletti LMM, Paschoal JAR, et al. Antioxidant activity of aqueous extract of passion fruit (*Passiflora edulis*) leaves: In vitro and in vivo study. Food Research International. 2013;**53**(2):882-890

[35] Valdés L, Cuervo A, Salazar N, Ruas-Madiedo P, Gueimonde M, González S. The relationship between phenolic compounds from diet and microbiota: Impact on human health. Food &Function [Internet]. 2015;**6**:2424-2439. Available from: www.rsc.org/foodfunction

[36] Gullon B, Pintado ME, Fernández-López J, Pérez-Álvarez JA, Viuda-Martos M. In vitro gastrointestinal digestion of pomegranate peel (*Punica granatum*) flour obtained from co-products: Changes in the antioxidant potential and bioactive compounds stability. Journal of Functional Foods [Internet]. 2015;**19**:617-628. DOI: 10.1016/j.jff.2015.09.056

[37] Quiñones M, Miguel M, Aleixandre A. Beneficial effects of polyphenols on cardiovascular disease. Pharmacological Research. 2013;**68**(1):125-131

[38] Slavin J. Fiber and prebiotics: Mechanisms and health benefits. Nutrients [Internet]. 2013;**5**(4):1417-1435. Available from: http://www.pubmedcentral.nih.gov/articlerender.fcgi?artid=3705355&tool=pmcentrez&rendertype=abstract

[39] Rodríguez-Cabezas ME, Camuesco D, Arribas B, Garrido-Mesa N, Comalada M, Bailón E, et al. The combination of fructooligosaccharides and resistant starch shows prebiotic additive effects in rats. Clinical Nutrition [Internet]. 2010;**29**(6):832-839. DOI: 10.1016/j.clnu.2010.05.005

[40] Savignac HM, Corona G, Mills H, Chen L, Spencer JPE, Tzortzis G, et al. Prebiotic feeding elevates central brain derived neurotrophic factor, N-methyl-d-aspartate receptor subunits and d-serine. Neurochemistry International [Internet]. 2013;**63**(8):756-764. DOI: 10.1016/j.neuint.2013.10.006

[41] Pyra KA, Saha DC, Reimer RA. Prebiotic fiber increases hepatic acetyl CoA carboxylase phosphorylation and suppresses glucose-dependent insulinotropic polypeptide secretion more effectively when used with metformin in obese rats. Journal of Nutrition. 2012;**142**(2):213-220

[42] Freitas KDC, Amancio OMS, Morais MB. High-performance inulin and oligofructose prebiotics increase the intestinal absorption of iron in rats with iron deficiency anaemia during the growth phase. British Journal of Nutrition [Internet]. 2012;**108**(6):1008-1016. Available from: http://www.ncbi.nlm.nih.gov/pubmed/22172251

[43] Gobinath D, Madhu AN, Prashant G, Srinivasan K, Prapulla SG. Beneficial effect of xylo-oligosaccharides and fructo-oligosaccharides in streptozotocin-induced diabetic rats. British Journal of Nutrition. 2010;**104**(1):40-47

[44] Macfarlane GT, Gibson GR, Cummings JH. Comparison of fermentation reactions in different regions of the human colon. Journal of Applied Bacteriology. 1992;**72**(1):57-64

[45] Moter A, Göbel UB. Fluorescence in situ hybridization (FISH) for direct visualization of microorganisms. Journal of Microbiological Methods. 2000;**41**(2):85-112

[46] Olson DW, Aryana KJ. Effect of prebiotics on *Lactobacillus acidophilus* growth and resulting pH changes in skim milk and a model peptone system. Journal of Microbial & Biochemical Technology. 2012;**4**:121-125

[47] Teixeira LS, Martim SR, Silva LSC, Kinupp VF, Teixeira MFS, Porto ALF. Efficiency of Amazonian tubers flours in modulating gut microbiota of male rats. Innovative Food Science and Emerging Technologies. 2016;**38**:1-6

[48] Tuohy KM, Probert HM, Smejkal CW, Gibson GR. Using probiotics and prebiotics to improve gut health. Drug Discovery Today. 2003;**8**(15):692-700

[49] Gómez B, Gullón B, Remoroza C, Schols HA, Parajó JC, Alonso JL. Purification, characterization, and prebiotic properties of pectic oligosaccharides from orange peel wastes. Journal of Agricultural and Food Chemistry. 2014;**62**(40):9769-9782

[50] Duarte FND, Rodrigues JB, da Costa Lima M, Lima MD, Pacheco MT, Pintado MM, de Souza Aquino J, de Souza EL. Potential prebiotic properties of cashew apple (*Anacardium occidentale* L.) agro-industrial byproduct on Lactobacillus species. Journal of Organizational Behavior. 2017;**31**:1-8

[51] Thuaytong W, Anprung P. Bioactive compounds and prebiotic activity in Thailand-grown red and white guava fruit (*Psidium guajava* L.). Food Science and Technology International. 2011;**17**(3):205-212

[52] Sousa S, Pinto J, Pereira C, Malcata FX, Pacheco MTB, Gomes AM, et al. In vitro evaluation of yacon (*Smallanthus sonchifolius*) tuber flour prebiotic potential. Food and Bioproducts Processing. 2015;**95**:96-105

[53] Condezo-Hoyos L, Mohanty IP, Noratto GD. Assessing non-digestible compounds in apple cultivars and their potential as modulators of obese faecal microbiota in vitro. Food Chemistry. 2014;**161**:208-215

[54] Gullon P, Gullon B, Tavaria F, Vasconcelos M, Gomes AM. In vitro fermentation of lupin seeds (*Lupinus albus*) and broad beans (Vicia faba): Dynamic modulation of the intestinal microbiota and metabolomic output. Food & Function [Internet]. 2015;**6**(10):3316-3322. Available from: http://www.ncbi.nlm.nih.gov/pubmed/26252418

[55] Oliveira DL, Costabile A, Wilbey RA, Grandison AS, Duarte LC, Roseiro LB. In vitro evaluation of the fermentation properties and potential prebiotic activity of caprine cheese whey oligosaccharides in batch culture systems. BioFactors. 2012;**38**(6):440-449

[56] Wichienchot S, Jatupornpipat M, Rastall RA. Oligosaccharides of pitaya (dragon fruit) flesh and their prebiotic properties. Food Chemistry [Internet]. 2010;**120**(3):850-857. DOI: 10.1016/j.foodchem.2009.11.026

[57] Gullon B, Pintado ME, Barber X, Fernández-López J, Pérez-Álvarez JA, Viuda-Martos M. Bioaccessibility, changes in the antioxidant potential and colonic fermentation of date pits and apple bagasse flours obtained from co-products during simulated in vitro gastrointestinal digestion. Food Research International [Internet]. 2015;**78**:169-176. DOI: 10.1016/j.foodres.2015.10.021

[58] Immerzeel P, Falck P, Galbe M, Adlercreutz P, Nordberg Karlsson E, Stålbrand H. Extraction of water-soluble xylan from wheat bran and utilization of enzymatically produced xylooligosaccharides by Lactobacillus, Bifidobacterium and Weissella spp. LWT—Food Science and Technology [Internet]. 2014;**56**(2):321-327. DOI: 10.1016/j.lwt.2013.12.013

[59] Liu Z, Wang W, Huang G, Zhang W, Ni L. In vitro and in vivo evaluation of the prebiotic effect of raw and roasted almonds (*Prunus amygdalus*). Journal of the Science of Food and Agriculture. 2016;**96**(5):1836-1843

[60] Shinohara K, Ohashi Y, Kawasumi K, Terada A, Fujisawa T. Anaerobe effect of apple intake on fecal microbiota and metabolites in humans. Anaerobe [Internet]. 2010;**16**(5):510-515. DOI: 10.1016/j.anaerobe.2010.03.005

[61] Macfarlane GT, Macfarlane S. Models for intestinal fermentation: Association between food components, delivery systems, bioavailability and functional interactions in the gut. Current Opinion in Biotechnology. 2007;**18**(2):156-162

[62] Han KS, Balan P, Molist Gasa F, Boland M. Green kiwifruit modulates the colonic microbiota in growing pigs. Letters in Applied Microbiology. 2011;**52**(4):379-385

[63] Cheesman SE, Neal JT, Mittge E, Seredick BM, Guillemin K. Epithelial cell proliferation in the developing zebrafish intestine is regulated by the Wnt pathway and microbial signaling via Myd88. Proceedings of the National Academy of Sciences. 2011;**108**(Suppl. 1):4570-4577

[64] Huang YL, Tsai YH, Chow CJ. Water-insoluble fiber-rich fraction from pineapple peel improves intestinal function in hamsters: Evidence from cecal and fecal indicators. Nutrition Research [Internet]. 2014;**34**(4):346-354. DOI: 10.1016/j.nutres.2014.03.001

[65] Samal L, Chaturvedi VB, Saikumar G, Somvanshi R, Pattanaik AK. Prebiotic potential of Jerusalem artichoke (*Helianthus tuberosus* L.) in Wistar rats: Effects of levels of supplementation on hindgut fermentation, intestinal morphology, blood metabolites and immune response. Journal of the Science of Food and Agriculture. 2015;**95**(8):1689-1696

[66] Yasmin A, Butt MS, Afzaal M, van Baak M, Nadeem MT, Shahid MZ. Prebiotics, gut microbiota and metabolic risks : Unveiling the relationship. Journal of Functional Foods [Internet]. 2015;**17**:189-201. DOI: 10.1016/j.jff.2015.05.004

[67] Al-Sheraji SH, Ismail A, Manap MY, Mustafa S, Yusof RM, Hassan FA. Prebiotics as functional foods: A review. Journal of Functional Foods [Internet]. 2013;**5**(4):1542-1553. Available from: http://linkinghub.elsevier.com/retrieve/pii/S175646461300193X

[68] den Besten G, van Eunen K, Groen AK, Venema K, Reijngoud D-J, Bakker BM. The role of short-chain fatty acids in the interplay between diet, gut microbiota, and host energy metabolism. Journal of Lipid Research [Internet]. 2013;**54**(9):2325-2340. Available from: http://www.jlr.org/cgi/doi/10.1194/jlr.R036012

[69] Salazar N, Dewulf EM, Neyrinck AM, Bindels LB, Cani PD, Mahillon J, et al. Inulin-type fructans modulate intestinal Bifidobacterium species populations and decrease fecal short-chain fatty acids in obese women. Clinical Nutrition [Internet]. 2015;**34**(3):501-517. Available from: http://linkinghub.elsevier.com/retrieve/pii/S0261561414001599

[70] McOrist AL, Miller RB, Bird AR, Keogh JB, Noakes M, Topping DL, et al. Fecal butyrate levels vary widely among individuals but are usually increased by a diet high in resistant starch. Journal of Nutrition. 2011;**141**(5):883-889

[71] Paturi G, Butts CA, Monro JA, Hedderley D, Stoklosinski H, Roy NC, et al. Evaluation of gastrointestinal transit in rats fed dietary fibres differing in their susceptibility to large intestine fermentation. Journal of Functional Foods [Internet]. 2012;**4**(1):107-115. DOI: 10.1016/j.jff.2011.08.005

[72] Hu Y, Le Leu RK, Christophersen CT, Somashekar R, Conlon MA, Meng XQ, et al. Manipulation of the gut microbiota using resistant starch is associated with protection against colitis-associated colorectal cancer in rats. Carcinogenesis. 2016;**37**(4):366-375

[73] de Moura NA, Caetano BFR, Sivieri K, Urbano LH, Cabello C, Rodrigues MAM, et al. Protective effects of yacon (*Smallanthus sonchifolius*) intake on experimental colon carcinogenesis. Food and Chemical Toxicology [Internet]. 2012;**50**(8):2902-2910. DOI: 10.1016/j.fct.2012.05.006

[74] Bränning C, Håkansson A, Ahrné S, Jeppsson B, Molin G, Nyman M. Blueberry husks and multi-strain probiotics affect colonic fermentation in rats. British Journal of Nutrition. 2009;**101**(6):859-870

[75] Adam CL, Williams PA, Dalby MJ, Garden K, Thomson LM, Richardson AJ, et al. Different types of soluble fermentable dietary fibre decrease food intake, body weight gain and adiposity in young adult male rats. Nutrition and Metabolism (London) [Internet]. 2014;**11**(1):36. Available from: http://nutritionandmetabolism.biomedcentral.com/articles/10.1186/1743-7075-11-36

[76] Massot-Cladera M, Costabile A, Childs CE, Yaqoob P, Franch À, Castell M, et al. Prebiotic effects of cocoa fibre on rats. Journal of Functional Foods [Internet]. 2015;**19**:341-352. DOI: 10.1016/j.jff.2015.09.021

[77] da Silva JK, Cazarin CBB, Bogusz Junior S, Augusto F, Maróstica Junior MR. Passion fruit (*Passiflora edulis*) peel increases colonic production of short-chain fatty acids in Wistar rats. LWT—Food Science and Technology [Internet]. 2014;**59**(2):1252-1257. Available from: http://www.sciencedirect.com/science/article/pii/S0023643814003193

[78] Fotschki B, Juśkiewicz J, Jurgoński A, Kołodziejczyk K, Milala J, Kosmala M, et al. Anthocyanins in strawberry polyphenolic extract enhance the beneficial effects of diets with fructooligosaccharides in the rat cecal environment. PLoS One. 2016;**11**(2):1-18

[79] Campos D, Betalleluz-Pallardel I, Chirinos R, Aguilar-Galvez A, Noratto G, Pedreschi R. Prebiotic effects of yacon (Smallanthus sonchifolius Poepp. & Endl), a source of fructooligosaccharides and phenolic compounds with antioxidant activity. Food Chemistry [Internet]. 2012;**135**(3):1592-1599. DOI: 10.1016/j.foodchem.2012.05.088

[80] Li T, Lu X, Yang X. Stachyose-enriched α-galacto-oligosaccharides regulate gut microbiota and relieve constipation in mice. Journal of Agricultural and Food Chemistry. 2013;**61**(48):11825-11831

[81] Cani PD, Possemiers S, Van de Wiele T, Guiot Y, Everard A, Rottier O, et al. Changes in gut microbiota control inflammation in obese mice through a mechanism involving GLP-2-driven improvement of gut permeability. Gut [Internet]. 2009;**58**(8):1091-1103. Available from: http://gut.bmj.com/cgi/doi/10.1136/gut.2008.165886

[82] Everard A, Lazarevic V, Derrien M, Girard M, Muccioli GG, Neyrinck AM, et al. Responses of gut microbiota and glucose and lipid metabolism to prebiotics in genetic obese and diet-induced leptin-resistant mice. Diabetes [Internet]. 2011;**60**(11):2775-2786. Available from: http://diabetes.diabetesjournals.org/cgi/doi/10.2337/db11-0227

[83] Franco BM, Latre ML, Esteban EMA, Ordovás JM, Casasnovas JA, Peñalvo JL. Soluble and insoluble dietary fibre intake and risk factors for metabolic syndrome and cardiovascular disease in middle-aged adults: The AWHS Cohort. Nutrición Hospitalaria. 2014;**30**(6):1279-1288

[84] Wu Y, Qian Y, Pan Y, Li P, Yang J, Ye X, et al. Association between dietary fiber intake and risk of coronary heart disease: A meta-analysis. Clinical Nutrition. 2015;**34**(4):603-611

[85] Barbalho SM, da Silva Soares de Souza M, de Paula e Silva JC, Mendes CG, de Oliveira GA, Costa T, et al. Yellow passion fruit rind (Passiflora edulis): An industrial waste or an adjuvant in the maintenance of glycemia and prevention of dyslipidemia? Journal of Diabetes Research and Clinical Metabolism. 2012;**1**(1):5

[86] Lenquiste SA, Batista AG, Marineli RDS, Dragano NRV, Maróstica MR. Freeze-dried jaboticaba peel added to high-fat diet increases HDL-cholesterol and improves insulin resistance in obese rats. Food Research International. 2012;**49**(1):153-160

[87] Amaya-Cruz DM, Rodríguez-González S, Pérez-Ramírez IF, Loarca-Piña G, Amaya-Llano S, Gallegos-Corona MA, et al. Juice by-products as a source of dietary fibre and antioxidants and their effect on hepatic steatosis. Journal of Functional Foods [Internet]. 2015;**17**:93-102. Available from: http://linkinghub.elsevier.com/retrieve/pii/S1756464615002340

[88] Wanchai K, Pongchaidecha A, Chatsudthipong V, Chattipakorn SC, Chattipakorn N, Lungkaphin A. Role of gastrointestinal microbiota on kidney injury and the obese condition. American Journal of the Medical Sciences [Internet]. 2017;**353**(1):59-69. Available from: http://www.ncbi.nlm.nih.gov/pubmed/28104104; http://linkinghub.elsevier.com/retrieve/pii/S0002962916306292

[89] Oriach CS, Robertson RC, Stanton C, Cryan JF, Dinan TG. Food for thought: The role of nutrition in the microbiota-gut-brain axis. Clinical Nutrition Experimental [Internet]. 2016;**6**:25-38. DOI: 10.1016/j.yclnex.2016.01.003

[90] Jose PA, Raj D. Gut microbiota in hypertension. Current Opinion in Nephrology and Hypertension. 2015;**24**(5):403-409

[91] Louis P, Hold GL, Flint HJ. The gut microbiota, bacterial metabolites and colorectal cancer. Nature Reviews Microbiology [Internet]. 2014;**12**(10):661-672. DOI: 10.1038/nrmicro3344

[92] Savignac HM, Couch Y, Stratford M, Bannerman DM, Tzortzis G, Anthony DC, et al. Prebiotic administration normalizes lipopolysaccharide (LPS)-induced anxiety and cortical 5-HT2A receptor and IL1-β levels in male mice. Brain, Behavior, and Immunity [Internet]. 2016;**52**:120-131. DOI: 10.1016/j.bbi.2015.10.007

[93] Schmidt K, Cowen PJ, Harmer CJ, Tzortzis G, Errington S, Burnet PWJ. Prebiotic intake reduces the waking cortisol response and alters emotional bias in healthy volunteers. Psychopharmacology (Berlin). 2015;**232**(10):1793-1801

[94] Vaziri ND, Liu SM, Lau WL, Khazaeli M, Nazertehrani S, Farzaneh SH, et al. High amylose resistant starch diet ameliorates oxidative stress, inflammation, and progression of chronic kidney disease. PLoS One. 2014;**9**(12):1-15

[95] de Sousa C, dos Santos EF, Sgarbieri VC. The importance of prebiotics in functional foods and clinical practice. Food and Nutrition Sciences [Internet]. 2011;**2**(2):133-144. Available from: http://www.scirp.org/journal/PaperInformation.aspx?PaperID=4536&#abstract

[96] Charalampopoulos D, Rastall RA. Prebiotics in foods. Current Opinion in Biotechnology [Internet]. 2012;**23**(2):187-191. DOI: 10.1016/j.copbio.2011.12.028

[97] Homayouni A, Amini A, Keshtiban AK, Mortazavian AM, Esazadeh K, Pourmoradian S. Resistant starch in food industry: A changing outlook for consumer and producer. Starch/Staerke. 2014;**66**(1-2):102-114

Chapter 13

Leveraging Bioactives to Support Human Health through the Lifecycle: Scientific Evidence and Regulatory Considerations

Deshanie Rai and Gyan Rai

Additional information is available at the end of the chapter

http://dx.doi.org/10.5772/intechopen.69836

Abstract

The identification of bioactive food components and understanding their role as adjunct therapeutic agents in disease management and prevention has become a significant area of research. Accumulating evidence suggests a link between certain bioactive food components and health outcomes, for example, lutein and zeaxanthin for visual performance and delaying age-related macular degeneration, probiotics for gastrointestinal outcomes related to irritable bowel syndrome or prebiotics for its potential programming of the microbiome in early life to influence later life outcomes.. This rapidly developing science has triggered discussions to determine if public health recommendations can be made on bioactive foods. However, regulatory guidance is necessary to guide the development the science, it's consideration for public health policy and the communication thereof to both healthcare professionals and consumers. This chapter will focus on the clinical and basic science supporting a role for lutein and pre- and probiotics in modulating several aspects of human health, including the gut microbiome through the human lifecycle. Opportunities to translate the science to consumers in a meaningful and accurate way will also be highlighted along with the regulatory landscape to shape the testing, communication and commercialization of these bioactives.

Keywords: bioactives, lutein, zeaxanthin, prebiotics, probiotics, gut microbiome, early life, development, programming, regulatory

1. Introduction

Interest in bioactive foods and ingredients is high among consumers. The 2013 Functional Foods Consumer Survey conducted by the International Food and Information Council showed that of the 1005 participants, 45% said they were very interested and 86% said they were very or somewhat interested in learning more about foods that have benefits beyond basic nutrition.

Although bioactive compounds typically occur in small amounts in foods, they are capable of influencing physiological and cellular activities, and in some cases can modify disease risk. Given the current focus on reducing risk of chronic diseases and programming health outcomes starting early in life, the potential beneficial role of bioactive compounds has garnered significant interest within the scientific community. The Office of Dietary Supplements at the NIH has defined bioactive compounds as constituents in foods or dietary supplements, other than those needed to meet basic human nutritional needs, which are responsible for changes in health status [1].

Over the past three decades, a variety of bioactive compounds have been identified that have potentially important health benefits. These include carotenoids, e.g., lutein; flavonoids, e.g., cocoa polyphenols; plant sterols, e.g., those found in soybeans; n-3 long chain polyunsaturated fatty acids (LCPUFA), e.g., docosahexaenoic acid (DHA); and more recently modulators of the gastrointestinal microbiota, e.g., prebiotics and probiotics. These compounds can act as antioxidants, enzyme inhibitors and inducers, inhibitors of receptor activities, and inducers and inhibitors of gene expression, among other actions [2, 3].

Currently, there are no public health recommendations for bioactive nutrients and bioactive containing foods as part of the everyday diet, including as part of supplementation or medical nutrition. An understanding of the strength of the established and emerging scientific evidence needs to be considered in order to establish such recommendations as well as to drive science-based communication strategies to consumers and healthcare professionals, within the guardrails of the regulatory environments.

Globally, healthcare costs are sky-rocketing and there is increased emphasis on reducing the prevalence of non-communicable diseases, e.g., heart disease, type II diabetes, obesity, and inflammatory bowel diseases. Not surprisingly, in most parts of the world today, the goal of healthcare is focused on disease prevention. Indeed, nutrition has an established role in "preventing, treating, mitigating, or curing a disease" but they are not drugs. However, there are obvious regulatory challenges to clinically testing and communicating the science of nutrients and bioactives to the general population. The FDA has an important role herein to guide the necessary framework that that supports development of the science on bioactives for relevant human health outcomes and its communication. We are certainly not there today.

Our goal with this review is to focus on the strengths and gaps in the science to help scientists, regulators and policymakers initiate dialogue on whether public health recommendations for key bioactive components, such as lutein and zeaxanthin, probiotics and/or prebiotics can be established. This review will focus on the published science supporting the role of these specific bioactives in influencing both early and later life outcomes. The focus on the lifecycle effect is highly relevant given the accumulating evidence linking early-life nutritional and microbiome exposures and health status or disease risk in later life. In particular, we will focus on: (1) lutein and zeaxanthin as it pertains to visual performance and macular degeneration, and (2) the role of probiotics and prebiotics as they pertain to beneficially modulating the gastrointestinal microflora, gastrointestinal symptom outcomes in the context of irritable bowel syndrome (IBS), as well as early programming effects.

2. Lutein and zeaxanthin—visual function

Lutein, zeaxanthin, and meso-zeaxanthin are the three major carotenoid-based xanthophylls found in the eye. They comprise the macular pigments and give the macula lutea its yellowish color. Humans are unable to synthesize lutein and zeaxanthin and hence rely on their food supply and/or dietary supplements. Meso-xanthin is a metabolite of lutein and also can be absorbed from the diet [4]. Lutein and zeaxanthin are found in highest amounts in green leafy vegetables, egg yolk, corn, citrus, and other fruits [5]. These two macular pigments are highly concentrated in the retina [6]. While they can also be detected in human serum, their levels here are ~2–3 fold lower vs. levels measured in the retina [7]. This preferential localization and concentration of lutein and zeaxanthin suggests a specific uptake and storage mechanism for these xanthophylls in the visual system and highlights their essential role in retinal function [8].

The localization of lutein and zeaxanthin within the retina and their ability to absorb light near 460 nm allows these carotenoids to filter out high energy blue light, typically within the short wavelength spectrum [9, 10]. As a result, lutein and zeaxanthin limits photochemical damage and simultaneously supports visual performance and increases contrast sensitivity [11–13].

In addition to blue light filtration, these macular pigments serve as effective antioxidants, capable of quenching singlet oxygen and triplet state photosensitizers, inhibiting peroxidation of membrane phospholipids, scavenging reactive oxygen species, and reducing lipofuscin formation [14]. Photoreceptors contain chromophores which are vulnerable to damage through oxidation and macular pigments can limit the compromising effects of lipid peroxidation within the retina by quenching reactive oxygen species [14, 15]. Moreover, long chain polyunsaturated fatty acids, especially docosahexaenoic acid (DHA), are also selectively concentrated in the rod outer segments and given its chemical structure, DHA is highly susceptible to lipid peroxidation and cellular damage. As an antioxidant, lutein can return singlet oxygen to the ground state limiting lipid peroxidation. Lutein auto-regenerates in the process and through this way, may work to be a more efficient quencher of singlet oxygen than other antioxidants such as alpha-tocopherol (vitamin E) [16].

Hence, macular pigments support visual function through multiple ways. The filtration of blue light results in reduced chromatic aberration and subsequently improved visual acuity and contrast sensitivity. Lutein and zeaxanthin also reduce discomfort glare and increase visual acuity, photo-stress recovery time, macular function, and neural processing speed. These are further discussed below.

2.1. Glare discomfort and disability glare

Glare discomfort is characterized by photophobia—a phenomenon that occurs when intense light enters the eye and the recipient experiences discomfort. Photosensitivity is an inherent mechanism to protect the eye from high energy wavelengths [17, 18]. Increased sensitivity to shorter wavelengths of light can trigger retinal damage with less energy compared to other wavelengths. Photophobic response studies have shown that subjects with higher macular pigment levels tolerated light better and have less glare [17]. Additionally, small increases in macular pigment were sufficient to increase photophobia thresholds and lessen visual discomfort [19]. These data support that macular pigment supplementation has a role in reducing discomfort associated with glare.

Bright light settings results in scattered light which subsequently causes decreased visual acuity. This phenomenon is commonly referred to as disability glare. Similar to data generated for glare discomfort, it has also been shown that subjects with higher macular pigment levels maintained acuity better than subjects with lower levels when exposed to both bright white light and short wavelength (blue) light. Additionally, lutein and zeaxanthin supplementation improved glare disability under these conditions [20].

2.2. Photo-stress recovery

The time required to recover vision after exposure to a bright light source is called photo-stress recovery. This visual performance parameter describes the time it takes for bleached photopigments to regenerate and it is affected by macular pigments. Similar to the data generated for glare, individuals with higher macular pigment levels had shorter photo-stress recovery time when tested with intense short wavelength and bright white light sources [21]. The mechanism for this benefit of the macular pigment appears to be related to the reduced photoreceptor exposure to short wavelength light in the foveal and parafoveal regions. Recovery time for the subject with the lowest macular pigment levels was twice as long as subjects with the highest macular pigment levels [22]. Moreover, supplementation with lutein and zeaxanthin significantly decreased photo-stress recovery time [20]. More specifically, supplementation with lutein (10 mg/d) and zeaxanthin at a dose of 2 mg/d over 3 months significantly increased serum levels of lutein and zeaxanthin, macular pigment optical density, and improved chromatic contrast and recovery from photo-stress [20].

2.3. Neural processing

It is not surprising that the brain is frequently referred to as the "window to the world" given the intimate relationship between the optical, neurological and physiological mechanisms underlying vision. In addition to the visual system, macular pigments are present in the brain [23, 24]. A reliable and commonly used proxy for macular pigment levels and hence lutein and zeaxanthin levels is macular pigment optical density (MPOD). MPOD correlates with processing speed and cognitive performance in healthy elderly subjects as well as those with mild cognitive impairment [25–27].

Consistent with data generated for visual function, higher macular pigment levels have been linked to improved critical flicker fusion frequency [28–31], higher concentrations in the visual cortex [53], and improvements in electroretinography responses [32, 33]. Bovier et al. found moderate but statistically significant improvements in both MPOD and cognitive function with lutein and zeaxanthin supplementation of young, healthy individuals considered to be at peak cognitive efficiency [34]. These studies suggest that both young, healthy adults and the elderly population can gain cognitive benefits from lutein and zeaxanthin supplementation.

2.4. Age-related macular degeneration (AMD)

Oxidative stress has been identified as a major contributing factor in the pathogenesis of AMD, a disease that is commonly associated with irreversible blindness in older people [35]. Given the selective localization of lutein and zeaxanthin within the retina and their potency as singlet oxygen scavengers to limit oxidative damage, there has been considerable scientific interest to identify if lutein and zeaxanthin can be used as a therapeutic approach to manage AMD.

Observational studies of dietary intakes of lutein and zeaxanthin, generally suggests that high intakes of these carotenoids in the diet are associated with lower risk of AMD. These studies were conducted globally and over multiple years of supplementation [36–40]. In regards to macular pigment levels and the risk of AMD, Bone et al. demonstrated that subjects with AMD had significantly lower levels of macular pigment and those with the highest quartile of lutein/zeaxanthin had a lower risk of having AMD compared those in the lowest quartile [41]. MPOD is positively correlated with dietary intake of lutein and zeaxanthin [31] and their serum levels [42, 43]. The CAREDS study, a prospective cohort analysis of nearly two thousand postmenopausal women, did not find a correlation between MPOD and AMD [44]. Several but not all trials have supported a lower MPOD in eyes with AMD, and several supplementation trials of AMD subjects have demonstrated reduced MPOD in those subjects not receiving supplementation [45–47].

Supplementation trials with lutein and zeaxanthin and reduced risk of AMD have yielded considerably consistent results compared to most other bioactive/nutrient studies as related to measure of MPOD and/or visual acuity. A meta-analysis performed by Liu et al. compared the results of seven randomized, double-blind, placebo-controlled trials, including the LAST, Weigert et al., Ma et al., CARMIS, LUTEGA, CLEAR, and CARMA studies [13, 47–52]. Out of these studies, four reported an increase in visual acuity with supplementation, and the benefit appeared more pronounced in those subjects with early AMD vs. late AMD. This may be due to a greater loss of macular photoreceptors in the late stage of the disease. A stronger effect was noted for studies using higher doses of supplements. Interestingly, a linear association of MPOD and an increase in visual acuity was also measured [53].

Most recently, the Age-Related Eye Disease 2 Study (AREDS2), a 5-year multicenter, double-blinded, placebo-controlled clinical trial involving 4203 participants with intermediate AMD or large drusen in 1 eye and advanced AMD in the fellow eye was completed. Participants were randomized to one of four groups: placebo, lutein (10 mg) and zeaxanthin (2 mg), omega-3 fatty acids (DHA 350 mg and EPA 650 mg), or a combination of lutein, zeaxanthin, and omega-3 fatty acids. Although the original analysis did not find significant effects from the lutein and zeaxanthin supplementation, a secondary analysis of the effects of xanthophyll supplementation demonstrated reduced AMD progression [54] but did not affect the development of geographic atrophy. Focusing the analyses to eyes with bilateral large drusen at baseline, the comparison of lutein/zeaxanthin vs. β-carotene showed even stronger effects for progression to late AMD and for neovascular AMD.

Collectively, the overall body of evidence supports that structural changes in the retina and improvements in visual acuity can be achieved with lutein and zeaxanthin supplementation. However, additional research is warranted to identify the optimal levels of supplementation in healthy individuals with compromised visual function as well as those with eye disease, e.g., AMD and cataracts, as well the role of early supplementation initiated before disease progression.

2.5. Visual development: role of lutein and zeaxanthin in the prenatal and postnatal periods

Although the placental transfer of carotenoids from mother to child *in utero* has not been directly studied through clinical supplementation trials, there is evidence of the deposition of carotenoids within the eye during the gestational period [14], with ratios of lutein: zeaxanthin: *meso*-zeaxanthin differing from the composition of serum [55]. It is likely that maternal carotenoid status during the gestational period may impact infant macular development, and prenatal supplementation may play a role in maximizing visual development.

Bernstein and colleagues demonstrated an age-dependent increase in MPOD in infants and children, but preterm infants in that cohort did not have measurable MPOD [56]. Interestingly, there were significant correlations between infant MPOD and infant serum zeaxanthin. Additionally, maternal serum zeaxanthin levels correlated with infant MPOD in term infants shortly after birth [57]. This suggests a potential role for maternal nutrition and macular development *in utero* and the opportunity for the mother to increase their lutein and zeaxanthin dietary intakes through food and/or supplements during pregnancy and the breastfeeding period.

The role of lutein in early maturation of the retina is further supported by data from non-human primate studies wherein xanthophyll-free diets resulted in the absence of macular pigmentation, more drusen-like bodies in the retinal pigment epithelial cells (cells that are crucial for nourishment of the retina), increased macular hyperfluorescence, and more retinal abnormalities [58].

Given the promising data on the use of lutein and zeaxanthin to delay macular degeneration, the potential role of these carotenoids in preventing oxidative damage in preterm infants is worthy of further study. There are preliminary data to suggest a potential protective role of carotenoids against oxidative stress during premature life, particularly in cases of retinopathy of prematurity (ROP) [59].

While there is at present no data in humans showing directly that lutein and zeaxanthin influence retinal/visual development, it is highly plausible that these bioactives are important for visual development given their involvement in three key aspects of the visual system: (1) influence input during a critical/sensitive period of visual development and/or (2) influence maturation and/or (3) protection of the retina during a period when it was particularly vulnerable [60].

2.6. Summary

Of all the carotenoids found in nature, only lutein and zeaxanthin are exclusively found in the retina and selectively concentrated in the macula. These macular pigments have been well documented through epidemiological, observational, and intervention studies to play a promising role in visual performance both in healthy individuals and those with macular degeneration. Preliminary data also suggest a relationship between lutein and zeaxanthin and visual development in infancy. Dietary intakes of lutein and zeaxanthin are dismally low among Americans with most adults and children not consuming intakes clinically demonstrated to be protective for eye health. Strategies need to be identified to increase dietary intake of these relevant bioactive nutrients and create awareness on their essentiality to the health of humans.

3. Probiotics, prebiotics and gastrointestinal microbiome

The gastrointestinal tract is best known for its role in the digestion of food and absorption of nutrients. It has the largest surface area in the body—it is ~9 m in length with a surface area of ~250–400 m^2, comparable to the size of a tennis court. It hosts a variety of immune cells making it the largest immunological organ in the body and equally interestingly, it contains a similar number of neurons as that found in the spinal cord—so in other words, the gastrointestinal tract has its own nervous system, the enteric nervous system. For this reason, the gastrointestinal tract is frequently referred to as the body's second brain. Additionally, the gastrointestinal tract also houses the greatest number and variety of bacteria in the body. There are 10×s as many bacteria in the gastrointestinal tract (GIT) as there are cells in the body. These bacteria have the unique ability to interact and communicate with the immune cells, intestinal cells, and the neurons in the body to influence digestive health, immune health and overall well-being. Certain lifestyle and environmental factors can influence the balance of the friendly vs. unfriendly bacteria in the gastrointestinal tract including diet, age, medication, stress, travel, and sleep.

At birth, the human gastrointestinal tract is relatively sterile but becomes rapidly colonized with a diverse microbial population comprising tens of trillions of bacteria and hundreds of different species by 3–5 years of age. The density and diversity increases exponentially from the stomach to the colon, where the microbial content is at its highest concentration. Although the phyla *Firmicutes* and *Bacteroidetes* dominate the human gut microbiota, it contains a core microbiome with shared functionality and shared mechanisms of action [61].

The abundance and diversity of the microbiota suggest an important physiological role for this "organ" within the gastrointestinal tract. Herein, this dynamic ecosystem facilitates multiple functions including the digestion of complex carbohydrates; shaping the immune system and modulating immune responses; contributing to the defense against pathogens by the mechanism of colonization resistance and fermentation of non-digestible carbohydrates. They produce metabolic products including short chain fatty acids such as acetate, propionate, and butyrate. These metabolites serve as a major energy source for intestinal epithelial cells wherein they can influence cell proliferation and differentiation, mucus secretion, intestinal motility, and barrier function; and may also exert anti-inflammatory and antioxidative activity [62].

A shift from a stable intestinal environment occurs when the gut microbiota community is temporarily or permanently altered and is termed "dysbiosis." Factors that may lead to dysbiosis include antibiotics, diet, host immune system, inflammation, and infectious gastroenteritis. Dysbiosis is observed in several gastrointestinal disorders including IBS, and Crohn's disease, and may play a key role in their pathogenesis and possibly in management. Some of the common features of dysbiosis in IBS and Crohn's disease are reduced microbial diversity, lower bifidobacteria, lower bacteroidetes to firmicutes ratio in IBS and in Crohn's disease, and decreased *Faecalibacterium prausnitzii* [63].

Current research efforts have focused on two main approaches to supporting and promoting the stability and diversity of the microbial community within the GIT: (1) offering specific substrates for fermentation by the colonic bacteria (prebiotics); and/or (2) introducing specific bacterial species or strains to the colonic microbiota (probiotics).

Probiotics and prebiotics have been evaluated in a number of clinical trials involving individuals at different stages of the lifecycle, including pregnancy, infants, children and adults and under different health conditions, e.g., infectious diarrhea, antibiotic-associated diarrhea, therapy and prevention of *Clostridium difficile* and other infections, inflammatory bowel disease, IBS, atopic dermatitis and allergic immune outcomes.

In this section, we review the role of probiotic and prebiotics in the context of gastrointestinal health, with a particular focus on IBS. IBS is a chronic functional disorder of the gastrointestinal system. Individuals experience abdominal pain and altered bowel habit, with either predominantly diarrhea (IBS-D), constipation (IBS-C), or both (IBS-M). It has an insidious onset, and frequently does not result in medical care. Irrespective of geography, IBS is a significant health care burden affecting around 11% of the population globally [64]. Recent studies suggest IBS may comprise ~20% of gastroenterology outpatient visits, and thus these statistics highlights the importance of identifying effective therapies to manage their symptoms and improve their quality of life.

Additionally, the role of prebiotics in influencing the gut microbiome composition and activity in early life and the subsequent long term benefits thereof will also be highlighted in this section.

3.1. Probiotics

The term "probiotic" as originally defined by FAO/WHO refers to "live microorganisms that, when administered in adequate amounts, confer a health benefit on the host" [65]. However, in order to be beneficial, probiotic bacteria must be able to survive along the gastrointestinal tract, to resist gastric acid, bile and pancreatic juice action and to demonstrate functional efficacy [66].

A meta-analysis involving 18 randomized-controlled trials including 1650 patients with IBS was conducted by Moayyedi et al. [67]. Although the review reported considerable heterogeneity among the studies, the analysis reported a preference toward probiotic treatment with statistically significant improvement of individual symptoms such as pain, flatulence and bloating. No side effects were reported and there was no significant differences detected between the various types of probiotics used in the studies, with three studies using *Lactobacillus* (n = 140 subjects), two trials using *Bifidobacterium* (n = 422 subjects), one trial using *Streptococcus* (n = 54 subjects), and four trials using a combination of probiotics (n = 319 subjects). The favorable safety profile reported in this meta-analysis are consistent with the findings of Hungin et al. who also showed several positive effects of probiotics on IBS symptoms and health-related quality of life measures. Their analysis involved 19 studies and 1807 patients [68].

Similar to the findings of Moayyedi et al., Clarke and coworkers also showed that despite significant studies heterogeneity in their analysis of 42 randomized-controlled trials, 34 studies reported beneficial effects on at least one pre-specified endpoint including improvement in abdominal pain/discomfort, improvement on abdominal bloating/distension compared to placebo [69]. Both *Bifidobacteria* and *Lactobacilli* were found effective in ameliorating IBS symptoms, while the beneficial effects of the multispecies lactic acid bacteria preparations, including the multi-strain preparation VSL#3, were less pronounced.

Another systematic review with meta-analysis has been recently published by Didari et al. which focused on a review of 15 studies involving 882 patients with IBS. Not surprising, significant study heterogeneity was observed given differences in the types of bacterial strains used, probiotic dosage, duration of either treatment or follow-up and endpoints/outcome. However, consistent with the other systematic reviews, probiotics were more effective than placebo in reducing abdominal pain after 8 and 10 weeks of treatment. Few adverse events were reported in both probiotics and placebo groups and this meta-analysis reconfirmed the safety profile of probiotic use [70].

3.1.1. Mechanisms of action of probiotics

Probiotics appear to exert their beneficial effects on gastrointestinal healthy through three general mechanisms: antimicrobial effects, mucosal barrier integrity, and immune modulation. Moreover, the important benefits of probiotics is based on their ability to metabolize complex carbohydrates and produce lactic acid and SCFAs such as butyrate [58, 59]. In the context of IBS, there is ample evidence to support the role of probiotics in managing the symptoms of IBS through positive changes in the composition and functionality of the intestinal bacteria, correcting intestinal motility, limiting visceral hypersensitivity, modulating immune responses and benefiting the gut-brain axis [71]. Indeed, more studies need to be conducted to further unravel the mechanisms through which probiotics beneficially influence the symptoms of IBS and thereby further enhanced focused and specific probiotic therapeutic modalities that can also be "personalized" based on an individual's needs.

3.2. Prebiotics

Prebiotics are selectively fermented ingredients that result in specific changes in the composition and/or activity of the gastrointestinal microbiota, thus conferring a benefit on the host. In order for a compound to be classified as a prebiotic, it has to fulfill three criteria: i] resistant to gastric acidity and hydrolysis by mammalian enzymes and gastrointestinal absorption; ii] can be fermented by intestinal microbiota; iii] selectively stimulates the growth and/or activity of the intestinal bacteria associated with health and wellbeing [72].

These non-digestible oligosaccharides, such as fructooligosaccharides (FOS), galactooligosaccharides (GOS), lactulose, and inulin, stimulate and nourish the growth of selective and beneficial gut bacteria, particularly lactobacilli and bifidobacteria [73]. Prebiotics have been clinically tested in a variety of settings for multiple health benefits, including improvement of intestinal function as measured by stool bulking, stool regularity, stool consistency, glucose and lipid metabolism, immune health including allergic outcomes, satiety and appetite regulation, and stimulation of mineral absorption and improvement of bone density. The majority of the studies have focused on inulin and FOS, whereby studies have consistently shown a benefit for overall digestive health, including an increase in the total bacterial mass, growth of beneficial bacteria, reduction in pathogenic bacteria, and production of numerous beneficial bacterial metabolites.

The proceeding paragraphs will highlight two areas of emerging evidence: (1) role of prebiotics in IBS and (2) programming effect of prebiotics when supplemented during the first 1000 days of life.

3.2.1. Prebiotics and IBS

Since IBS is generally categorized by an imbalance of bacteria, the mechanisms through which prebiotics work suggest that they could potentially be used as a therapy, either alone, or in combination with probiotics, to manage IBS and its related symptoms. To date however, there have only been a handful of randomized control trials investigating the effect of prebiotics on IBS. As summarized in the literature, two studies in adults with IBS at doses of 6 g/d of oligofructose and 20 g/d of inulin showed no improvement in symptom or stool output measures. Another trial showed an improvement in composite symptom score with 5 g/d of short-chain FOS in the per-protocol population, but this was not analyzed intention to treat, with a high non-compliance rate and only 50/105 being included in the per protocol analysis. Separately, a 12-week parallel cross-over trial, which used a β-GOS, showed a dose-dependent stimulation of bifidobacteria at 3.5 and 7.0 g/d. Global symptom relief scores were significantly improved in the prebiotic group vs. the placebo, including for flatulence, bloating, and stool consistency [74, 75].

These preliminary data suggest that prebiotics may offer promise as a therapeutic option in the dietary management of IBS but more studies certainly need to be conducted to confirm the benefit of prebiotics for this population, including the optimal type and dose. These factors need to be first addressed prior to prebiotics being considered as therapy option in individuals with IBS.

3.2.2. *Gut microbiome in the first 1000 days and the "programming" effects of prebiotics*

The first 1000 days of a child's life is now well recognized as a critical timeframe for health into adulthood, wherein nutrition plays a key role. Additionally, a robust link between nutrition and gut microbiota composition with health outcomes has been documented. It is intriguing to consider that events early in life may determine the activity of our gut microbiota for the rest of our life. It is equally fascinating that the gut microbiota in early life can determine our risk of later life heath outcomes.

Colonization of the infant gut contributes to the intestinal homeostasis and mucosal barrier function, that both are essential for our health, at the start of life and apparently also in adulthood. In this regard several studies have demonstrated that the mode of delivery affects the composition of the newborn's microbiota wherein caesarean section birth is associated with a lower total microbial diversity and delayed colonization. Other factors influencing this composition include infant hospitalization and antibiotic use, antibiotic use in the pregnant mother, solid-feeding practices and day care attendance. Alterations of the development of the gut microflora during infancy has been linked to altered immune system development and thus increased risk of allergic immune outcomes, as well as altered metabolic profiles and increased obesity risk [76].

A new exciting development is the role of the gut microbiome as an epigenetic regulator wherein sequencing of DNA methylomes of pregnant women revealed an association between bacterial predominance and epigenetic profiles. Epigenetics comprise genomic modifications that occur due to environmental factors and do not change the nucleotide sequence. In the context of cardiovascular disease and obesity, different methylation status of gene promoters have been correlated with specific gut microbiota signatures, with either *Firmicutes* or *Bacteroidetes* represented as a dominant group. These observations parallel previous studies linking higher levels of *Firmicutes* to obesity. Additionally, an elegant study by Paul HA and colleagues showed that consumption of prebiotics during pregnancy and lactation improves metabolism in diet-induced obese rats and limits the detrimental nutritional programming of offspring associated with maternal obesity. More specifically, there was a reduction in gestational weight gain, increased circulating concentrations of satiety hormones and abundance of *Bifidobacterium* spp. in the gut. These effects were accompanied by an attenuation of increased adiposity in both dams and offspring at weaning [77].

Over the past decade, studies have investigated the effect of specific mixtures of prebiotics, for example short chain GOS + long chain FOS, on the composition of the intestinal microbiota in preterm, term, and weaning infants and have consistently shown that prebiotic supplementation influences early microbial pattern similar that of human milk with an intestinal microbiota dominated by Bifidobacterium and Lactobacillus [78–82].

Studies have also shown that changes in early-life microbial composition by such prebiotics parallels metabolic production of the microbiota, including increased short-chain fatty acid production, lactate and a reduced pH [83, 84]. These favorable metabolic changes induced by prebiotics have been associated with increased colonization resistance to pathogens and this characteristic is supported by in-vitro data [85]. Moreover, the modulation of early-life microbiota by prebiotics correlates with improved immune system maturation. More specifically dietary supplementation with short chain GOS + long chain FOS has been positively associated with increased production of secretory IgA. Additionally, there are preclinical data supporting the role of such prebiotics in modulating systemic immune responses through direct binding of specific receptors on immune cells and/or through short-chain fatty acid production [86, 87].

Given the accumulating evidence supporting the association between the infant's gut microbiota composition and health in later life, the potential for gut microbe-based modulation including prebiotics, may be a promising approach to improve health during prenatal life, infancy, childhood and thus, later life outcomes.

4. Bioactive foods and the regulatory environment

The functional food components discussed in this chapter can be commercialized under several of the FDA categories that researchers and manufacturers need to consider carefully prior to launch. FDA's authority to regulate a product as a food, supplement, device, or a drug, depend on the product presentation, intended uses, target population, and claims they make about their product. This "intended use" criterion also defines the materials that can be used in the formulation of the product. Together, these dictate the appropriate regulations applicable and regulatory agencies responsible for regulating them. Most important among the claims is whether the product is intended to be used to diagnose, cure, mitigate, treat, or prevent a disease. Although the intended uses of Drugs and Devices may also be applicable, only the dietary regulations are covered in this subsection given the focus on nutritional bioactives.

The FDA regulates claims in four categories [88]:

- Nutrient Content Claims: characterize the amount of nutrients present in the product,
- Health Claims: describe a relation between a nutrient and a disease based on Significant Scientific Agreement (SSA),
- Qualified Health Claims: provide for health claims based on less scientific evidence than SSA standard as long as the claims do not mislead the consumers, and
- Structure/Function Claims: relate the role of a nutrient to the normal structure or function in humans and do not make reference to a disease.

The food labels and messaging are controlled through several federal regulations and agencies such as the Federal Food, Drug, and Cosmetic Act (FFDCA), the Nutrition Labeling and Education Act (NLEA), the FDA, and the Federal Trade Commission (FTC), the false advertising litigations permitted under state laws and section 43(a) of the Lanham Act, and the consumer protection laws in general [89]. The National Advertising Division (NAD) of the Council of Better Business Bureaus, Inc. (CBBB) is another active player in regulating and shaping the food industry communication, including the dietary supplements category, where most of these products are today placed. The NAD is an industry-funded body that reviews nationally disseminated advertising for truth and accuracy [90].

4.1. Dietary supplement

The Dietary Supplement Health and Education Act (DSHEA) of 1994 a "dietary supplement is a product intended for ingestion that contains a "dietary ingredient" intended to add further nutritional value to (supplement) the diet. A "dietary ingredient" may be one, or any combination, of the following substances:

- A vitamin
- A mineral
- An herb or other botanical
- An amino acid
- A dietary substance for use by people to supplement the diet by increasing the total dietary intake
- A concentrate, metabolite, constituent, or extract

Dietary supplements may be found in many forms such as tablets, capsules, softgels, gelcaps, liquids, or powders. Some dietary supplements can help ensure that you get an adequate dietary intake of essential nutrients; others may help you reduce your risk of disease." [91]

Ingredients used in dietary supplements must either demonstrate evidence of use prior to 1994, or that they were used in food in the present form.

4.2. Conventional food

Congress passed the FFDCA in 1938, which grants the FDA the power to ensure that "foods are safe, wholesome, sanitary, and properly labeled." Section 201(f) of the FD&C Act (21 U.S.C. 321(f)) defines a food as "(1) articles used for food or drink for man or other animals, (2) chewing gum, and (3) articles used for components of any such article" and a drug to include "articles (other than food) intended to affect the structure or any function of the body" and "intended for use in the diagnosis, cure, mitigation, treatment, or prevention of disease."

4.3. Food for special dietary uses (FSDU)

FSDU are defined as food "(i) used for supplying particular dietary needs which exist by reason of a physical, physiological, pathological or other condition, including but not limited to the conditions of diseases, convalescence, pregnancy, lactation, allergic hypersensitivity to food, underweight, and overweight; (ii) uses for supplying particular dietary needs which exist by reason of age, including but not limited to the ages of infancy and childhood; (iii) uses for supplementing or fortifying the ordinary or usual diet with any vitamin, mineral, or other dietary property. Any such particular use of a food is a special dietary use, regardless of whether such food also purports to be or is represented for general use" [92].

4.4. Medical food

In 1988, with the Orphan Drug Act Amendments, Congress recognized the need to encourage development of "medical foods" for the management of disease and health conditions and defined "medical food" as "food which is formulated to be consumed or administered enterally under the supervision of a physician and which is intended for the specific dietary management of a disease or condition for which distinctive nutritional requirements, based on recognized scientific principles, are established by medical evaluation." (Orphan Drug Act –1988; 21 U.S.C. §360ee(b)(3), 5(b); FFDCA §528)

The FDA later clarified this definition into the five criteria used to define medical food listed below [92]:

(i) it is a specially formulated and processed product (as opposed to a naturally occurring foodstuff used in its natural state) for the partial or exclusive feeding of a patient by means of oral intake or enteral feeding by tube;

(ii) it is intended for the dietary management of a patient who, because of therapeutic or chronic medical needs, has limited or impaired capacity to ingest, digest, absorb, or metabolize ordinary foodstuffs or certain nutrients, or who has other special medically determined nutrient requirements, the dietary management of which cannot be achieved by the modification of the normal diet alone;

(iii) it provides nutritional support specifically modified for the management of the unique nutrient needs that result from the specific disease or condition, as determined by medical evaluation;

(iv) it is intended to be used under medical supervision; and

(v) it is intended only for a patient receiving active and ongoing medical supervision wherein the patient requires medical care on a recurring basis for, among other things, instructions on the use of the medical food.

All ingredients used in conventional foods, FSDU, or Medical Foods, must be either Generally Recognized as Safe (GRAS) or pre-approved by the FDA as additives. Further, conventional and FSDU must conform to all nutrient and health claims provisions in 21 CFR Subpart A 101.13 and 101.14 along with the specific requirements for the claims in 21 CFR subpart D (Specific Requirements for Nutrient Content Claims) and subpart E (Specific Requirements for Health Claims). Medical foods must follow many of the same labeling requirements of conventional foods except only Medical Food is exempt from the nutritional and health claims labeling of food [92].

Among the above categories, none is most strife for abuse as the Medical Foods category, primarily because of its claims exemption requirements. This it is also the most controlled as the "FDA considers the statutory definition of medical foods to narrowly constrain" to the definition [93]. The FDA has consistently applied a standard that food are "articles consumed primarily for taste, aroma, or nutritive value" but "used as a drug for some other physiological effect" [94].

The two most important hurdles to overcome in meeting the regulatory requirements for Medical Food are [95]:

(i) Distinctive Nutritional Requirements, and

(ii) Cannot be achieved by the Modification of the Diet Alone.

Nutrient intake requirements assessed by Institute of Medicine (now National Academy of Medicine) are based on population estimates of estimated average requirements. What is interesting to note is that these are based on estimated average intake levels correlated with measure of inadequacy (on the lower end) and risk of adverse events (on a higher end) [96]. The bio-functional molecules covered in this category are currently considered by regulatory agencies as non-essential and therefore ineligible for dietary reference intakes (DRI) estimates. The "nutritive value" today is not a function of DRI, estimated average requirements (EAR) or daily values (DV), but a complex biochemical function derived through genome, epigenetics, nutrigenomics, and the microbiome. In the roundtable and workshop on obesity report by the National Academies of Science [97], the early origins of obesity can be traced to metabolic programming that starts pre-conception and defines the individual's predisposition for a nutrient uptake and metabolism.

Simple biochemical statistics suggest that uptake or utilization of cellular molecules, including nutrients, is multimodal kinetics. For illustration purposes only, the Distinct Nutritional Requirements for an individual can be depicted using a sigmoidal curve nutrient uptake and utilization by the following model (**Figure 1**). In any individual, the individual's diet,

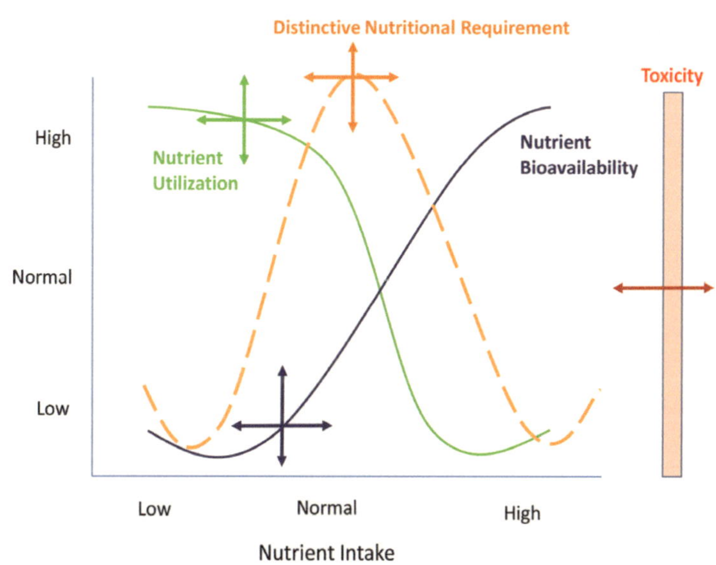

Figure 1. Distinct nutritional requirements for an individual can be depicted using a sigmoidal curve nutrient uptake and utilization of the multimodal kinetics model.

metabolic rate, or any situation-specific external and internal requirements, will dictate the bioavailability curve that in turn will drive the nutrient requirements. Simply because the nutrient shares the cellular network with a pharmacological function should not dictate its classification as a drug. Preponderance of evidence now suggest that a modified view of nutrients as bio-functional components are mandatory and the old way of nutrient intake should give way to the new scientific knowledge. Current FDA regulations that surround definition of Food and Dietary Supplements, not only do not provide consideration for individualistic or disease specific intake (or utilization) of the nutrient, but specifically prohibit such interpretation. The FDA has to acknowledge the complex role of a nutrient in the health and wellbeing of an individual is simply not relatable to the DRI.

In any individual, it is the nutrient bioavailability curve that will dictate the biochemical utilization of that nutrient, all conditions deemed equal. However, depending on the individual's diet, metabolic rate, genetics, epigenetics, or any situation-specific requirements, the availability curve can be right or left shifted. So also, it is not unusual that the nutrient utilization for its cellular or metabolic function can similarly be right or left shifted, again depending on their own cellular availability of nutrient requirements, metabolites concentration, or any other external or internal factors. Rapid net catabolism of body protein occurring in major trauma, burns and sepsis patients have a higher resting energy and protein requirements. In these examples, the utilization curve is right-shifted along the X-axis and without proper balancing of the nutrient bioavailability (and thus intake) curve also right shifted, patients would not recover well. Thus, both the availability and utilization functions can be moved along the Y-axis where a minimum threshold need to be met before the nutrient is available for its cellular functions. Thus, the "distinctive nutrition requirements" for that individual can be the equilibrium function of the two biochemical curves that can also move along the two axes depending on the "conditional" needs of that individual. Similarly, the toxicity function can slide along the X-axis depending on the individual's needs. For example, Lofenalac was specifically formulated for patients with phenylketonuria (PKU) unable to adequately metabolize phenylalanine and is considered FSDU. In this case, the toxicity function would be to the extreme left. This would also be true of other allergic diseases for nutrients where depending on the individual's tolerance to that allergen, the toxicity function can be anywhere along the X-axis. Since the mechanism of action for most nutrients share the same cellular pathway as their pharmacological counterparts, just because the nutrient takes part in that pathway is not a sufficient criteria to qualify as a drug. Hydrolyzed protein epitopes of an allergen are a classic example. Maternal consumption of peanut during pregnancy reduced peanut allergy sensitization in infants born to these mothers [99]. Similarly, maternal serum zeaxanthin levels correlated with infant MPOD in term infants shortly after birth [57] thereby providing opportunity for maternal diet supplementation during pregnancy and the breastfeeding period. Neither of these examples fit the static EAR definition and yet meet all the classic requirements for a nutrient and their role in disease without being a drug. This dynamic nature of the nutrient are necessary and sufficient conditions for "distinctive nutritional requirements."

In the context of bioactives for human health support, the FDA has to take a holistic view of human health where food, drug, and supplements, and alternative therapies, all have a role. The healthcare costs are sky-rocketing and nutrition has an established role in "preventing, treating, mitigating, or curing a disease" and they are not drugs. Manufacturers should be able to provide functional nutrients to consumers provided the claims are well substantiated. There are obvious challenges to impose drug clinical study design on the substantiation of dietary ingredients since they are not a single chemical entity or are easily achievable by a double blind placebo controlled multi-center trials like drugs [98]. However, a reasonable study design to measure clinical outcomes is still necessary and the FDA should exercise regulatory responsibility to provide the necessary framework that takes into consideration the developing science and the practical limits of the diet. This will considerably help the conscientious industry players as well as to control mavericks trying to circumvent the food category utilization for product placement and claims. When it comes to policy making, Nutrition, Diet, or Food manufacturers are conspicuously absent from the stakeholder list. The Nutrition industry is a necessary partner in the healthcare discussion [100].

5. Concluding remarks: where do we go from here?

There is increasing interest by consumers, researchers, and regulators into the roles that certain bioactive compounds, such as lutein, zeaxanthin, prebiotics and probiotics, can play in health maintenance and promotion, as well as potentially programming health outcomes starting in early life. The state of the science for these bioactives and their benefits to health and wellbeing appear to be sufficiently mature to bring together key stakeholders including policymakers, regulators, and toxicologists, to initiate dialogue on advancing the process for establishing recommended intakes and its communication to the public.

These collaborative dialogues will need to address difficult and controversial questions, e.g., (1) what constitutes sufficient evidence and do we have to adopt an evidence-based medicine model focusing only on randomized controlled trials; (2) should the evidence focus on demonstrating the bioactive is health-promoting or are the study participants performing better than baseline?; (3) availability of reliable and validated biomarkers for both exposure and effect and their relation to health outcomes, especially in vulnerable populations, e.g., pregnancy, postnatal, and early childhood; (4) limited databases for bioactives such as lutein and prebiotics—without such databases in place, intakes of these compounds by groups and populations cannot be evaluated in food consumption surveys; (5) methods to standardize and measure bioactive components.

From a regulatory perspective, the FDA should take a holistic view on the process to regulate and hence commercialize nutritional bioactives. The current health care environment where consumers are taking more ownership for their healthcare starts with laying the foundation where consumers are encouraged to find accurate and reliable information from both the industry and the government. A proper regulatory framework that allows for nutritional benefits of the dietary ingredients and their role in diseases to be conveyed to the consumers will be beneficial for society at large.

Moreover, the process for communicating the science on bioactives to the public including healthcare professionals certainly lags behind commercialization. The accumulating and promising scientific evidence for lutein, zeaxanthin and pro- and prebiotics, warrants guidance and alignment from key stakeholders on approaches to help educate and communicate the benefits of these bioactives in a manner that is science-based, meaningful, accurate and not misleading. Platforms such continuing medical education programs, webinars, and conference proceedings can be leveraged to disseminate scientific information. This will empower consumers to leverage these self-care strategies in a responsible and compliant manner.

Author details

Deshanie Rai[1]* and Gyan Rai[2]

*Address all correspondence to: deshanierai@gmail.com

1 Tufts U, Mountain Lakes, NJ, USA

2 IU School of Medicine, Mountain Lakes, NJ, USA

References

[1] NIH, Office of Dietary Supplements. Federal Register. Vol. 69 No. 179 FR Dec 04-20892, Sept. 16, 2004

[2] Kris-Etherton PM, Hecker KD, Bonanome A, Coval SM. Bioactive compounds in foods: Their role in the prevention of cardiovascular disease and cancer. The American Journal of Medicine. 2002;**113**:71S-88S

[3] Weaver CM. Bioactive foods and ingredients for health. Advances in Nutrition. 2014;**5**: 306S-311S

[4] Rasmussen HM, Muzhingi T, Eggert EMR, Johnson JE. Lutein, zeaxanthin, meso-zeaxanthin content in egg yolk and their absence in fish and seafood. Journal of Food Composition and Analysis. 2012;**27**:139-144

[5] Sommerburg O, Keunen JEE, Bird AC, van Kuijk FJGM. Fruits and vegetables that are sources for lutein and zeaxanthin: the macular pigment in human eyes. British Journal of Ophthalmology. 1998;**82**:907-910

[6] Bernstein S, Khachik F, Carvalho LS, Muir DJ. Identification and quantitation of carotenoids and their metabolites in the tissues of the human eye. Experimental Eye Research. 2001;**72**:215-223

[7] Handelman GJ, Snodderly DM, Adler AJ, Russett MD. Measurement of carotenoids in human and monkey retinas. Methods in Enzymology. 1992;**213**:220-230

[8] Granado F, Olmedilla B, Blanco I. Nutritional and clinical relevance of lutein in human health. British Journal of Nutrition 2003;**90**:487-502

[9] Snodderly DM, Auran JD, Delori FC. The macular pigment. II. Spatial distribution in primate retinas. Investigative Ophthalmology and Visual Science. 1984;**25**:674-685

[10] Ham WT Jr., Mueller HA, and Sliney DH. Retinal sensitivity to damage from short wavelength light. Nature. 1976;**260**(5547):153-155

[11] Kim CBY, Mayer MJ. Foveal flicker sensitivity in healthy aging eyes. II. Cross-sectional aging trends from 18 through 77 years of age. Journal of Optical Society of America A: Optics and Image Science, and Vision. 1994;**11**:1958-1969

[12] Renzi LM, Hammond BR. The effect of macular pigment on heterochromatic luminance contrast. Experimental Eye Research. 2010;**91**:896-900

[13] Richer S, Stiles W, Statkute L. Double-masked, placebo-controlled, randomized trial of lutein and antioxidant supplementation in the intervention of atrophic age-related macular degeneration: The Veterans LAST study (Lutein Antioxidant Supplementation Trial). Optometry. 2004;**75**:216-230

[14] Bone RA, Landrum JT, Fernandez L, Tarsis SL. Analysis of the macular pigment by HPLC: Retinal distribution and age study. Investigative Ophthalmology and Visual Science. 1988;**29**:843-849

[15] Beatty S, Koh HH, Phil M, Henson D. The role of oxidative stress in the pathogenesis of age-related macular degeneration. Survey of Ophthalmology. 2000;**45**:115-134

[16] Fukuzawa K, Inokami Y, Tokumura A, Terao J. Rate constants for quenching singlet oxygen and activities for inhibiting lipid peroxidation of carotenoids and α-tocopherol in liposomes. Lipids. 1998;**33**:751-756

[17] Stringham JM, Fuld K, Wenzel AJ. Action spectrum for photophobia. Journal of the Optical Society of America A: Optics, Image Science Vision. 2003;**20**:1852-1858

[18] Stringham JM, Fuld K, Wenzel AJ. Spatial properties of photophobia. Investigative Opthalmology and Visual Science. 2004;**45**: 3838-3848

[19] Wenzel AJ, Fuld K, Stringham JM, Curran-Celentano J. Macular pigment optical density and photophobia light threshold. Vision Research. 2006;**46**:4615-4622

[20] Stringham JM, Garcia PV, Smith PA, McLin LN. Macular pigment and visual performance in glare: Benefits for photostress recovery, disability glare, and visual discomfort. Investigative Ophthalmology and Visual Science. 2011;**52**:7406-7415

[21] Stringham JM, Hammond BR Jr. The glare hypothesis of macular pigment function. Optometry and Vision Science. 2007;**84**:859-864

[22] Stringham JM, Hammond BR. Macular pigment and visual performance under glare conditions. Optometry and Vision Science. 2008;**85**: 82-88

[23] Johnson JE, Stringham JM, Hammond BR Jr, Yeum K-J. Relation among serum and tissue concentrations of lutein and zeaxanthin and macular pigment density. The American Journal of Clinical Nutrition. 2000;**71**:1555-1562

[24] Vishwanathan R, Iannaccone A, Scott TM. Macular pigment optical density is related to cognitive function in older people. Age and Ageing. 2014;**43**:271-275

[25] Feeney J, Finucane C, Savva GM. Low macular pigment optical density is associated with lower cognitive performance in a large, population-based sample of older adults. Neurobiology of Aging. 2013;**34**:2449-2456

[26] Johnson EJ, McDonald K, Caldarella SM, Chung H-Y. Cognitive findings of an exploratory trial of docosahexaenoic acid and lutein supplementation in older women. Nutritional Neurosciene. 2008;**11**:75-83

[27] Renzi LM, Dengler MJ, Puente A, Miller LS. Relationships between macular pigment optical density and cognitive function in unimpaired and mildly cognitively impaired older adults. Neurobiology of Aging. 2014;**35**:1695-1699

[28] Hammond BR Jr, Wooten BR. CFF thresholds: Relation to macular pigment optical density. Ophthalmic and Physiological Optics. 2005;**25**:315-319

[29] Brown ED, Micozzi MS, Craft NE. Plasma carotenoids in normal men after a single ingestion of vegetables or purified β-carotene. The American Journal of Clinical Nutrition. 1989;**49**:1258-1265

[30] Chasan-Taber L, Willett WC, Seddon JM. A prospective study of carotenoid and vitamin A intakes and risk of cataract extraction in US women. The American Journal of Clinical Nutrition. 1999;**70**:509-516

[31] Hammond BR Jr, Wooten BR, Snodderly DM. Density of the human crystalline lens is related to the macular pigment carotenoids, lutein and zeaxanthin. Optometry and Vision Science. 1997;**74**:499-504

[32] Berrow EJ, Bartlett HE, Eperjesi F, Gibson JM. The effects of a lutein-based supplement on objective and subjective measures of retinal and visual function in eyes with age-related maculopathy—A randomised controlled trial. British Journal of Nutrition. 2013;**109**:2008-2014

[33] Ma L, Dou H-L, Huang Y-M. Improvement of retinal function in early age-related macular degeneration after Lutein and zeaxanthin supplementation: A randomized, double-masked, placebo-controlled trial. American Journal of Ophthalmology. 2012;**154**:625.e1-634.e1

[34] Bovier ER, Renzi LM, Hammond BR. A double-blind, placebo-controlled study on the effects of lutein and zeaxanthin on neural processing speed and efficiency. PLoS One. 2014;**9**:e108178

[35] Congdon N, O'Colmain B, Klaver CC. Causes and prevalence of visual impairment among adults in the United States. Archives of Ophthalmology. 2004;**122**: 477-485

[36] Ma L,. Dou H-L, Wu Y-Q. Lutein and zeaxanthin intake and the risk of age-related macular degeneration: A systematic review and meta-analysis. British Journal of Nutrition. 2012;**107**:350-359

[37] Seddon JM, Ajani UA,. Sperduto RD. Dietary carotenoids, vitamins A, C, and E, and advanced age-related macular degeneration. JAMA. 1994;**272**:1413-1420

[38] Tan JSL, Wang JJ, Flood V, Rochtchina E. Dietary antioxidants and the long-term incidence of age-related macular degeneration: The Blue Mountains Eye study. Ophthalmology. 2008;**115**:334-341

[39] SanGiovanni JP, Chew EY,Clemons TE. The relationship of dietary carotenoid and vitamin A, E, and C intake with age-related macular degeneration in a case-control study: AREDS report No. 22. Archives of Ophthalmology. 2007;**125**:1225-1232

[40] Ho L, van Leeuwen R, Witteman JCM. Reducing the genetic risk of age-related macular degeneration with dietary antioxidants, zinc, and ω-3 fatty acids: The Rotterdam study. Archives of Ophthalmology. 2011;**129**: 758-766

[41] Bone RA, Landrum JT, Tarsis SL. Preliminary identification of the human macular pigment. Vision Research. 1985;**25**;1531-1535

[42] Hammond BR Jr, Curran-Celentano J, Judd S. Sex differences in macular pigment optical density: Relation to plasma carotenoid concentrations and dietary patterns. Vision Research. 1996;**36**:2001-2012

[43] Nolan JM, Kenny R, O'Regan C. Macular pigment optical density in an ageing Irish population: The Irish Longitudinal study on ageing. Ophthalmic Research. 2010;**44**:131-139

[44] LaRowe TL, Mares JA, Snodderly DM, Klein ML. Macular pigment density and age-related maculopathy in the carotenoids in age-related eye disease study. An ancillary study of the women's health initiative. Ophthalmology. 2008;**115**:876-883

[45] Beatty S, Murray IJ, Henson DB, Carden D. Macular pigment and risk for age-related macular degeneration in subjects from a northern European population. Investigative Ophthalmology and Visual Science. 2001;**42**:439-446

[46] Bernstein PS, Zhao D-Y, Wintch SW, Ermakov IV. Resonance Raman measurement of macular carotenoids in normal subjects and in age-related macular degeneration patients. Ophthalmology. 2002;**109**:1780-1787

[47] Beatty S, Chakravarthy U, Nolan J. Secondary outcomes in a clinical trial of carotenoids with coantioxidants versus placebo in early age-related macular degeneration. Ophthalmology. 2013;**120**: 600-606

[48] Dawczynski J, Jentsch S, Schweitzer D, Hammer M. Long term effects of lutein, zeaxanthin and omega-3-LCPUFAs supplementation on optical density of macular pigment in AMD patients: The LUTEGA study. Graefe's Archive for Clinical and Experimental Ophthalmology. 2013;**251**: 2711-2723

[49] Ma L, Yan S-F, Huang Y-M. Effect of lutein and zeaxanthin on macular pigment and visual function in patients with early age-related macular degeneration. Ophthalmology. 2012;**119**:2290-2297

[50] Murray IJ, Makridaki M, van der Veen RLP, Carden D. Lutein supplementation over a one-year period in early AMD might have a mild beneficial effect on visual acuity: The CLEAR study. Investigative Ophthalmology and Visual Science. 2013;**54**:1781-1788

[51] Piermarocchi S, Saviano S, Parisi V. Carotenoids in age-related maculopathy Italian study (CARMIS): Two-year results of a randomized study. European Journal of Opthalmology. 2012;**22**: 216-225

[52] Weigert G, Kaya S, Pemp B. Effects of lutein supplementation on macular pigment optical density and visual acuity in patients with age-related macular degeneration. Investigative Ophthalmology and Visual Science. 2011;**52**: 8174-8178

[53] Liu R, Wang T, Zhang B. Lutein and zeaxanthin supplementation and association with visual function in age-related macular degeneration. Investigative Ophthalmology and Visual Science. 2014;**56**:252-258

[54] Chew EY, Clemons TE, Sangiovanni JP. Secondary analyses of the effects of lutein/zeaxanthin on age-related macular degeneration progression: AREDS2 report no. 3. JAMA Ophthalmology. 2014;**132**:142-149

[55] Bone RA, Landrum JT, Hime GW, Cains A. Stereochemistry of the human macular carotenoids. Investigative Ophthalmology and Visual Science 1993;**34**:2033-2040

[56] Bernstein PS, Sharifzadeh M, Liu A, Ermakov I. Blue-light reflectance imaging of macular pigment in infants and children. Investigative Ophthalmology and Visual Science. 2013;**54**:4034-4040

[57] Henriksen BS, Chan G, Hoffman RO, Sharifzadeh M. Interrelationships between maternal carotenoid status and newborn infant macular pigment optical density and carotenoid status. Investigative Ophthalmology and Visual Science. 2013;**54**:5568-5578

[58] Malinow MR Feeney-Burns L Peterson LH, et al. Diet-related macular anomalies in monkeys. Investigative Ophthalmology and Visual Science. 1980;**19**:857-863

[59] Manzoni P, Guardione R, Bonetti P, Priolo C. Lutein and zeaxanthin supplementation in preterm very low-birth-weight neonates in neonatal intensive care units: a multicenter randomized controlled trial. American Journal of Perinatology. 2013;**30**:25-32

[60] Hammond BR, Jr. Possible role for dietary lutein and zeaxanthin in visual development. Nutrition Reviews. 2008;**66**:695-702

[61] Rutten NB, Gorissen DM, Eck A, Niers LE. Long term development of gut microbiota composition in atopic children: Impact of probiotics. PLoS One. 2015;**10**:e0137681

[62] Lynch SV, Pedersen O. The human intestinal microbiome in health and disease. The New England Journal of Medicine. 2016;**375**:2369-2379

[63] Linares DM, Ross P, Stanton C. Beneficial Microbes: The pharmacy in the gut. Bioengineered. 2016;**7**:11-20

[64] Canavan C, West J, Card T. The epidemiology of irritable bowel syndrome. Journal of Clinical Epidemiology. 2014;**6**:71-80

[65] Hill C, Guarner F, Reid G, Gibson GR. Expert consensus document. The International Scientific Association for Probiotics and Prebiotics consensus statement on the scope and appropriate use of the term probiotic. Nature Reviews Gastroenterology & Hepatology. 2014;**11**:506-514

[66] Food and Agriculture Organization/World Health Organization. Evaluation of Health and Nutritional Properties of Probiotics in Food, Including Powder Milk with the Live Lactic acid Bacteria. Report of a Joint FAO/WHO Expert Consultation on evaluation of health and nutritional properties of probiotics in food including powder milk with live lactic acid bacteria. Available from: http://www.who.int/foodsafety/publications/fs_management/en/probiotics.pdf [Accessed: 27-02-2017]

[67] Moayyedi P, Ford AC, Talley NJ, Cremonini F. The efficacy of probiotics in the treatment of irritable bowel syndrome: A systematic review. Gut 2010;**59**:325-332

[68] Hungin AP, Mulligan C, Pot B, Whorwell P. Systematic review: Probiotics in the management of lower gastrointestinal symptoms in clinical practice—An evidence-based international guide. Alimentary Pharmacology & Therapeutics. 2013;**38**:864-886

[69] Clarke G, Cryan JF, Dinan TG, Quigley EM. Review article: Probiotics for the treatment of irritable bowel syndrome—Focus on lactic acid bacteria. Alimentary Pharmacology & Therapeutics. 2012;**35**:403-413

[70] Didari T, Mozaffari S, Nikfar S, Abdollahi M. Effectiveness of probiotics in irritable bowel syndrome: Updated systematic review with meta-analysis. World Journal of Gastroenterology. 2015;**21**:3072-3084

[71] Quigley EM. Probiotics in irritable bowel syndrome: The science and the evidence. Journal of Clinical Gastroenterology. 2015;**49**:S60-S64

[72] Gibson GR, Probert HM, Loo JV, Rastall RA. Nutrition Research Reviews. 2004;**17**:259-275

[73] Manning TS, Gibson GR.Microbial-gut interactions in health and disease. Prebiotics. Best Practice and Research Clinical Gastroenterology. 2004;**18**:287-298

[74] Silk DBA, Davis A, Vulevic J, Tzortzis G. Clinical trial: The effects of a trans-galactooligosaccharide prebiotic on faecal microbiota and symptoms in irritable bowel syndrome. Alimentary Pharmacology & Therapeutics. 2009;**29**:508-518

[75] Wilson B, Whelan K. Prebiotic inulin-type fructans and galacto-oligosaccharides: Definition, specificity, function, and application in gastrointestinal disorders. Journal of Gastroenterology and Hepatology. 2017;**32**:64-68

[76] Marques TM, Wall R, Ross RP, Fitzgerald GF, Ryan CA, Stanton C.Current Opinion in Biotechnology. 2010;**21**:149-156

[77] Paul HA, Bomhof MR, Vogel HJ, Reimer RA. Diet-induced changes in maternal gut microbiota and metabolomic profiles influence programming of offspring obesity risk in rats. Scientific Reports. 2016;**6**:20683

[78] Boehm G, Moro G. Structural and functional aspects of prebiotics used in infant nutrition. The Journal of Nutrition. 2008;**138**:1818S-1828S

[79] Schmelzle H, Wirth S, Skopnik H, Radke M. Randomized double-blind study of the nutritional efficacy and bifidogenicity of a new infant formula containing partially hydrolyzed protein, a high beta-palmitic acid level, and nondigestible oligosaccharides. Journal of Pediatric Gastroenterology and Nutrition. 2003;**36**:343-351

[80] Moro G, Minoli I, Mosca M, Fanaro S. Dosage-related bifidogenic effects of galacto- and fructooligosaccharides in formula-fed term infants. Journal of Pediatric Gastroenterology and Nutrition. 2002;**34**:291-295

[81] Boehm G, Lidestri M, Casetta P, Jelinek J. Supplementation of a bovine milk formula with an oligosaccharide mixture increases counts of faecal bifidobacteria in preterm infants. Archives of Disease in Childhood. Fetal and Neonatal Edition. 2002;**86**:F178

[82] Knol J, Boehm G, Lidestri M, Negretti F. Increase of faecal bifidobacteria due to dietary oligosaccharides induces a reduction of clinically relevant pathogen germs in the faeces of formula-fed preterm infants. Acta Paediatrica. Supplement. 2005;**94**:31-33

[83] Koenig JE, Spor A, Scalfone N, Fricker AD. Succession of microbial consortia in the developing infant gut microbiome. Proceedings of the National Academy of Sciences of the United States of America. 2011;**108**:4578-4585

[84] Klaassens ES, Boesten RJ, Haarman M, Knol J. Mixed-species genomic microarray analysis of fecal samples reveals differential transcriptional responses of bifidobacteria in breast- and formula-fed infants. Applied and Environmental Microbiology. 2009;**75**:2668-2676

[85] Oozeer R, van Limpt K, Ludwig T, Ben Amor K. Intestinal microbiology in early life: Specific prebiotics can have similar functionalities as human-milk oligosaccharides. The American Journal of Clinical Nutrition. 2013 ;**98**:561S-571S

[86] Scholtens PA, Alliet P, Raes M, Alles MS. A dietary fiber mixture versus lactulose in the treatment of childhood constipation: A double-blind randomized controlled trial. The Journal of Nutrition. 2008;**138**:1141-1147

[87] Georgi G, Bartke N, Wiens F, Stahl B. Functional glycans and glycoconjugates in human milk. The American Journal of Clinical Nutrition. 2013;**98**:578S-585S

[88] Guidance for Industry: A Food Labeling Guide. Available from: https://www.fda.gov/Food/GuidanceRegulation/GuidanceDocumentsRegulatoryInformation/LabelingNutrition/ucm064908.htm [Accessed 29-03-2017]

[89] Roller S, Pippins R. Marketing nutrition & health-related benefits of food & beverage products: Enforcement, litigation and liability issues. Food and Drug Law Journal. 2010; **65**:447-469

[90] Regulation of Dietary Supplement Advertising: Current Claims of Interest to the Federal Trade Commission, Food and Drug Administration and National Advertising Division. Available from: https://www.ftc.gov/tips-advice/business-center/guidance/dietary-supplements-advertising-guide-industry [Accessed 29-03-2017]

[91] What is a dietary supplement? Available from: https://www.fda.gov/AboutFDA/Transparency/Basics/ucm195635.htm [Accessed 29-03-2017]

[92] CFR—Code of Federal Regulations Title 21. Available from: https://www.gpo.gov/fdsys/granule/CFR-2008/CFR-2008-title21-vol2-sec105-3 [Accessed 29-03-2017]

[93] Regulation of medical foods, advance notice of proposed rulemaking. Federal Register. **61**(231):60661-60671; GPO website. www.gpo.gov/fdsys/pkg/FR-1996-11-29/pdf/96-30441.pdf [Accessed 29-03-2017]

[94] Guidance for Clinical Investigators, Sponsors, and IRBs Investigational New Drug Applications (INDs)—Determining Whether Human Research Studies Can Be Conducted Without an IND (September 2013, stayed version October 2015). FDA website. http://www.fda.gov/downloads/Drugs/Guidances/ UCM229175.pdf [Accessed 29-03-2017]

[95] Guidance for Industry: Frequently Asked Questions about Medical Foods. 2nd ed. May 1997; May 2007; May 2016. FDA website. https://www.fda.gov/Food/GuidanceRegulation/GuidanceDocumentsRegulatoryInformation/ucm054048.htm [Accessed 29-03-2017]

[96] Nutrient Recommendations: Dietary Reference Intakes (DRI). Available from: https://ods.od.nih.gov/Health_Information/Dietary_Reference_Intakes.aspx [Accessed 29-03-2017]

[97] National Academies of Sciences, Engineering, and Medicine. Obesity in the Early Childhood Years: State of the Science and Implementation of Promising Solutions: Workshop Summary. Washington, DC: The National Academies Press; 2016. DOI: 10.17226/23445

[98] Guidance for Industry: Evidence-Based Review System for the Scientific Evaluation of Health Claims Final. Available from: https://www.fda.gov/Food/GuidanceRegulation/GuidanceDocumentsRegulatoryInformation/LabelingNutrition/ucm073332.htm [Accessed 28-03-2017]

[99] Sicherer SH. Maternal consumption of peanut during pregnancy is associated with peanut sensitization in atopic infants. Journal of Allergy and Clinical Immunology. 2010;**126**:1191-1197

[100] National Academies of Sciences, Engineering, and Medicine. Biomarker Tests for Molecularly Targeted Therapies: Key to Unlocking Precision Medicine. Washington, DC: The National Academies Press; 2016. DOI: 10.17226/21860

Chapter 14

Diet Quality Indices for Nutrition Assessment: Types and Applications

Maria Luisa Poyatos Guerrero and
Fernando Pérez-Rodríguez

Additional information is available at the end of the chapter

http://dx.doi.org/10.5772/intechopen.69807

Abstract

According to the World Health Organization, the proportion of noncommunicable diseases (NCD) burden is foreseen to increase to 57% in 2020. Consumption patterns have a positive effect on healthy growth and development during childhood and adolescence, and on health problems in adulthood. Diet quality indices are mathematical algorithms used for nutritional epidemiology, aimed at quantifying the degree of adequacy between actual intakes of nutrients or food groups within a population and the reference intakes, which are established based on scientific facts assuring an optimal state of health while preventing consumers from chronic diseases. Similarly, indexes allow to analyse dietary pattern of target population and its consumption trends. In general the terms, DQI (Diet Quality Index), HDI (Healthy Diet Indicator) and MDS (Mediterranean Diet Score), are referred to three internationally recognized diet indexes, which several indices have been adapted from. This chapter includes an extensive review of existing diet indexes, 1) providing a brief description of the most relevant ones, 2) highlighting the weaknesses and strengths and 3) defining the suitable scope of application of each index.

Keywords: diet quality, dietary quality index, food habits, lifestyles, diet variety

1. Introduction

According to the World Health Organisation, the burden of chronic diseases is rapidly increasing worldwide. In 2001, chronic diseases accounted for around 60% of the total reported deaths in the world and 46% of the global burden of disease [1]. Almost half of total chronic disease deaths are attributable to cardiovascular diseases. Obesity and diabetes are also showing worrying trends, not only because they already affect a large proportion of the population, but also because they have started to appear earlier in life [2]. The relevance of diet quality in the prevention and management of disease and premature death caused by noncommunicable diseases (NCDs) is scientifically supported by epidemiological data. Eating patterns may have a positive impact on healthy growth and development throughout childhood and adolescence [3], and on the mitigation of health problems in adults [4].

However, "diet quality" remains a somewhat imprecise term, due to the heterogeneous and multidimensional nature of the concept itself, whose definition should ideally reflect aspects relevant to a number of fields, including nutrition, toxicology, economics and the food industry [5]. Although consensus has yet to be reached on the concept, a high-quality diet may be defined as one which is hygienically safe; nutritious and balanced and adapted to individual requirements in order to prevent disease and ensure a good state of health as well as optimal development and growth [6].

Methods for assessing diet quality have attracted growing interest since Patterson published the first dietary quality index in 1994 [7]. While the concept itself is undoubtedly heterogeneous and multidisciplinary, dietary quality indices (DQIs) are no more than mathematical algorithms aimed at quantifying the extent to which real food and nutrient intake complies with the reference intake values recommended in national dietary guidelines, or at analysing dietary patterns in the population and weighting those components whose consumption has been linked, in scientific studies, either to the appearance of the disease or to the preservation of an optimal state of health [8].

In the first global review of DQIs, published in 1996, Kant assigned each index to one of three groups depending on the items it comprised [9]:

Indices based solely on an analysis of nutrient intake. Examples include the index developed by Cusatis et al. for use as a tool in the Penn State Young Women's Health Study [10].

1.1. Indices based on food groups

Indices based on a combination of food groups and nutrient intake. This is the most widespread approach, given that—since people do not consume foods or nutrients singly—research into the potential link between food/nutrient intake and disease should focus not on a single food or nutrient but rather on the diet as whole [9].

Later, in 2015, Gil defined a new category: indices comprising items intended to assess a range of specific behaviour patterns: eating habits, physical activity and rest, and certain sociocultural or lifestyle habits. Gil labelled these items "healthy lifestyle indicators" (HLIs) [8].

Dietary quality indices traditionally focused on the extent to which real dietary intake complied with nutritional recommendations, and on variations in the intake of basic food groups [9]; current indices tend to include additional items relating to life style and physical exercise [8].

2. Reference values: ideal consumption of food groups or nutrients

There is clear disagreement regarding the servings or amounts of different food/nutrient groups that should ideally be consumed. The determination of nutritional needs is a complex matter. Each country, drawing on the findings of expert groups, has issued its own recommendations, reflecting the characteristics of the population, specific recommendations are made for target groups as a function of age, sex and physiological status (e.g. pregnant or breastfeeding women). Recommendations are also taken into account the extent to which a given nutrient is used, its bioavailability, the existence of precursors, potential interactions with other substances [11] and potential loss or alteration during food transport, storage, processing and preparation [12]. According to Bolzetta, the factors to be borne in mind when establishing the Recommended Intake (RI) values for essential nutrients, which scientific research has shown to be sufficient to meet the nutritional needs of practically all healthy people, can be divided into three major categories [13]:

1. Person-related factors, that is, those which govern inter-individual variability.
2. Environment-related factors.
3. Diet-related factors, that is, those linked to food intake.

On this basis, the Food and Nutrition Board of the American Institute of Medicine (FNB-IOM) established Dietary Reference Intakes (DRIs) for North America (United States and Canada) comprising a number of parameters such as estimated average requirement (EAR), recommended dietary allowances (RDA), adequate intake (AI) and tolerable upper intake level (UL). These guidelines, which replaced the RIs in force until 1997, provide reference values for the nutrients that a diet should contain in order to prevent deficiency diseases, reduce the burden of chronic disease and achieve optimal health by making maximum use of each nutrient. If there is sufficient confirmed scientific evidence, an EAR[1] is set; this value, after further testing, shall be considered as RDA.[2] If testing is inconclusive, an estimated value is provided, known as the IA.[3] When sufficient data is available, ULs[4] are also established for nutrients. Since these values are not readily understood by the layman, they are used as the basis for Food-Based Dietary Guidelines (FBDG) which express nutrient requirements in the form of servings of different food groups; the language used is straightforward and the information is often additionally expressed graphically, for example, in the form of a pyramid, a diamond-shape or a wheel. The most commonly used graphic worldwide was the pyramid: foods to be consumed most frequently were placed at the base, and foods

[1]EAR, estimated average requirement: a nutrient intake value that is estimated to meet the requirement of half the healthy individuals in a group of people of a given age and sex.
[2]RDA, recommended dietary allowances: the average daily dietary intake level that is sufficient to meet the nutrient requirement of nearly all (97–98%) healthy individuals in a group of people of given age and sex.
[3]AI, adequate intake, a value based on observed or experimentally determined approximations of nutrient intake by a group (or groups) of healthy people—used when an RDA cannot be determined.
[4]UL, tolerable upper intake level: the highest level of daily nutrient intake that is likely to pose no risk of adverse health effects to almost all individuals in the general population.

to be avoided or consumed only occasionally at the apex [14]. In 2011, the United States Department of Agriculture (USDA) replaced the pyramid with a simpler design, where the proportions of the different food groups to be consumed in the course of the day are shown as servings on a plate. "My plate" promotes fruit and vegetables, which together occupy half the plate; grains and proteins each occupy a quarter of the plate. Dairy products are present in the form of a glass of milk beside the plate [15].

Although the Dietary Guidelines for Americans—published every 5 years since 1980, and now in their eighth edition (2015–2020), accompanied since 1992 by a pyramid graphic—have long served as an international benchmark, several countries have developed their own guidelines, taking into account specific national dietary requirements. Most national guidelines advocate greater variety in the diet, increased intake of plant-based foods—especially fruit and vegetables—and lower consumption of solid fats, salt and sugar [16, 17].

3. Methods for collecting food consumption data

Consumption data provide the essential basis for any assessment of nutritional status and for the drawing up of nutritional guidelines for the population as a whole. The type of application for which data are used is largely determined by the method of collection, which also accounts for certain limitations.

Since 1940, FAO has regularly published national food balance sheets, which provide data on food potentially available for human consumption during a specified reference period. The balance sheets provide a picture of the overall food supply situation in a country, but give no indication of the diet consumed by different population groups depending on their socioeconomic status or geographical location; or do they provide information on seasonal variations in food consumption (i.e. the distribution of national consumption at different times of year) [18, 19].

Household surveys are bound by the same constraints. Available foods are weighed at the beginning and end of the reference period; any incoming food is added daily to the supply figure, while any food consumed other than by the survey participants is subtracted. The total amount of food consumed by the household over the period is then divided equally between its members. While certain survey groups tend to be fairly homogeneous—for example, school dining-rooms or old people's room—the composition of a household may vary considerably. Such surveys thus provide an overall view of the group, but do not record the real intake of each household member. However, they do enable identification of groups at risk due to inappropriate intake, which require closer investigation.

Individual dietary surveys can be carried out prospectively, that is, studying current intake, or retrospectively, focussing on past intake. Most dietary quality indices (DQIs) are based on retrospective studies, since these enable measurement of intake in the immediate, recent or distant past, and also provide data for epidemiological studies, by relating past diet to present disease. The main limitation is that data collection requires participants to remember their past diet and their reports may be influenced by their current diet [19]. The respondent's diet history may comprise various sections and may take several forms:

1. Record of food consumed over 2 or 3 days or, failing this, a 24 h-food recall, in order to gain an idea of diet type and eating habits [20].
2. Food frequency questionnaire (daily, weekly and monthly) covering the last month. This is a structured, organised list, broken down by meals: breakfast, lunch and dinner (first course, second course and dessert). The amount consumed is estimated by approximate measurement at home or by reference to photograph collections showing different serving sizes of the same food or dish [21].
3. Specific questions relating to the study aims.

4. Dietary quality index applications

DQIs can be classed depending on their application, as health assessment indices or risk assessment indices. Health assessment indices focus on dietary quality in terms of compliance with dietary guidelines, and thus provide a practical way of assessing the health status of the population. The findings can be used to draw up specific effective prevention strategies. Indices of this type include, for example, those designed to assess compliance with the Mediterranean diet, whose consumption by various groups has been associated with lower rates of chronic disease, myocardial infarction, arthritis, various neoplasms (including breast, bowel and prostate cancers), diabetes, other oxidative-stress-related pathologies, childhood asthma and rhinitis. When repeated at intervals, the findings of such indices can be used to chart changes in eating patterns and to compare dietary quality in different groups and populations, thus providing a useful basis for the design of nutritional intervention policies. Risk assessment indices measure the risk of developing certain diseases, by examining overall diet composition and nutrient intake. A number of indices have been used, for example, to investigate cardiovascular risk and cancer mortality [22–27].

Kant was the first to review global diet quality indices, noting that—not unexpectedly—the definition of diet quality depended on attributes selected by the investigators [9]. In a second review, published 8 years later, she reviewed the literature on dietary patterns, both empirically derived and theoretically defined—in relation to health outcomes [28]. Later, Waijers et al. reviewed 20 DQIs, and found that existing indices did not predict disease or mortality significantly better than individual dietary factors, although they may be useful for measuring the extent to which individuals adhere to dietary guidelines [29]. Arvaniti and Panagiotakos also reviewed 23 DQIs, most of which overlapped with those reported by Waijers [30]. Bach et al. revised a number of DQIs developed for the general adult population based on the Mediterranean diet and their correlation with health outcomes [31]. More recently, Alkerwi highlighted the complexities involved in defining and quantifying the concept of *diet quality*, and recommended an integrated approach that combines not only nutritional characteristics but also other facets of diet quality, including food safety, organoleptic and sociocultural aspects for which there are currently no established thresholds or criteria, with a view to dispelling the confusion generated by multiple DQIs [4]. Similarly, in 2015 Gil hailed DQIs as an important tool for assessing diet quality within specific populations, in terms not only of nutrient intake but also of diversity and moderation, although advocating a more global concept taking into account, in addition to food groups and nutrients, factors such as certain sociocultural habits, physical activity, sedentariness and rest create healthy lifestyle indicators (HLIs) [8].

Other reviews have focused on the use of indices in children: in 2011, Lazarou and Newby [32] examined 90 indices used in developed countries, while in 2014 Marshall et al. [33] conducted a similar review at world level. Both acknowledged the value of dietary indices, regardless of the individual methodology used but noted that, in general, higher *diet-quality* scores were associated with more favourable nutrient and food intakes, better lifestyles, lower chronic disease risk factors, more favourable body composition, less obesity and fewer asthma-related conditions. However, certain measures or techniques used in data processing need to be more clearly defined in order to increase the robustness of estimations regarding diet quality.

5. Main dietary indexes

The first Diet Quality Index, published by Patterson in 1994, was based on an epidemiological study analysing the diet of 5484 adults in the United States. Her index comprised eight items, six of which related to specific nutrients (total fat, saturated fatty acids (SFA), cholesterol, protein, sodium and calcium), while the remaining two assessed only intake of food groups (fruit and vegetables, grain and legumes). She found that low index scores correlated positively with vitamin and mineral intake, and negatively with fat intake [7].

Taking this index as a basis, Haines et al. [34] produced the Diet Quality Index Revised (DQI-R), which sought to reflect subsequent changes in the recommendations for the American population. The major new feature was the incorporation of dietary diversity and moderation as specific items. Diversity was scored using 23 subcategories of the 4 food-group categories established in the Food Guide Pyramid: 7 for grains, 7 for vegetables, 2 for fruits and 7 for meat/dairy components. Diversity was recorded over two survey days. Moderation was measured in terms of added sugars, discretionary fat, sodium intake and alcohol intake. Scoring criteria were taken from the Food Guide Pyramid and adjusted for Kcal intake per individual; AI was used for calcium and RDA for iron [34]. The index thus analysed both nutrient and food-group intake. The most recent review was published by Fung in 2005 [35].

In 1995, the United States Department of Agriculture (USDA) published the Healthy Eating Index (HEI), designed by Kennedy et al. with a view to monitoring changes in the quality of American diets and to developing and evaluating nutritional intervention strategies [36]. The original index comprised 10 variables covering nutrient intake, food-group intake and dietary variation. The first five items were based on the five major food groups included in the American Food Guide Pyramid (1992), which was later replaced by MyPyramid (2005) [37]: grains, vegetables, fruits milk and meat. A further four items (fat, SFA, cholesterol and sodium) reflected the intake values recommended by national dietary guidelines. The tenth item, dietary variety, was a measure of the variety in food choices, scored as intake of 16 items over an arbitrary 3-day period. Each item was a discrete variable scored from 0 to 10, giving a maximum HEI score of 100 points. Two subsequent revisions were carried out by Guenther et al. (HEI-2005) [38] and in 2010 (HEI-2010) [39], which included an additional food group item—fish and seafood—in the design. The revised versions comprised 12 components, and additionally evaluated energy density, adequacy and dietary moderation.

Using the variety component of the original HEI designed by Kennedy et al. [36], Cox et al. developed the Food Variety Index for Toddlers (VIT) to assess diet in children aged between 24 and 36 months. Ideal intake for each food item was as given in the Food Guide Pyramid, adapted for this age range [40].

Feskanich et al. also used the HEI as the basis for assessing diet quality in children aged 9–14 whose parents were participants in US Nurses Health Study II cohort. The resulting Young Healthy Eating Index (YHEI) includes both eating habits and intake of food groups associated with "empty calorie" intake. The Index comprises 13 items. The first 7 items (intake of wholemeal grains, vegetables, fruits, dairy products, meat, snacks and soft drinks) have a maximum score of 10, while the remaining 6 (intake of multivitamins, margarine and butter, fried foods outside the home, visible animal fat, eating breakfast and dinner with parents) had a maximum of 5 points. The score for the overall Index thus ranged from 0 to 100. The HEI score was highly correlated with total energy intake ($r = 0.67$), and as inversely associated with time spent in inactive pursuits ($r = -0.27$) [41].

Huijbregts et al. were the first to devise a method for simplifying calculations and extracting total scores from an index. Using the World Health Organisation (WHO) guidelines for the prevention of chronic diseases, they conducted a longitudinal cohort study of a random sample of men aged between 1950 and 1970 in 1970; a total of five cohorts were followed up in Finland, Italy and the Netherlands. Findings were used to develop the Healthy Diet Indicator (HDI) [42], in which a dichotomous variable was generated for each food group or nutrient that was included in the WHO guidelines, thus making calculation easier than in earlier indices. If a person's intake was within the recommended range this variable was coded as 1; otherwise it was coded as 0. The HDI was the sum of values for nine variables: SFA, polyunsaturated fatty acids (PUFA), cholesterol, protein, complex carbohydrates, monosaccharides and disaccharides, dietary fibre, fruits and vegetables, pulses, nuts and seeds. The overall score therefore ranged between 0 and 9. Although we still assess nutrient and food group intake, the HDI did so in a more specific manner than earlier indices. Instead of evaluating total fat intake, its origin was taken into account, whit the result that a single item in earlier studies (total fat) was broken down into three items (SFA, PUFA and cholesterol), enabling more detailed analysis. A further modification was the assessment of dietary fibre intake and nut intake as separate components.

Later, in order to compare diet quality between populations with different eating habits and evaluate their current stage of nutrition transition, Kim et al. developed the Diet Quality Index-International (DQI-I), which was initially applied to China and the United States [43]. The DQI-I focused on three major aspects of diet: adequacy, moderation and overall balance, and total index scores ranged from 0 to 100. A new feature was the assessment of Vitamin C intake. Other items were similar to those used in the DQI-R, although scoring was completely different and somewhat arbitrary, as shown in **Table 1**. Within the block of items assessing dietary moderation, the alcohol item was included under "empty calories". Since European countries follow the Mediterranean diet, and well-known characteristic variables for the European region were not taken into account in designing this index, the DQI-I was modified by Tur et al. to assess diet quality in Andalusia: fat intake guidelines were increased by 10% for the Mediterranean region, largely to reflect olive oil consumption [44].

Variety	**0–20 points**	
Overall food group variety (meat/poultry/fish/eggs; dairy/beans; grain; fruit; vegetable)	>1 serving from each food group/day	15 points
	Any 1 food group missing/day	12 points
	Any 2 food groups missing/day	9 points
	Any 3 food groups missing/day	6 points
	>4 food groups missing/day	3 points
	None from any food groups	0 points
Within-group variety for protein source (meat, poultry, fish, dairy, beans, eggs)	>3 different sources/day	5 points
	Two different sources/day	3 points
	From 1 source/day	1 points
	None	0 points
Adequacy	0–40 points	
Vegetable group	3–5 servings/day	5 points
	0 servings/day)	0 points
Fruit group	2–4 servings/day	5 points
	0 servings/day	0 points
Grain group	6–11 servings/day	5 points
	0 servings/day	0 points
Fibre	20–30 g/day	5 points
	0 g/day	0 points
Protein	10% of energy/day	5 points
	0% of energy/day	0 points
Iron	100% RDA (AI)/day	5 points
	0% RDA (AI)/day	0 points
Calcium	100% AI/day	5 points
	0% AI/day	0 points
Vitamin C	100% RDA (RNI)/day	5 points
	0% RDA (RNI)/day	0 points
Moderation	0–30 points	
Total fat	<20% of total energy/day	6 points
	20–30% of total energy/day	3 points
	>30% of total energy/day	0 points
Saturated fat	>7% of total energy/day	6 points
	7–10% of total energy/day	3 points
	10% of total energy/day	0 points

Variety	0–20 points	
Cholesterol	<300 mg/day	6 points
	300–400 mg/day	3 points
	>400 mg/day	0 points
Sodium	<2400 mg/day	6 points
	2400–3400 mg/day	3 points
	>3400 mg/day	0 points
Empty calorie foods	<3% of total energy/day	6 points
	3–10% of total energy/day	3 points
	>10% of total energy/day	0 points

Table 1. Scoring and items included in the Diet Quality Index international (DQI).

6. Indices for the Mediterranean diet

The Seven Countries Study carried out by Dr Ancel Keys from 1958 onwards was the first to systematically examine the links between diet, lifestyle, risk factors and rates of coronary disease and cerebrovascular accident [45]. A total of 12,763 men aged between 40 and 59 were recruited into 16 cohorts in 7 different countries: Finland, Italy, Netherlands, Greece, Yugoslavia, United States and Japan. Participants were given standardised tests relating to lifestyles and cardiovascular risk factors: they were tested at the start of the study (baseline data) and then after 5 and 10 years' follow-up.

One major conclusion of the study was that cardiovascular diseases can be prevented and are strongly influenced by the fatty composition of the habitual diet. The study also suggested that there may be other and important protective elements in the diet and lifestyles of Crete and Japan [46]. The healthy nature of the Greek diet and more particularly the Cretan diet moved Keys to label it the Mediterranean Diet and to note that it is characterised by high intake of fruit and vegetables, pulses, nuts and grains and, especially, olive oil, together with moderate consumption of fish, eggs and dairy products—preferably yoghurt and cheese—and lower intakes of meats and animal fats [47].

It is regarded as the prototype of a healthy diet, since it ensures a supply of calories and nutrients in sufficient amounts and adequate proportions, and also contributes to the prevention of cardiovascular disease, hypertension, diabetes and cancer, and generally to increased life expectancy [48–54].

In view of its many benefits, numerous indices have been designed to assess compliance with this diet. The original Mediterranean Diet Score (MDS) was developed by Trichopoulou et al. to assess adherence to the Mediterranean diet consumed by the Greek population, in view of the apparent beneficial effects of this diet on health and longevity [55]. The MDS comprised eight components: seven food groups (vegetables, legumes, fruit and nuts, dairy products, cereals, meat and meat products and alcohol) and the MUFA:SFA ratio. A value of 0 or 1 was assigned to each item, using the median value for each sex as a cut-off point. The total score thus ranged from 1 to 8. A value of 1 was assigned for a daily intake of 10-50g of alcohol for men and 5-25 g of alcohol for women. In general terms, a score of 4 or more was taken to indicate satisfactory compliance. Food frequencies were adjusted to daily intakes of 2500 kcal for men and 2000 kcal for women, so estimations reflected variations in median energy intake. A later revision by Trichopoulou et al. included a ninth item—fish intake—bringing the total maximum score to 9 [56]. The MDS index has been widely used in studies relating the Mediterranean diet to disease, in some cases incorporating modifications to reflect the specific purpose of the study. Bach et al. have reviewed studies applying MDS or variations of it [31]. One limitation of this index is the use of median intake for the population as a cut-off point, rather than the intake deemed suitable for that population. To address this issue, Schröder et al. developed a rapid 14-point screening questionnaire, the Mediterranean Diet Adherence Screener (MEDAS) for assessing adherence to the Mediterranean diet [57]. Twelve items related to food consumption frequency, and the remaining two food intake habits are considered as characteristic of the Spanish Mediterranean diet. Each question was scored 0 or 1, so that overall scores ranged from 0 to 14. A score of less than 9 was considered indicative of poor adherence, and a score of more than 9 indicative of satisfactory adherence. This index was used in the 2013 PREDIMED Prevention with Mediterranean Diet (PREDIMED) study [58]. The criterion used are set out in **Table 2**.

One variant is the KidMed index used in the "enKid" study to assess adherence to the Mediterranean diet in younger age groups. It was developed in 2003 to evaluate the eating habits of 3850 Spanish children, adolescents and young adults (age 2–24) [59]. The questionnaire com-

Olive oil as the principal source of fat for cooking	+1
Four or more tablespoons—1 tablespoon 13.5 g— of olive oil/day (including that used in frying, salads, meals eaten away from home)	+1
Two or more serving of vegetables/day	+1
Three or more pieces of fruit/day	+1
<1 Serving of red meat or sausages/day	+1
<1 Serving of animal fat/day	+1
<1 Cup (1 cup = 100 ml) of sugar-sweetened beverages/day	+1
Seven or more servings of red wine/week	+1
Three or more servings of fish/week	+1
Three or more serving of nuts/week(30g/serving)	+1
<2 Commercial pastries/week	+1
Three or more pieces of fruit/day	+1
Two or more servings/week of a dish with a traditional sauce of tomatoes, garlic, onion, or leeks sautéed in olive oi.	+1
Preferring white meat over red meat	+1

Table 2. Scoring and items included in the Mediterranean Diet Adhere Screener (MEDAS).

prises 16 items: affirmative responses to questions denoting a negative connotation with regard to the Mediterranean diet were assigned a value of −1, while affirmative responses to questions denoting a positive connotation were assigned a value of +1. The scoring range for the index was therefore from 0 to 16. For purposes of interpretation, the sum of the values were classified into three levels: ≤3—very low diet quality; from 4 to 7—improvement needed to adjust intake to Mediterranean patterns; ≥8—optimal Mediterranean diet (**Table 3**). A number of studies have used the KidMed index to assess diet quality in children and adolescents. More recent research using this index is reported by Toktas and Yildiz [60], Mistretta et al. [61] and Idelson et al. [62].

As indicated earlier, the Mediterranean diet involves more than just healthy eating habits. In order to benefit fully from the diet, a number of cultural and lifestyle elements need to be borne in mind: moderation in calorie intake; regarding eating as a social act in which the act of cooking the food to be eaten plays a major role; physical activity and rest. Awareness of these additional elements has led, over the last few years, to the development of healthy lifestyle indicators that focus on these variables as well as on food intake [63]. This more holistic approach gave rise to the Mediterranean Lifestyle Index (MEDLIFE), based on the current guidelines reflected in the Spanish Mediterranean diet pyramid: in the base, food items that should sustain the diet, with recommendations concerning the composition and serving size of main meals; in the upper levels, foods to be eaten in moderate amounts. But the pyramid also contains cultural and social elements characteristic of the Mediterranean way of life in a broader sense [64].

Takes a fruit or fruit juice every day	+1
Has a second fruit every day	+1
Has fresh or cooked vegetables regularly once a day	+1
Has fresh or cooked vegetables more than once a day	+1
Consumes fish regularly (at least 2–3 times per week)	+1
Goes more than once a week to a fast-food (hamburger) restaurant	−1
Likes pulses and eats them more than once a week	+1
Consumes pasta or rice almost every day (5 or more times per week)	+1
Has cereals or grains (bread, etc) for breakfast	+1
Consumes nuts regularly (at least 2–3 times per week)	+1
Uses olive oil at home	+1
Skips breakfast	−1
Has a dairy product for breakfast (yogurt, milk, etc.)	+1
Has commercially baked goods or pastries for breakfast	−1
Takes two yogurts and/or some cheese (40 g) daily	+1
Takes sweets and candy several times every day	−1

Table 3. Scoring and items used in KidMed of adherence to the Mediterranean diet for childhood.

The pyramid was used as the basis for the MEDLIFE index [65], which assess adherence to the Mediterranean lifestyle. MEDLIFE comprises 28 items covering 3 separate aspects of Mediterranean lifestyle: 15 items assess consumption of various food groups; 7 items focus on traditional Mediterranean eating habits and physical activity and the remaining 6 examine physical activity, rest and social interaction. Compliance with each item was assigned 1 point and incompliance 0 points scored. The total score for the index as a whole thus ranged from 0 to 28 points. The index is viewed as a more holistic tool to measure adherence to the Mediterranean lifestyle in epidemiological studies.

In line with this broad-based approach to lifestyle indices, the E-KINDEX developed by Lazarou et al. for use with children [66] comprises three blocks of items. The first block, designated Foods E-KINDEX, assesses intake of eight food-groups ensuring a varied diet (bread, cereals and grain foods, fruits and fruit juices, vegetables, legumes, milk, fish and meat); three additional items relate to cooking technique (smoked or salted meats, fried food, grilled food) and to others to empty calorie intake (sweets and snacks, and soft drinks). The second block, designated Dietary Behavior E-KINDEX consists of eight statements regarding the child's attitude to his own diet. The third and final block, labelled Dietary Habits E-KINDEX, comprises nine items designed to assess eating habits, for example, number of meals per day, meals eaten outside the home, eating alone or with parents. Scoring on the E-KINDEX ranges from 1 (the lowest) to 87 (the highest). This index is of particular interest, since in addition to actually quantifying diet quality and eating habits in the child, it detects potential problems of nutritional education through the questions in Block 2. For validation of the E-KINDEX, multiple linear and logistic regression analyses were applied, taking as dependent outcomes various body composition indices of 1140 children from the CYKIDS study, aged 9–13 [67]. In all models, adjustments were made for age, gender, physical activity level, TV viewing time, socioeconomic status, breastfeeding and parental obesity status. The highest E-KINDEX category (60 points) was associated with 85% less likelihood of a child being obese or overweight and 86% less likelihood of having a waist circumference ≥ 75th percentile.

Another index used to assess children's lifestyle is the Preschoolers Diet-Lifestyle Index (PDL-Index) designed by Manios et al. [68] and validated using a sample of 2287 Greek preschool children (from the GENESIS study). The index comprises 11 items: fruits, fish and seafood, sweets, grains, unsaturated fats, vegetables, red meat and meat products, white meat and legumes, dairy products, physical activity and hours of TV viewing. The scoring interval for each item was from 0 to 4, so the total index score ranged from 0 to 44. In the absence of dietary guidelines for preschool children in Greece, components were selected on the basis of the USDA's Food Guide Pyramid, Canada's Food Guide, and the dietary recommendations of the American Heart Association and the American Academy of Pediatrics (AAP). Higher scores indicated greater adherence to dietary-lifestyle guidelines. Overweight and obese children are more likely to have cardiovascular disease risks (hypertension, type 2 diabetes mellitus) and to be overweight or obese as adults [68, 69].

7. Variety as a criterion for dietary quality in childhood

As indicated in earlier sections, most current diet quality indices were developed for the adult population and based on US dietary guidelines (**Figure 1**). Many have subsequently

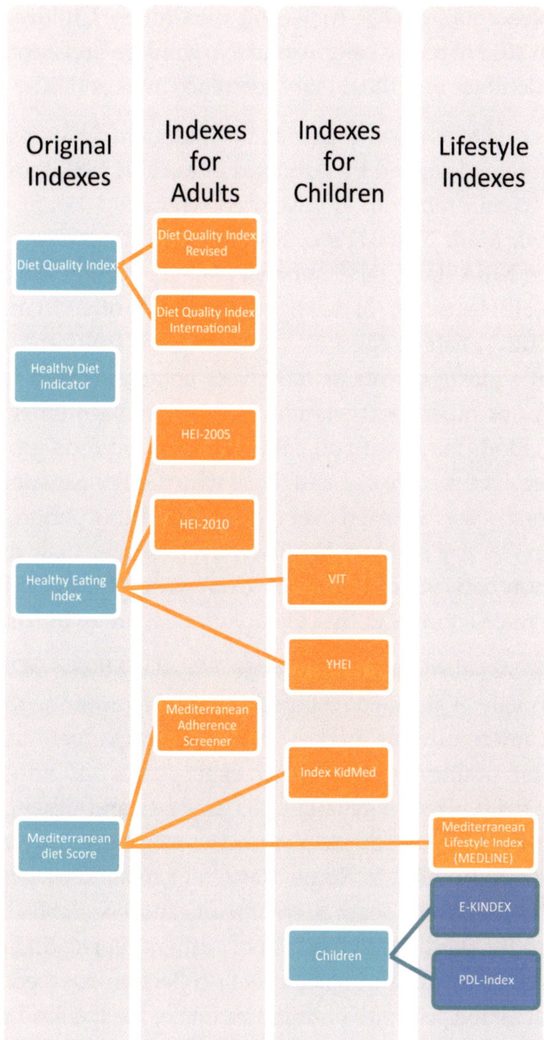

Figure 1. Scheme representing the relationship between original and revised Diet Quality/Lifestyle Indexes for Adults and Children.

been developed or modified for use with children and adolescents in the United States (age 2–18) [70–73], Australia (age 4–16) [73–75], Germany (age 0–17) [76–78], Finland (age 1–6) [79], Spain (age 2–24) [80–82], Canada (age over 3) [83, 84] and a number of Asian countries [85, 86]. The scoring for these indices was based on the assumption that a diet containing an adequate variety of food groups was equally adequate in nutritional terms [86–89]. The items contained in the indices are often arbitrary, reflecting the importance of regional foodstuffs. examples include INCAP Papers Dietary Diversity Score [90], Saibul's Food Variety Score for indigenous Malaysians [91] and Roche's 2008 Dietary Diversity Score (DDS) for the Awajun in the Peruvian Amazon [92]. In some cases, no explanation is given for food classifications [93–95]. Some studies refer to country-specific guidelines, including the New Zealand Diet

Quality Index for Adolescents (NZDQI-A) [96, 97], the Chinese Children Dietary Index developed by Cheng et al. in 2013 to assess calorie intake in children and adolescents with reference to Chinese Dietary Guidelines and the dietary reference intakes (DRIs) for Chinese [85, 86].

The DDS most widely used to assess diet quality in children in developing countries are based on the diet quality surveys designed by Arimond et al. 2004 [88] to assess dietary diversity and nutritional status in children from 11 developing countries (Benin, Cambodia, Colombia, Ethiopia, Haiti, Malawi, Mali, Nepal, Peru, Rwanda and Zimbabwe) using the seven food groups included in the MEASURE DHS surveys [98]: (1) starchy staples (foods made from grains, roots or tubers); (2) legumes; (3) dairy products (milk other than breast milk, cheese or yoghurt); (4) meat, poultry, fish or eggs; (5) vitamin A-rich fruits and vegetables (pumpkin; red or yellow yams or squash; carrots or red sweet potatoes; green leafy vegetables; fruits such as mango, papaya or other local vitamin A-rich fruits); (6) other fruits and vegetables (or fruit juices) and (7) foods made with oil, fat or butter. Food/food groups that the child had consumed regularly received a score of 1 and those who did not consume regularly received a score of 0. Dietary diversity was assessed over an arbitrary 3-day period, and terciles of dietary diversity were used to classify children into low, average and high diversity. The findings suggested an association between child dietary diversity and nutritional status that is independent of socioeconomic factors, and that dietary diversity may indeed reflect diet quality.

Minor modifications were subsequently introduced to adapt these DDS to different research purposes. In 2007, Kennedy et al. [86] developed a variant to examine the correlation between dietary diversity and micronutrient intake. Ten food groups were used to calculate DDS: cereals and tubers; meat, poultry and fish; dairy; eggs; pulses and nuts; vitamin A-rich fruits and vegetables; other fruit; other vegetables; oils and fats and other). The choice of the 10 food groups was based on the outcome of discussions held during a workshop on validation methods for dietary diversity held in Rome, Italy in October 2004 [98]. Modifications with regard to the original included the separation of fruits and vegetables, the treatment of eggs as a separate item, and the addition of a group of "others" consisting of sugar, non-juice or dairy beverages, and condiments and spices. Data collection covered a 24-h period. An all inclusive DDS was calculated without a minimum intake for the food group. A second DDS was calculated applying a 10-g minimum intake for all food groups (DDS 10 g) except fats and oils.

Using Kennedy's 10 g consumption criterion, in 2004 Moursi et al. [99] developed 4 variations on the DDS to study dietary diversity in 1667 children aged between 6 and 23 months in the districts of Sahalava and Antsororokavo, Fianarantsoa, Madagascar, as part of the Nutrimad project [100], and to confirm statistically the correlation between intake of 10 g of each food group and dietary micronutrient density. Two DDS (DDS8 and DDS8-R) covered a total of eight possible food groups: grains, roots and tubers; legumes and nuts; dairy products; flesh foods (meat, fish, poultry and liver/organ meats); eggs; vitamin A-rich fruits and vegetables (>130 retinol equivalents/100 g); other fruits and vegetables; and fats and oils. A score of 1 was assigned if a child ate 1 or more foods from a given food group and 0 if not. For DDS8-R, a food group was counted only if at least 10 g were consumed, except for fats and oils, for which the cut-off of ≥1 g was used. Two additional scores were calculated after excluding the fats and oils group (DDS7 and DD7-R) using the 1 g and 10 g minimum cut-offs (range 0–7).

All DDS scores correlated positively with dietary micronutrient density. These results support the growing evidence of the usefulness of dietary diversity to predict dietary quality, and among infants and young children more specifically.

8. Dietary Quality Index Items

Generally speaking, most indices include fruit and vegetables, and grains or cereals, as food groups. However, the criteria used for their classification vary considerably. The DQI places fruits and vegetable in the same group, but treats wholemeal cereals as a separate item. DQI-R scores fruit separately from vegetables, but—unlike DQI, HEI, MDS and HDI—places grains in a single category, without treating wholemeal grains as a specific item. Although HDI does not treat cereals as a single group, it addresses this design defect by evaluating their intake in the form of nutrients, measuring complex carbohydrate, monosaccharide and disaccharide and fibre intake. HEI increases the number of food groups to be assessed, introducing milk, meat, cereals fruit and vegetables as separate components. In contrast to other indices, "Nuts" are treated as a specific category in HDI, but are grouped with fruits in MDS. Pulses and olive oil feature as items in MDS, similar to fish intake.

In terms of the nutrients assessed, the greatest disparity is in relation to the treatment of data on fat intake. DQI, DQI-I and DQI-R assess intake of total fat, saturated fate and cholesterol, while HDI addresses only intake of SFA and PUFA and MDS deals only with the MUFA:SFA ratio. Quantification of vitamin and mineral intake is arbitrary: although most indices include calcium and sodium, DQI-R is one of the few indices to include iron intake, and DQI-I is among the few to assess Vitamin C intake.

The disparity is even more marked in child-related dietary quality indices. These are based on food groups rather than nutrients, and there is little consensus regarding nomenclature. The E-Kindex, for example, separates bread intake from cereal intake, and also takes into account cooking techniques which may lead to harmful eating habits, such as consumption of fried foods, and of smoked or salted meats. The PDL-Index places white meat and vegetables in the same category, but treats fish as an independent component. Reflecting current lifestyle trends, most indices used to assess dietary quality in developed countries include items to evaluate empty calorie intake and time devoted to physical activity. By contrast, the DDS—aimed at developing countries—dispense with these variables to focus on food intake and, in all cases, food structure: starchy staples (foods made from grains, roots or tubers); legumes; dairy products (milk other than breast milk, cheese or yoghurt); meat, poultry, fish or eggs; vitamin A-rich fruits and vegetables (pumpkin; red or yellow yams or squash; carrots or red sweet potatoes; green leafy vegetables; fruits such as mango, papaya or other local vitamin A-rich fruits; other fruits and vegetables (or fruit juices)) and foods made with oil, fat or butter. However, some indices treat certain components separately: the DDS developed by Kennedy in 2006, for example, separates fish and eggs from the protein group.

The components used by each index are shown in **Table 4**.

Nutrients	
Total fat	DQI, DQI-R, DQI-I, HEI, HEI-2005
SFA	DQI, DQI-R, DQI-I, HEI, HEI-2005, HDI, MDQI, The Chinese Children Dietary Index
Ratio of MUFA or PUFA to SFA	DQI-I, MDS, MDS-f, MDS-a I, MDS-a III
PUFA	HDI, PDL-Index
Protein	DQI, DQI-I, HDI
Carbohydrate	DQI
Complex carbohydrates	DQI, HDI
(Cereal) fibre	DQI-I, HDI, The Chinese Children Dietary Index
Mono- and disaccharides	DQI, HDI
Sucrose o sucrose added	DQI-a I
Cholesterol	DQI, DQI-R, DQI-I, DQI-a I – III, HEI, HEI-2005, HDI,
Alcohol	MDS, MDS-f, MDS-a I, III, IV
Sodium	DQI, DQI-I, DQI-a II, HEI, HEI-2005
Calcium	DQI, DQI-R, DQI-I
Iron	DQI-R, DQI-I
Vitamin C	DQI-I
Vit A	The Chinese Children Dietary Index
Ratio of carbohydrates to protein to fat	DQI-I
Energy balance	The Chinese Children Dietary Index
Foods	
Fruit and vegetables	DQI, MDQI, MDS-a I, HDI, VIT, YHEI, DDS8, DDS Eneman
Fruits (and nuts)	DQI-R, DQI-I, HEI, HEI-2005, AHEI, MDS, MDS-f, MDS-a II – IV, HuSKY , NZDQI-A, E-KINDEX, PDL-Index, GINI-plus/LISA-plus Studies, KIDMED, DDS Mirmiram, DDS Kennedy, FVS Saibul, DDS Rah, DDS Torheim, DDS Steyn, The Chinese Children Dietary Index
Vegetables	DQI-R, DQI-I, HEI, HEI-2005, AHEI, MDS, MDS-f, MDS-a II – IV, NZDQI-A, HuSKY, UFCS, E-KINDEX, PDL-Index, GINI-plus/LISA-plus Studies, KIDMED, DDS Mirmiram, DDS Kennedy, DDS Rah, DDS Torheim, DDS Steyn, The Chinese Children Dietary Index
Legumes (and nuts and seeds)	MDS, MDS-f, MDS-a I–IV, HDI, E-KINDEX, PDL-Index, DDS Kennedy, DDS Rah, DDS8, DDS Eneman, DDS Torheim, DDS Steyn
Nuts (and soya)	MDS-a II, MDS-a III, The Chinese Children Dietary Index
(Whole) cereals or grains	DQI-R, DQI-I, HEI, HEI-2005, all MDS, VIT, YHEI, E-KINDEX, PDL-Index, KIDMED, DDS Mirmiram, DDS Kennedy, DDS Torheim, The Chinese Children Dietary Index

Nutrients	
Meat (and meat products)	HEI, HEI-2005, MDS, MDS-f, MDQI, MDS-a I – IV, VIT, YHEI, NZDQI-A, E-KINDEX, PDL-Index, DDS Mirmiram, DDS Kennedy, FVS Saibul, DDS Rah, DDS8, DDS Eneman,, DDS Torheim , DDS Steyn, The Chinese Children Dietary Index
Eggs	DDS Kennedy, FVS Saibul, DDS8, DDS Eneman, DDS Torheim, DDS Steyn, The Chinese Children Dietary Index
White meat	PDL-Index
Red and processed meat	MDS-a III
Poultry	MDS-a IV, DDS Kennedy, FVS Saibul, DDS Rah, DDS8, DDS Eneman, DDS Steyn
Fish	MDS-f, MDS-a II–IV, E-KINDEX, DDS Kennedy, FVS Saibul, DDS Rah, DDS8, DDS Eneman, DDS Torheim, DDS Steyn, The Chinese Children Dietary Index
Milk (and dairy products)	HEI, HEI-2005, MDS, MDS-a I, VIT, YHEI, NZDQI-A, E-KINDEX, DDS Mirmiram, FVS Saibul, DDS Rah, DDS8, DDS Eneman, DDS Torheim, DDS Steyn
High fat dairy	MDS-a II, IV
Oil	MDS-a IV, KIDMED, DDS Kennedy, DDS8, DDS Torheim, DDS Steyn
Potatoes	MDS-a IV, DDS8, DDS Eneman
Cheese	KIDMED
Red wine	MDS-a III
Butter, margarine, animal fat	YHEI, DDS8, DDS Eneman, DDS Steyn
Sweets/sweet beverages	E-KINDEX, KIDMED, DDS Torheim
Dietary variety	DQI-R, DQI-I, HEI, HEI-2005,The Chinese Children Dietary Index
Dietary moderation	DQI-R
Behaviour	
Multivitamin use never	YHEI
Snack foods	YHEI, E-KINDEX
Sweetened beverages	YHEI
Sugary sodas	E-KINDEX
Fried food outside home	YHEI, E-KINDEX, KIDMED
Beverages	The Chinese Children Dietary Index
Breakfast and quality	YHEI, KIDMED, The Chinese Children Dietary Index
Dinner at home	YHEI, The Chinese Children Dietary Index
Physical activity	PDL-Index

Table 4. Overview of items included in means quality index.

9. Uses of quality indices by the food companies

Food industry could uses DQIs to develop healthier menus intended to be consumed by different specific population groups so as to ensure the correct nutritional inputs in terms of macro and micronutrients according to the physiological characteristics of each group. In this way, not to evaluate food products or prepared meals, but within a more global vision such as the field of collective catering (hospitals, nursing homes, schools, colleges and universities). DQIs would be a useful and easy-to-manage tool for assessing the quality of the diet as a whole [101, 102].

10. Conclusions

Indices are a useful tool for epidemiological research, for the development and application of nutritional strategies, for charting changes in eating habits within a given population, and for measuring adherence to dietary guidelines. In the same way, DQIs can be used by the food industry to offer healthier menus within the scope of collective catering. However, it is very difficult to compare findings due to the marked disparity between the variables used in each index. There is clearly a need for an easily-applied index which scores intake of each item on a simple 0/1 basis and adopts a more holistic approach, embracing not only healthy eating habits but also healthy lifestyle choices such as physical activity and a less sedentary way of life.

Author details

Maria Luisa Poyatos Guerrero and Fernando Pérez-Rodríguez*

*Address all correspondence to: b42perof@uco.es

Food Science and Technology Department, University of Córdoba, Córdoba, Spain

References

[1] WHO Library Cataloguing-in-Publication Data, editor. Global Action Plan for the Prevention and Control of Noncommunicable Diseases 2013-2020. Switzerland: World Health Organization 2013; 2013. 55 p

[2] WHO Library Cataloguing-in-Publication Data, editor. Global Status Report on Noncommunicable Diseases 2014. Switzerland: World Health Organization 2014; 2014. 298 p

[3] Assman KE, et al. Prospective association of animal protein intake during puberty with body composition in young adulthood. Obesity (Silver Spring). 2013;**21**(12):E782-E789. DOI: 10.1002/oby.20516

[4] Alkerwi A. Diet quality concept. Nutrition. 2014;**30**(6):613-618. DOI: 10.1016/j.nut.2013.10.001

[5] Elmadfa I, Meyer AL. Diet quality, a term subject to change over time. International Journal for Vitamin and Nutrition Research. 2012;**82**(3):144-147. DOI: 10.1024/0300-9831/a000104

[6] Pelletier JE, Laska MN, MacLehose R, Nelson TF, Nanney MS. Evidence-based policies on school nutrition and physical education: Associations with state collaboration, obesity, and socio-economic indicators. Preventive Medicine. 2017;**99**:87-93. DOI: 10.1016/j.ypmed2017.02.005

[7] Patterson RE, Haines PS, Popkin BM. Diet quality index: Capturing a multidimensional behavior. Journal of the American Dietetic Association. 1994;**94**(1):57-64

[8] Gil A, Martinez de Victoria E, Olza J. Indicators for the evaluation of diet quality. Nutrición Hospitalaria. 2015;**31**(Suppl. 3):128-144

[9] Kant AK. Indexes of overall diet quality: A review. Journal of the American Dietetic Association. 1996;**96**(8):785-791. DOI: 10.1016/S0002-8223(96)00217-9

[10] Cusatis DC, Chinchilli VM, Johnson-Rollings N, Kieselhorst K, Stallings VA, Lloyd T. Longitudinal nutrient intake patterns of US adolescent women: The Penn State Young Women's Health Study. Journal of Adolescent Health. 2000;**26**(3):194-204

[11] Martin C, Zhang Y, Tonelli C, Petroni K. Plants, diet, and health. Annual Review of Plant Biology. 2013;**64**:19-46. DOI: 10.1146/annurev-arplant-050312-120142

[12] Pellegrini N, Fogliano V. Cooking, industrial processing and caloric density of foods. Current Opinion in Food Science. 2017;**14**:89-102

[13] Bolzetta F, et al. Are the recommended dietary allowances for vitamins appropriate for elderly people? . Journal of the Academy of Nutrition and Dietetics. 2015;**115**(11):1789-1797. DOI: 10.1016/j.jand.2015.04.013

[14] García A. Nutrient intakes: Concepts and international recommendations (fits part). Nutrición Hospitalaria. 2006;**21**(3):291-299

[15] USDA. Choosemyplate [Internet]. 25/01/2007 [Updated: 25/01/2017]. Available from: https://www.choosemyplate.gov/ten-tips-choose-myplate. [Accessed: 23-03-2017]

[16] García A. Nutrient intakes: Concepts and international recommendations (part two). Nutrición Hospitalaria. 2006;**21**(4):437-447

[17] Aranceta J, Pérez C, Serra-Majem L. Objetivos nutricionales para la población española. Consenso de la Sociedad Española de Nutrición Comunitaria 2011. Revista Española de Nutrición Comunitaria . 2011;**17**(4):178-199. ISSN: 1135-3074

[18] FAO. In: editors. Food Balance Sheets: A Handbook, Rome: Food and Agriculture Organization of the 48 United Nations; 2001

[19] Wilson MA, Reedy J, Krebs-Smith SM. American diet quality: Where it is, where it is heading, and what it could be. Journal of the Academy of Nutrition and Dietetics. 2016;**116**(2):302-310. DOI: 10.1016/j.jand.2015.09.020

[20] Poslusna K, Ruprich J, de Vries JH, Jakubikova M, van't Veer P. Misreporting of energy and micronutrient intake estimated by food records and 24 hour recalls, control and adjustment methods in practice. British Journal of Nutrition. 2009;**101**(Suppl 2):S73-S85. DOI: 10.1017/S0007114509990602

[21] Kobayashi S, et al. Both comprehensive and brief self-administered diet history questionnaires satisfactorily rank nutrient intakes in Japanese adults. Journal of Epidemiology. 2012;**22**(2):151-159. DOI: 10.2188/jea.JE20110075

[22] George SM, Ballard R, Shikany JM, Crane TE, Neuhouser ML. A prospective analysis of diet quality and endometrial cancer among 84.415 postmenopausal women in the Women's Helth Initiative. Annals of Epidemiology. 2015;**25**(10):788-793. DOI: 10.1016/j.annepidem.2015.05.009

[23] Freedman MR, Fulgoni VL, 3rd. Canned vegetable and fruit consumption is associated with changes in nutrient intake and higher diet quality in children and adults: National health and nutrition examination survey 2001-2020. Journal of the Academy of Nutrition and Dietetics. 2016;**116**(6):940-948. DOI: 10.1016/j.jand.2015.10.013

[24] Schwingshackl L, Hoffmann G. Diet quality as assessed by the Healthy Eating Index, the Alternate Healthy Eating Index, the Dietary Approaches to Stop Hypertension score, and health outcomes: A systematic review and meta-analysis of cohort studies. Journal of the Academy of Nutrition and Dietetics. 2015;**115**(5):780-800. DOI: 10.1016/j.jand.2014.12.009

[25] Byung-Joon K, et al. Diet quality and diet patterns in relation to circulating cardiometabolic biomarkers. Clinical Nutrition. 2016;**35**(2):484-490. DOI: 10.1016/j.clnu.2015.03.022

[26] Bail J, Meneses K, Demark-Wahnefried W. Nutritional status and diet in cancer prevention. Seminars in Oncology Nursing. 2016;**32**(3):206-214. DOI: 10.1016/j.soncn.2016.05.004

[27] Nissensohn M, Román-Viñas B, Sánchez-Villegas A, Piscopo S, Serra-Majem L. The effect of the mediterranean diet on hypertension: A sistematic review and meta-analysis. Journal of Nutrition Education and Behavior. 2016;**48**(1):42-53. DOI: 10.1016/j.jneb.2015.08.023

[28] Kant AK. Dietary patterns and health outcomes. Journal of the American Dietetic Association. 2004;**104**(4):615-635. DOI: 10.1016/j.jada.2004.01.010

[29] Waijers PM, Feskens EJ, Ocké MC. A critical review of predefined diet quality scores. British Journal of Nutrition. 2007;**97**(2):219-231. DOI: 10.1017/S0007114507250421

[30] Arvaniti F, Panagiotakos DB. Healthy indexes in public health practice and research: A review. Critical Reviews in Food Science and Nutrition. 2008;**48**(4):317-327. DOI: 10.1080/10408390701326268

[31] Bach A, et al. The use of indexes evaluating the adherence to the Mediterranean diet in epidemiological studies: A review. Public Health Nutrition. 2006;**9**(1a):132-146. DOI: 10.1079/PHN2005936

[32] Lazarou C, Newby PK. Use of dietary indexes among children in developed countries. Advances in Nutrition. 2011;**2**(4):295-303. DOI: 10.3945/an.110.000166

[33] Marshall S, Burrows T, Collins CE. Systematic review of diet quality indices and their associations with health-related outcomes in children and adolescents. Journal of Human Nutrition and Dietetics. 2014;**27**(6):577-598. DOI: 10.1111/jhn.12208

[34] Haines PS, Siega-Riz AM, Popkin BM. The Diet Quality Index revised: A measurement instrument for populations. Journal of the American Dietetic Association. 1999;**99**(6):697-704. DOI: 10.1016/S0002-8223(99)00168-6

[35] Fung TT, et al. Diet-quality scores and plasma concentrations of markers of inflammation and endothelial dysfunction. American Journal of Clinical Nutrition. 2005;**82**(1):163-173

[36] Kennedy ET, Ohls J, Carlson S, Fleming K. The Healthy Eating Index: Design and applications. Journal of the American Dietetic Association. 1995;**95**(10):1103-1108. DOI: 10.1016/S0002-8223(95)00300-2

[37] Britten P, Marcoe K, Yamini S, Davis C. Development of food intake patterns for the MyPyramid Food Guidance System. Journal of Nutrition Education and Behavior. 2006;**38**(6 Suppl):S78-S92. DOI: 10.1016/j.jneb.2006.08.007

[38] Guenther PM, Reedy J, Krebs-Smith SM. Development of the Healthy Eating Index-2005. Journal of the American Dietetic Association. 2008;**108**(11):1896-1901. DOI: 10.1016/j.jada.2008.08.016

[39] Guenther P, et al. Update of the Healthy Eating Index: HEI-2010. Journal of the Academy of Nutrition and Dietetics. 2013;**116**(4):170-175. DOI: 10.1016/j.jand.2012.12.016

[40] Cox DR, Skinner JD, Carruth BR, Moran J, Houck KS. A Food Variety Index for Toddlers (VIT): Development and application. Journal of the American Dietetic Association. 1997;**97**(12):1382-1386; quiz 1387-8

[41] Feskanich D, Rockett HR, Colditz GA. Modifying the Healthy Eating Index to assess diet quality in children and adolescents. Journal of the American Dietetic Association. 2004;**104**(9):1375-1383. DOI: 10.1016/j.jada.2004.06.020

[42] Huijbregts P, et al. Dietary pattern and 20 year mortality in elderly men in Finland, Italy, and The Netherlands: Longitudinal cohort study. BMJ. 1997;**315**(7099):13-17

[43] Kim S, Haines PS, Siega-Riz AM, Popkin BM. The Diet Quality Index-International (DQI-I) provides an effective tool for cross-national comparison of diet quality as illustrated by China and the United States. The Journal of Nutrition. 2003;**133**(11):3476-3484

[44] Tur JA, Romaguera D, Pons A. The Diet Quality Index-International (DQI-I): Is it a useful tool to evaluate the quality of the Mediterranean diet? British Journal of Nutrition. 2005;**93**(3):369-376

[45] Keys A, editor. Seven Countries. A Multivariate Analysis of Death and Coronary Heart Disease. 1st ed. Cambridge: Harvard University Press; 1980. 381 p

[46] Keys A, et al. The diet and 15-year death rate in the seven countries study. American Journal of Epidemiology. 1986;**124**(6):903-915

[47] Trichopoulou A, Vasilopoulou E. Mediterranean diet. Encyclopedia of Food and Health. 2016;711-714. DOI: 10.1016/B978-0-12-384947-2.00454-2

[48] Kastorini CM, et al. Metabolic syndrome, adherence to the Mediterranean diet and 10-year cardiovascular disease incidence: The ATTICA study. Atherosclerosis. 2016;**246**:87-93

[49] Arós F., Estruch R. Dieta mediterránea y prevención de la enfermedad cardiovasular. Rev Esp Cardiol. 2013; **66**(10): 771-4. DOI: 10.1016/j.recesp.2013.04.026

[50] Guallar-Castillón P, et al. Major dietary patterns and risk of cohoronary heart disease in middle-aged persons from a Mediterranean country: The EPIC- Spain country study. Nutrition, Metabolism, and Cardiovascular Diseases. 2012;**22**(3):192-199. DOI: 10.1016/j.numecd.2010.06.004

[51] Cueto-Galán R, et al. Changes in fatty liver index after consuming a Mediterranean diet: 6-year follow-up of the PREDIMED- Malaga trial. Medicina Clínica. 2017;**148**(10):435-443. DOI: 10.1016/j.medcli.2016.11.032

[52] Ruiz-Cabello P, et al. Influence of the degree of adherence to the Mediterranean diet on the cardiometabolic risk in peri and menopausal women. The Flamenco project. Nutrition, Metabolism, and Cardiovascular Diseases. 2017;**27**(3):217-224. DOI: 10.1016/j.numecd.2016.10.008

[53] Della Corte C, et al. Good adherence to the Mediterranean diet reduces the risk for NASH and diabetes in pediatric patients with obesity: The results of an Italian Study. Nutrition. 2017;**39-40**:8-14

[54] Izadi V, et al. Adherence to the DASH and Mediterranean diets is associated with decreased risk for gestational diabetes mellitus. Nutrition. 2016;**32**(10):1092-1096. DOI: 10.1016/j.nut.2016.03.006

[55] Trichopoulou A, et al. Diet and overall survival in elderly people. BMJ. 1995;**311**(7018): 1457-1460

[56] Trichopoulou A, Costacou T, Bamia C, Trichopoulos D. Adherence to a Mediterranean diet and survival in a Greek population. New England Journal of Medicine 2003;**348**:2599-2608. DOI: 10.1056/NEJMoa025039

[57] Schröder H, et al. A short screener is valid for assessing Mediterranean diet adherence among older Spanish men and women. The Journal of Nutrition. 2011;**141**(6):1140-1145. DOI: 10.3945/jn.110.135566

[58] Martínez-González M, et al. Benefits of the mediterranean diet: Insights from the PREDIMED study. Progress in Cardiovascular Diseases. 2015;**58**(1):50-60. DOI: 10.1016/j.pcad.2015.04.003

[59] Serra-Majem L, et al. Food, youth and the Mediterranean diet in Spain. Development of KIDMED, Mediterranean Diet Quality Index in children and adolescents. Public Health Nutrition. 2004;**7**(7):931-935

[60] Toktas N, Yildiz Y. Assessment of nutritional status of 10-14 years old adolescents using Mediterranean Diet Quality Index (kidmed). Procedia-Social and Behavioral Sciences. 2013;**106**:512-518. DOI: 10.1016/j.sbspro.2013.12.057

[61] Mistretta A, et al. Mediterranean diet adherence and body composition among Southern Italian adolescents. Obesity Research & Clinical Practice. 2017;**11** (2):215-226. DOI: 10.1016/J.ORCP.2016.05.007

[62] Idelson P, Scalfi L, Valerio G. Adherence to the Mediterranean diet in children and adolescents: A systematic review. Nutrition, Metabolism, and Cardiovascular Diseases. 2017;**27**(4):283-299. DOI: 10.1016/j.numecd.2017.01.002

[63] Bonaccio M, Bes-Rastrollo M, de Gaetano G, Iacoviello L. Challenges to the Mediterranean diet at a time of economic crisis. Nutrition, Metabolism, and Cardiovascular Diseases. 2016;**26**(12):1057-1063. DOI: 10.1016/j.numecd.2016.07.005

[64] Rizza W, De Gara L, Incalzi RA, Pedone C. Prototypical versus contemporary Mediterranean Diet. Clinical Nutrition ESPEN. 2016;**15**:44-48. DOI: 10.1016.j.clnesp.2016.06.007

[65] Sotos-Prieto M, Moreno-Franco B, Ordovás JM, León M, Casasnovas JA, Peñalvo JL. Design and development of an instrument to measure overall lifestyle habits for epidemiological research: the Mediterranean Lifestyle (MEDLIFE) index. Public Health Nutrition. 2015;**18**(6):959-967. DOI: 10.1017/S1368980014001360

[66] Lazarou C, Panagiotakos DB, Spanoudis G, Matalas AL. E-KINDEX: A dietary screening tool to assess children's obesogenic dietary habits. Journal of the American College of Nutrition. 2011;**30**(2):100-112

[67] Lazarou C, Karaolis M, Matalas A, Panagiotakos DB. Dietary patterns analysis using data mining method. An application to data from the CYKIDS study. Computer Methods and Programs in Biomedicine. 2012;**108**(2):706-714. DOI: 10.1016/j.cmpb.2011.12.011

[68] Manios Y, Kourlaba G, Grammatikaki E, Androutsos O, Moschonis G, Roma-Giannikou E. Development of a diet–lifestyle quality index for young children and its relation to obesity: The Preschoolers Diet–Lifestyle Index. Public Health Nutrition. 2010;**13**(12):2000-2009. DOI: 10.1017/s1368980010000698

[69] Manios Y, Moschonis G, Grammatikaki E, Anastasiadou A, Liarigkovinos T. Determinants of childhood obesity and association with maternal perceptions of their children's weight status: The "GENESIS" study. Journal of the American Dietetic Association. 2010;**110**(10):1527-1531. DOI: 10.1016/j.jada.2010.07.004

[70] Marshall S, Burrows T, Collins CE. Systematic review of diet quality indices and their associations with health-related outcomes in children and adolescents. Journal of Human Nutrition and Dietetics. 2014;**27**(6):557-598. DOI: 10.1111/jhn.12208

[71] Ramsay S, Shriver L, Taylor C. Variety of fruit and vegetables is related to preschoolers' overall diet quality. Preventive Medicine Reports. 2017;**5**:112-117. DOI: 10.1016/j.pmedr.2016.12.003

[72] Spence A, McNaughton S, Lioret S, Hesketh K, Crawford D, Campbell K. A health promotion intervention can affect diet quality in early childhood. The Journal of Nutrition. 2013;**143**(10):1672-1678. DOI: 10.3945/jn.113.177931

[73] Scott JA, Chih TY, Oddy WH. Food variety at 2 years of age is related to duration of breastfeeding. Nutrients. 2012;**4**(10):1464-1474. DOI: 10.3390/nu4101464

[74] Marshall S, Watson J, Burrows T, Guest M, Collins CE. The development and evaluation of the Australian child and adolescent recommended food score: A cross-sectional study. Nutrition Journal 2012;**11**:96. DOI: 10.1186/1475-2891-11-96

[75] Collins L, Lacy K, Campbell K, McNaughton S. The predictors of diet quality among Australian children aged 3.5 years. Journal of the Academy of Nutrition and Dietetics. 2016;**116**(7):1114-1126. DOI: 10.1016/j.jand.2015.12.014

[76] Kleiser C, Mensink GB, Scheidt-Nave C, Kurth BM. HuSKY: A healthy nutrition score based on food intake of children and adolescents in Germany. British Journal of Nutrition. 2009;**102**(4):610-618. DOI: 10.1017/s0007114509222689

[77] Alexy U, Sichert-Hellert W, Kersting M, Lausen B, Schöch G. Development of scores to measure the effects of nutrition counselling on the overall diet: A pilot study in children and adolescents. European Journal of Nutrition. 1999;**38**(4):196-200

[78] Kohlboeck G, et al. Food intake, diet quality and behavioral problems in children: Results from the GINI-plus/LISA-plus Studies. Annals of Nutrition and Metabolism. 2012;**60**:247-256. DOI: 10.1159/000337552

[79] Kyttälä P, et al. Finnish Children Healthy Eating Index (FCHEI) and its associations with family and child characteristics in pre-school children. Public Health Nutrition. 2014;**17**(11):2519-2527. DOI: 10.1017/S1368980013002772

[80] Maiz E., Balluerka N. Nutritional status and Mediterranean diet quality among Spanish children and adolescents with food neophobia. Food Quality and Preference. 2016;**52**:113-142. DOI: 10.1016/j.foodqual.2016.04.011

[81] Royo-Bordonada MA. Greater dietary variety is associated with better biochemical nutritional status in Spanish children: The Four Provinces Study. NMCD. 2003;**13**(6):357-364. DOI: 10.1016/S0939-4753(03)80004-2

[82] Pérez-Fainós N, et al. The ALADINO study: A national study of prevalence of overweight and obesity in Spanish children in 2011. BioMed Research International. 2013;**2013**:163687. DOI: 10.1155/2013/163687

[83] Fung C, McIsaac J, Kuhle S, Kirk S, Veugelers P. The impact of a population-level school food and nutrition policy on dietary intake and body weights of Canadian children. Preventive Medicine. 2013;**57**(6):934-940. DOI: 10.1016/j.ypmed.2013.07.016

[84] Chu Y, Storey K, Veugelers P. Involvement in meal preparation at home is associated with better diet quality among Canadian children. Journal of Nutrition Education and Behavior. 2014;46(4):304-308

[85] Cheng G, Duan R, Kranz S, Libuda L, Zhang L. Development of a dietary index to assess overall diet quality for Chinese school-aged children: The Chinese children dietary index. Journal of the Academy of Nutrition and Dietetics. 2016;**116** (4):608-617. DOI: 10.1016/J.JAND.2015.11.010

[86] Kennedy GL, Pedro MR, Seghieri C, Nantel G, Brouwer I. Dietary diversity score is a useful indicator of micronutrient intake in non-breast-feeding Filipino children. The Journal of Nutrition. 2007;**137**(2):472-477

[87] Falciglia GA, Troyer AG, Couch SC. Dietary variety increases as a function of time and influences diet quality in children. Journal of Nutrition Education and Behavior. 2004;**36**(2):77-83

[88] Arimond M, Ruel MT. Dietary diversity is associated with child nutritional status: evidence from 11 demographic and health surveys. The Journal of Nutrition. 2004;**134**(10):2579-2585

[89] Hoddinott J, Yohannes Y. In: Academy for Educational Development, editor. Dietary Diversity as a Food. Food Consumption and Nutrition Division Discussion Paper 136, Washington: Food and Nutrition Technical Assistance Project (FANTA); 2002. pp. 1-54

[90] Onyango A, Koski KG, Tucker KL. Food diversity versus breastfeeding choice in determining anthropometric status in rural Kenyan toddlers. International Journal of Epidemiology. 1998;**27**(3):484-489

[91] Saibul N, Shariff ZM, Lin KG, Kandiah M, Ghani NA, Rahman H. Food variety score is associated with dual burden of malnutrition in Orang Asli (Malaysian indigenous peoples) households: Implications for health promotion. Asia Pacific Journal of Clinical Nutrition. 2009;**18**:412-422

[92] Roche ML, Creed-Kanashiro HM, Tuesta I, Kuhnlein HV. Traditional food diversity predicts dietary quality for the Awajún in the Peruvian Amazon. Public Health Nutrition. 2008;**11**(5):457-465. DOI: 10.1017/S1368980007000560

[93] Enneman A, Hernández L, Campos R, Vossenaar M, Solomons NW. Dietary characteristics of complementary foods offered to Guatemalan infants vary between urban and rural settings. Nutrition Research. 2009;**29**(7):470-479. DOI: 10.1016/j.nutres.2009.06.007

[94] Eckhardt CL, Suchindran C, Gordon-Larsen P, Adair LS. The association between diet and height in the postinfancy period changes with age and socioeconomic status in Filipino youths. The Journal of Nutrition. 2005;**135**(9):2192-2198

[95] Rah JH, et al. Low dietary diversity is a predictor of child stunting in rural Bangladesh. European Journal of Clinical Nutrition. 2010;**64**(12):1393-1398. DOI: 10.1038/ejcn.2010.171

[96] Wong JE, Parnell WR, Black KE, Skidmore PM. Reliability and relative validity of a food frequency questionnaire to assess food group intakes in New Zealand adolescents. Nutrition Journal. 2012;**11**:65. DOI: 10.1186/1475-2891-11-65

[97] Wong J, Parnell W, Howe A, Black K, Skidmore P. Development and validation of a food-based diet quality index for New Zealand adolescents. BMC Public Health. 2013;**13**:562. DOI: 10.1186/1471-2458-13-562

[98] WHO Multicentre Growth Reference Study Group. Complementary feeding in the WHO Multicentre Growth Reference Study. Acta Paediatrics. 2006;**450**(Suppl):27-37

[99] Moursi MM, Arimond M, Dewey KG, Trèche S, Ruel MT, Delpeuch F. Dietary diversity is a good predictor of the micronutrient density of the diet of 6- to 23-month-old children in Madagascar. The Journal of Nutrition. 2008;**138**(12):2448-2453. DOI: 10.3945/jn.108.093971

[100] IRD/GRET/LABASAN. Nutrimad: Programme de lutte contre la malnutrition a` Madagascar, une composante du programme Nutridev GRET/IRD [Internet]. Available from: http://www.rencontres-montblanc.coop/node/1151 [Accessed: 22-03-2017]

[101] De Mateo B, et al. Diseño y aplicación de un cuestionario de calidad dietética de los menús escolares. Nutrición Hospitalaria. 2015;**31**(1):225-235

[102] Morán L, et al. La planificación del menú escolar; resultado de un sistema de asesoría dietética para la adecuación a las recomendaciones nutricionales. Nutrición Hospitalaria. 2013;**28**(4):1145-1150